I0814757

Cities of Repetition
Jason F. Carlow, Christian J. Lange

Cities of Repetition
Hong Kong's Private Housing Estates

Jason F. Carlow, Christian J. Lange

ar+d
publishing

Table of Contents

10 Estates

Analysis

Interview

Photographs

[fig. 1] This diagram shows the floor plan outlines of the 529 tower blocks across the ten main estates presented and analyzed in the main section of the book in chronological order from upper left to lower right. The black colored tower blocks represent unique types, while the gray plans denote copies. The drawing reveals the high degree of formal similarity across private estates in the latter part of the 20th Century, as well as the use of repetition as an architectural device for urban development.

Foreword The Cliché and the Archetype

by Cole Roskam

Hong Kong and an Architectural Theory of the Cliché and the Archetype

In 1961, the architect Eric Cumine (1905–2002) described Hong Kong's rapidly expanding built environment and its architecture as "cliché-ridden."[1] Cumine explained that in the British-controlled territory, "novel features introduced into any building are repeated in others ad nauseum," resulting in an urban landscape of unvarying predictability. Cumine did not necessarily see the situation as exceptionally problematic, however. Through Hong Kong's excess of conventionality, in fact, an unusual form of originality had paradoxically taken shape. "Such repetition in itself is partially, however, responsible for a place producing its own style," surmised Cumine, "or a style with its own peculiar clichés."[2]

Cumine's reflections capture a consequential moment in the city's architectural history, at a time in which tens of thousands of refugees fleeing China's civil war sought shelter and opportunity in the British colony. Their arrival triggered a humanitarian crisis that resulted in the establishment of Hong Kong's public housing program, which was deployed to effectively discipline a destabilized population and landscape. Not only was a certain aesthetic and formal repetition inevitable to the mission at hand; it was an essential, core component justified in the name of the colony's social, political, and financial well-being. As the Honorable David Ronald Holmes, founder of the Hong Kong Resettlement Department, explained in 1955, "These things are done primarily for the community as a whole, not for the individuals affected."[3]

Cumine was not alone in this assessment; other observers who lived or spent time in the colony during the 1950s and 1960s came to similar conclusions, both with respect to the city's tendency to repeat ad nauseum, and to the point that through such replication, something idiosyncratic had materialized.[4] What distinguishes Cumine's reflections and makes them relevant to the work of *Cities of Repetition*, however, is the extent to which he identified the cliché's potential as a foundation for some new kind of architectural expression specific to Hong Kong—namely, an archetype.

Both the terms cliché and archetype evoke notions of patterning and repetition, albeit in different ways, and with different results. To identify something as clichéd is to remark on its overuse to the point of meaninglessness—a word or image that requires nothing from either its user or its audience, both of whom already understand its meaning.[5] The archetype, by contrast, retains its meaning and thus its cultural significance over time. If the cliché is a copy of some fundamental truth circulated and recirculated over time, it may be understood to have been originally descended from the archetype.

More than a simple acknowledgement of the repetitive act of copying on display, Cumine gestured toward the ways in which such reiteration could paradoxically produce something original. Other theorists have extended the meanings of cliché and the archetype in similar ways, borrowing lessons from language and literature, applying them to readings of technology and the physical environment, and transforming our understanding of how architecture shapes our perceptions of the social and political systems that structure our lives in the process. For example, the philosopher Marshall McLuhan, in his 1970 book From the Cliché to the Archetype, argued that our cities depend upon services with physical forms—what he called "service environments"—that alternate between cliché and archetype depending upon the extent to which they are seen and aestheticized.[6]

The study of clichés, archetypes, and their various architectural and social implications resonate, both explicitly and implicitly, through the pages of *Cities of Repetition*. As the book's introduction suggests, Hong Kong's public housing program evokes Le Corbusier's archetypal urban proposals for Paris, both in form and aesthetic. Ideologically, meanwhile, Le Corbusier's legacy also surfaces in the initiative's underlying sense of urgency, particularly his calls to militarize architecture through the rapid deployment of a plan, as one might do during war.[7] Other influences need also be taken into account, however, including Britain's New Towns program, contemporaneous examples of high-density, low-cost housing in Japan, Singapore, Puerto Rico, and the Federation of Malaya, and architectural responses to refugee crises reverberating through India and Pakistan during the early to late 1950s.[8]

Relocated to Hong Kong, these influences took on a space and time of their own, shaped in large part by the city's population density, distinctive topography, and unique geopolitical identity. Over time, the urgencies and contingencies at work in the city's resettlement and public housing programs informed its private housing development and arguably its architectural production more generally. Arguably, the city's modern housing environments became increasingly constrained rather than enabled by the original ethos of crisis and urgency at their core. Mundanity gradually extended from the architecture itself to the conventional, oft-circulated conclusions typically used to explain it, including an inescapable desire for efficiency and the simple transactional burdens of the property market.[10]

Yet as *Cities of Repetition* suggests, these projects nevertheless remain significant and worthy of architectural study, with broader lessons regarding how and why design ideas can repeat, again and again, in the face of an uncertain future. As McLuhan himself observed, "in a world where all is change, creativity requires the conservation of mutations."[11] Hong Kong is a city made and reproduced, again and again, through constant cycles of change, and the kind of research captured in this book draws our attention to the ways in which architectural responses to change are capable of producing both cliché and archetype depending upon how, when, and where one sees them, now and in the future.

[1] Eric Cumine, "Hong Kong," in J.M. Richards, ed., NEW BUILDINGS IN THE COMMONWEALTH (London: The Architectural Press, 1961), 219.

[2] Ibid.

[3] Hon. Dr. Holmes, "The Problem of Resettlement," in HONG KONG BUSINESS SYMPOSIUM: A COMPILATION OF AUTHORITATIVE VIEWS ON THE ADMINISTRATION, COMMERCE, AND RESOURCES OF BRITAIN'S FAR EAST OUTPOST, comp. J.M. Braga (Hong Kong: South China Morning Post, 1957), 370.

[4] See Lionel Weaver, "Hong Kong," THE BUILDER 196 (1959), 772; J.A. Prescott, BUILDING BUT LITTLE ARCHITECTURE, South China Morning Post, 29 May 1964, 8.

[5] See Jakob Norberg, "The Political Theory of the Cliché," CULTURAL CRITIQUE 76 (Fall 2010), 77.

[6] Marshall McLuhan with Wilfred Watson, FROM CLICHÉ TO ARCHETYPE, ed. W. Terrence Gordon, 2nd edition (New York: Viking, 1970; New York: Gingko Press, 2011).

[7] Kenneth Frampton, LE CORBUSIER: ARCHITECT AND VISIONARY (London: Thames & Hudson, 2001), 119. Quoted in Gregory Clancey, "Toward a Spatial History of Emergency: Notes from Singapore," in BEYOND DESCRIPTION: SINGAPORE, SPACE, HISTORICITY, ed. Ryan Bishop, John Phillips, and Wei-Wei Yeo (London: Routledge, 2004), 34.

[8] See "Study of Low Cost Housing," SOUTH CHINA MORNING POST, 24 August 1954; "World Housing Plans," SOUTH CHINA MORNING POST, 8 June 1954, 5.

[9] See, for example, Li Shiqiao, "Hong Kong: City of Maximum Quantities," in William S. W. Lim, ed. ASIAN ALTERITY (Singapore: World Scientific, 2008), 28-36; Hendrik Tieben, "The Origin of Hong Kong Building Types," in HONG KONG TYPOLOGY (Zurich: gta publishers, 2010), 33-56.

[10] Marshall McLuhan, with Wilfred Watson, FROM CLICHÉ TO ARCHETYPE, ed. W. Terrence Gordon, 2nd ed. (New York: Viking Press, 1970; New York: Gingko Press, 2011), 54.

[fig. 2]

1950
MARK I
Resettlement Block Plan
[1953]
STANDARD BLOCK
Twin Tower
[1970]
1960
STANDARD BLOCK
Single H
[1977]
BLOCK 1-3
Mei Foo Sun Chuen
[1965]
STANDARD BLOCK
New Slab (Double Wing)
[1977]
1970
BLOCK 7/9/10/14/15
Taikoo Shing
[1977]
STANDARD BLOCK
Trident 2 (35 Story)
[1979]
BLOCK 3
Heng Fa Chuen
[1979]
BLOCK 1-8
Kornhill
[1980]
1980
STANDARD BLOCK
Lineral L1
[1979]
STANDARD BLOCK
Single I (21 Story)
[1980]
BLOCK 29
Whampoa Gardens
[1985]
BLOCK 37
City One
[1981]
STANDARD BLOCK
Slab (Link E=135)
[1984]
BLOCK 1
Belvedere Gardens
[1987]
STANDARD BLOCK
New Cruciform
[1986]
1990
STANDARD BLOCK
Harmony 1 (Option 10)
[1993]
STANDARD BLOCK
Harmony 2 (Option 3)
[1993]
BLOCK 2
Laguna City
[1991]
BLOCK 29
South Horizons
[1991]
BLOCK 1
Kingswood Villas
[1991]
2000

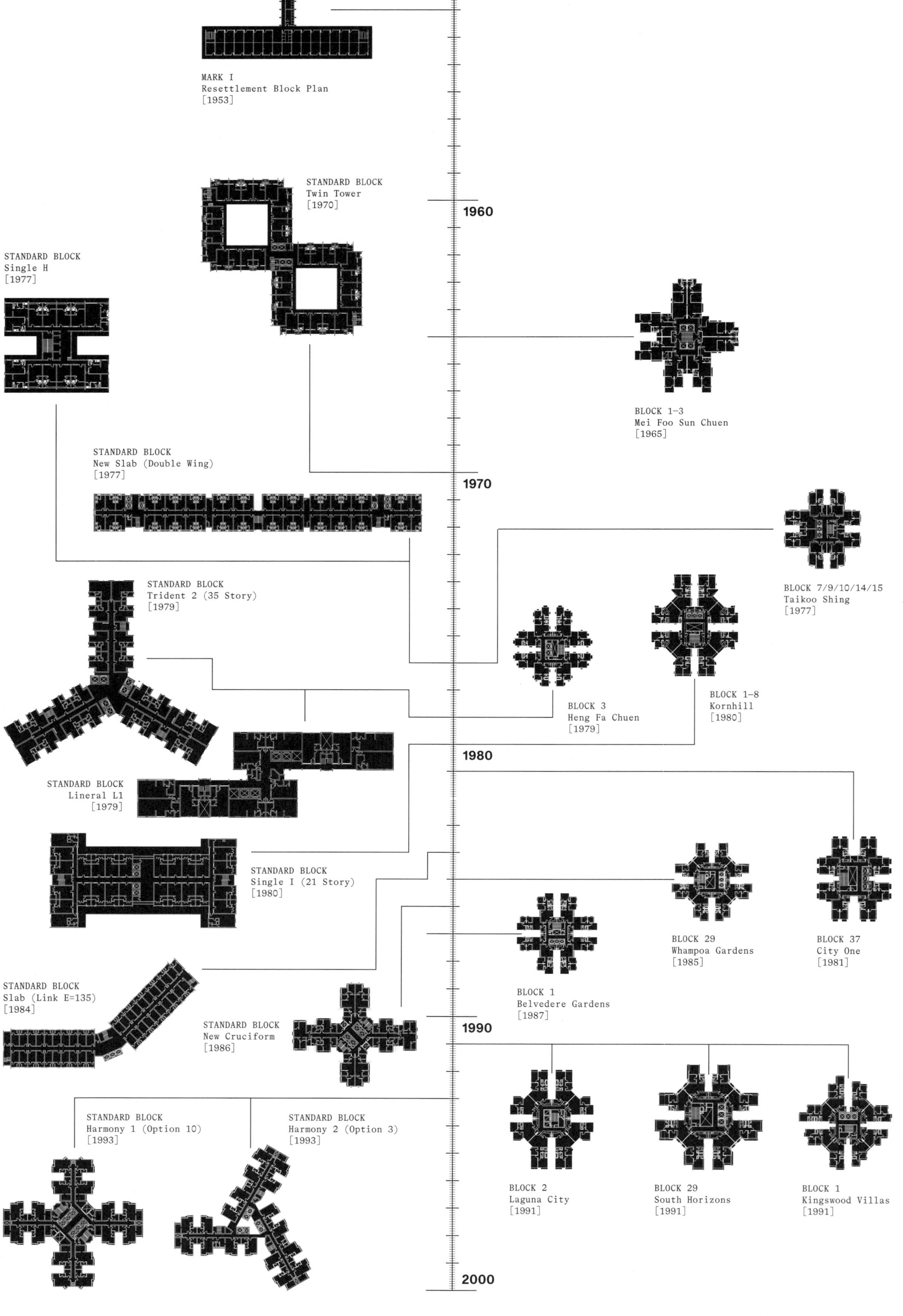

Hong Kong Housing History, Influences, and Impacts

Overview

Housing is one of the most fundamental elements of urban growth, and Hong Kong has for decades hosted some of the most intense built environments on the planet. The city's urbanization has produced unparalleled living conditions in terms of building scale and density. As a socio-political microcosm, Hong Kong has been dealing with the impacts of hyper-dense urban environments since the mid-twentieth century. Due to lack of space, topographical constraints, political conditions, and extremely high population density, the city has become an incubator for the development of housing models and tower typologies for high-density living.

This book provides a comprehensive graphic documentation of privately developed housing estates and tower typologies that have been designed and built from the late 1960s through the year 2000. At its core is the analysis of the ten largest private housing estates and their dozens of identical towers. The book is conceived as an atlas, giving insight into one extraordinary segment of the housing universe of Hong Kong. The plans, drawings, and diagrams in this graphic treatise illustrate the ultra-dense, mass-produced, highly repetitive built environments in which tens of thousands of Hong Kong residents live. Drawings not only display the immense scale of the housing estates within the city but present the hundreds of similarly planned apartment units and tower types with their subtle differences. Detailed analyses compare graphical and statistical information to show how the planning of these massive estates, driven by efficiency and building regulations, has become more uniform over time. Photographs reveal the spatial realities, design differences, and urban monotony of some of the most densely populated urban environments ever built.

The primary ten estates presented in this book were chosen for study and analysis based on their enormous population size, number of towers, and urban scale. They operate as miniature cities, each made up of thousands of similar units with similar floor plans. Despite being designed and built by several architecture firms and developers, they all resort to almost identical formal, spatial, and organizational systems. These ten "cities of repetition" are examples of how the pressures of strict building code regulation and extremely high land prices can disincentivize architectural experimentation and diversity.

As authors, we have approached this project of redrawing, modeling, and cataloging a highly repetitive and standardized built environment as architects and not as urban planners or social scientists. Therefore, the project focuses largely on building design as opposed to urban form or social conditions. The initial intent of the project was to simply study the organization of the buildings in relationship to the rules that govern and regulate architecture. A research methodology of cataloging and mapping tower types, in an approach similar to encyclopedic modes of cataloging species of flora or fauna, reveals an architectural relentlessness toward uniformity—a copy and paste attitude toward urban development and architecture. A method of color-coding in the main section of the book was utilized to disclose the differences and commonalities within the systems. Overall, the book presents an investigation and analysis of how hundreds of residential towers in Hong Kong, built in the final decades of the twentieth century, were shaped by regulatory and economic forces that radically standardized the city and limited architectural specificity in relation to context.

Public Housing in Hong Kong

To understand the forces that impact private, residential, real estate development in Hong Kong, one must first look back at the historical development of public housing in the city. Hong Kong's British colonial government formed a public housing strategy in the 1950s in response to waves of immigration into the territory from neighboring mainland China. In response to war, instability, and political change in China in the 1940s and 1950s, tens of thousands of

[fig. 3]
[fig. 4]

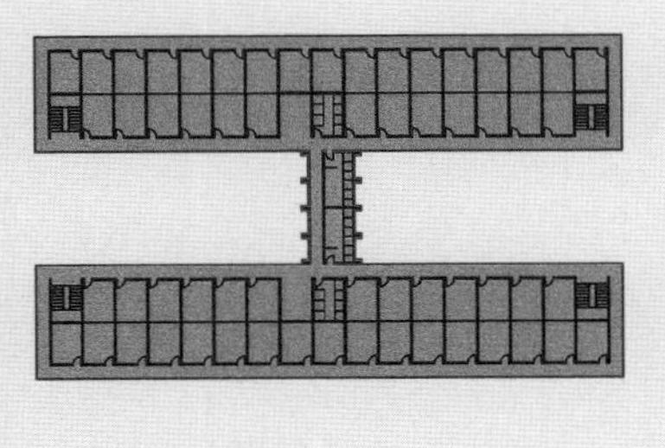

[fig. 2] Public and Private Housing Types. Timeline showing the relative diversity of public housing types developed by the Hong Kong Housing Authority as compared with the extremely homogenous plan layouts of typical towers in the ten largest, privately developed, housing estates in Hong Kong from 1950–2000. Drawings by authors.

[fig. 3] Redevelopment of Shek Kip Mei Estate in June 1958. Courtesy of the Hong Kong Government Records Service.

[fig. 4] Floor Plan of the Mark I housing block.

Chinese refugees immigrated to the then British colony of Hong Kong. Due to a lack of housing availability for the swelling population, many groups of immigrants constructed and lived in squatter settlements on the hillsides of the city. In 1953, a fire on Christmas day blazed through a large squatter area in the district of Shek Kip Mei, leaving more than fifty thousand immigrants homeless in the course of a single night.[1] Faced with a significant humanitarian crisis, the British colonial government of Hong Kong, under the leadership of Governor Alexander Grantham, acted to develop publicly funded housing for the residents displaced by the fire.

The Hong Kong government took several steps to address the housing crisis in 1954. By February, temporary housing in the form of small, two-story bungalows were built to house displaced Shek Kip Mei residents.[2] A "Resettlement Department" was formed, charged with housing the homeless refugees in Hong Kong through the construction of new "resettlement" estates. New public housing was quickly built to safely accommodate thousands of new residents. Other government-funded housing institutions were also established in this year. The government provided funding to a new "Hong Kong Housing Society," a volunteer-based institution, to develop rental housing. A semi-independent "Hong Kong Housing Authority" was also set up by the government in 1954 to develop "low-cost housing that provided a better living environment."[3]

The Resettlement Department speedily designed new mass housing prototypes for the displaced Shek Kip Mei community. The Mark I housing type [fig. 3, fig. 4] was the first design, built in response to the crisis, and the starting point of an evolutionary process of typology production for the public building sector in Hong Kong. The very basic, six- to seven-story building featured an H-shaped floor plan that organized single-room dwelling units along an exterior corridor on each of the outside wings. Communal bathroom and kitchen facilities were designed in the central connecting core of the "H." Compact, 120-square-foot flats were designed to provide a minimum of 24 square feet per adult and 12 square feet per child resident. This translated to space for five adults per unit, or three adults and four children.[4] Twenty-six Mark I housing types were built in total through the year 1962. Mass production in design and construction was implemented as standard practice in reaction to a housing crisis.[5] The buildings, and units within them, were densely packed and repetitive in nature.

The now globally established, modernist practice of standardization was employed to provide these first public housing units in Hong Kong as quickly and efficiently as possible. A philosophy of minimum space requirements and repetitive architectural and urban forms was adopted by the British colonial government at the very beginning of Hong Kong's public housing timeline. Policies built around minimum standards for space, light, and air were also built into Hong Kong building codes developed around the same time.

During the latter half of the twentieth century, the influx of new immigrants to Hong Kong continued, and more public housing types were designed and built across Hong Kong Island, Kowloon, and the New Territories. Mark II blocks, a variation on the Mark I type, were first introduced in the Tung Tau Tseun district in 1961. By 1972, Mark I and Mark II blocks housed over 500,000 residents—more than one-eighth of Hong Kong's entire population.[6] Resettlement Department housing variations (the Mark III, IV, V, and VI Blocks) were introduced during the 1960s and early 1970s until the administrative reorganization of Hong Kong's various housing departments and authorities in 1973. In that year, the Hong Kong Housing Authority was re-established under the administration of a newly formed Housing Department—assembled from the Resettlement Department and Housing Division of the Urban Services Department—to manage the design, development, and maintenance of public housing in the territory. Highly repetitive resettlement housing units—and their serial deployment across the rapidly growing city—set a precedent for housing policy and delivery in Hong Kong for decades to come. Housing types were designed as prototypical constructions, focused on efficiencies in floor plan and conceptually independent of urban or environmental context.

Since the extremely dense and standardized building types developed for the resettlement Mark I—Mark VI housing programs, planners and architects working with the Housing Authority since 1973 have actually developed a broad spectrum of housing types [fig. 2]. These typologies experimented with building form and organization and have grown increasingly diverse and elaborate through the years. In comparison with many cities across the world, the HK Housing Authority's public housing program has been one of the most responsive in terms of management and successful in terms of the overall provision of housing to citizens.[7] Progressive planning and housing management policies have helped to provide housing and long-term housing strategies for the city's neediest residents as Hong Kong developed into a thriving financial and economic hub in the latter half of the twentieth century. However, in comparison with the Housing Authority solutions of the 1970s and 80s that were offering a new range of residential models, Hong Kong's privately developed housing typologies of the same period utilized a narrow set of design strategies and became increasingly more standardized and repetitive.

Repetition and Privately Developed Housing in Hong Kong

While the history and architecture of public housing estates and housing typologies in the city has been well researched and documented, relatively little has been done to trace the evolution of Hong Kong's privately developed housing models. In many other developed nations in the world, when housing is left to the private sector, the results are typically diverse. Interestingly, when mass housing has been developed by the private sector in Hong Kong, the resultant architecture has proven to be highly formulaic. Over the last decades of the twentieth century, the common approach to mass housing in Hong Kong has been reduced to standardized development strategies that maximize spatial efficiency—and therefore profit for developers. Much of the city is built with repetitive canons utilizing fixed layouts and building designs that conform to building code restrictions without regard to site or environmental factors.

In Hong Kong during the years of the Resettlement Department, standardized architectural elements and forms enabled the government to increase the scale and pace of construction. Changes in housing policy and a shift of responsibility for housing from governmental agencies to the private sector were other tools to house Hong Kong's rapidly growing population. In the 1970s, the government of Hong Kong instituted two schemes to increase the amount of housing and the level of home ownership in the territory. The first scheme, the Ten-Year Housing Program, announced by the Hong Kong Government in 1972, set a goal to create housing for 1.8 million people by the year 1982.[8] The program was a reaction to a growing manufacturing economy and the continued high levels of immigration into the territory. Over time, the Hong Kong Housing Authority, under pressure from the private sector and especially the HK Real Estate Developers Association, began to design policy that encouraged the supply of new housing from the private sector through interest-free home-purchase loans.[9]

In 1976, the Housing Authority introduced a Home Ownership Scheme (HOS) that encouraged middle class families to move out of government rental housing and into flats that they could purchase for discounted prices with favorable lending terms. One year later, in 1977, the Housing Authority created incentives for private real estate developers to provide property for sale under the Private Sector Participation Scheme (PSPS).[10] By encouraging private developers to participate in the construction of low- to middle-income housing, the government's burden to provide housing for 1.8 million people under the Ten-Year Housing Program was mitigated.

Starting with the housing policies of the 1970s, the Hong Kong government has increasingly relied upon, and incentivized, the private sector to develop more housing to keep pace with demand. The strategy of mass standardization, initially established by Hong Kong's government agencies with inherent social responsibilities, was adopted and repurposed by the profit-driven real estate developers of the private sector.

[fig. 5]

[fig. 5] Aerial photograph of the Mei Foo Sun Chuen estate developed between 1965–78 (on left) compared with a model of Le Corbusier's Plan Voisin (on right). Image Credits: Wong Tung & Partners Limited, Mei Foo Investments Limited (Broadway-Nassau Investments Limited). 'Copy aerial photographs of construction, Mei Foo Sun Chuen (circa 1965–1978), Hong Kong', [circa 1970]. digital print, 3 items. M+, Hong Kong. Gift of Wong Tung & Partners Limited, 2013. [CA25/1/12]. © Wong Tung & Partners Ltd. Model of Plan Voisin, Paris, by Le Corbusier © F.L.C. / ADAGP, Paris, 2021.

[fig. 6] The incredibly dense figure-ground plan of the Mei Foo Sun Chuen estate in Kowloon (by authors) with ninety nine tower blocks, compared at the same scale with Le Corbusier's 1925 Plan Voisin for Paris. Original drawing of Le Corbusier's Plan Voisin by Stuart E. Cohen and Steven W. Hurtt, from their M.Arch thesis, "Le Corbusier: The Architecture of City Planning", Cornell University, 1967. © Stuart E. Cohen and Steven W. Hurtt.

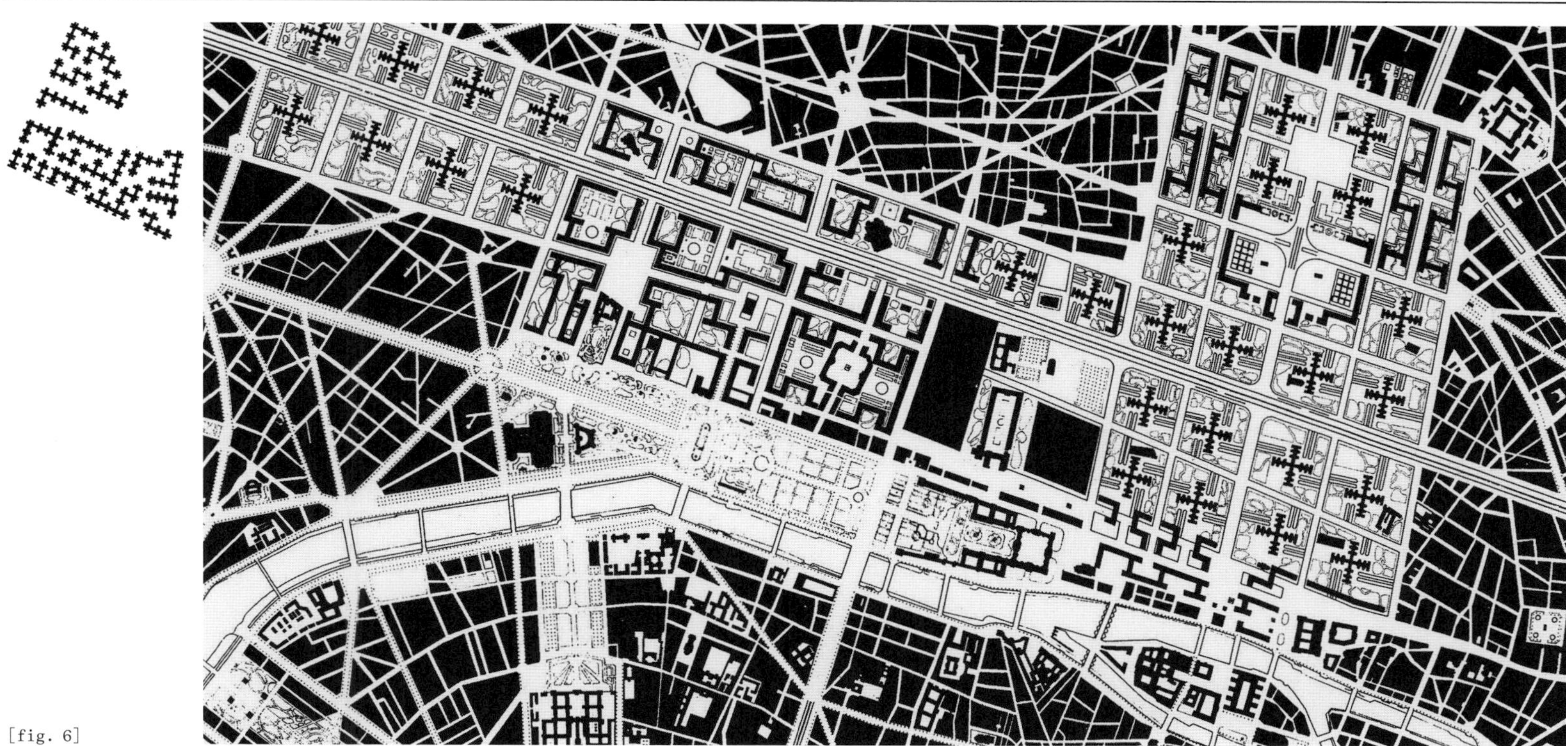

[fig. 6]

Hong Kong and Global Legacies of Standardization

The movement away from hand-crafted, or one-off construction methods, to factory made and industrially standardized units is not new and not exclusive to Hong Kong. The mass production of architectural elements and standardization of building components and techniques dates back to the development of cast-iron construction systems for train stations and large-span glass houses in the nineteenth century.[11] As leading architects, developers, and industrial visionaries around the globe sought to adapt the standardized, production-line techniques of the Ford automobile factory in the twentieth century to housing, variation and site-specificity were often reduced or completely lost. Mass production, scientific management, and the benefits of efficiency were elucidated by engineer Frederick Winslow Taylor and were looked upon by many modern architects as a solution to bring affordable architecture to the masses by embracing industrial production.[12] Walter Gropius, founder of the famous Bauhaus school, remarked:

> "There can be no doubt that the systematic application of standardization to housing would effect enormous economies, so enormous, indeed, that it is impossible to estimate their extent at present. Standardization is not an impediment to the development of civilization, but, on the contrary, one of its immediate prerequisites."[13]

As shown in the following sections of this book, the extreme standardization of architectural layouts, units, elements, and systems in the privately developed housing estates of Hong Kong has led to a profound standardization of domestic space and therefore the social structures of the communities they house.

Neighborhoods of regularly spaced high-rise towers are the predominant urban development model for one hundred of Hong Kong's largest, privately developed housing estates [pp. 15–21]. This model for housing is closely related to the urban proposals of one of the most influential architects and urbanists of the twentieth century, the Swiss-French modernist, Charles-Édouard Jeanneret, known as Le Corbusier. By the early 1920s, Le Corbusier had drawn up several proposals for new contemporary cities for millions of inhabitants. Plans for a Contemporary City of Three Million People, his Radiant City, and his Plan Voisin for central Paris all consisted of a core of high-rise commercial towers surrounded by mid-rise residential blocks, all situated in a large, park-like landscape. The projects were designed in oppositional reaction to the dense and congested urban areas of Paris around the tenets of modern rapid transportation, mixed-use programming, orderly street layouts, high speed motorways, and an urban structure that would provide residents with an efficient and healthy environment.

However, many of the realized housing developments and urban designs that imitated Le Corbusier's unbuilt proposals were misinterpretations of his plans.[14] Housing developments in cities around the world utilized high-rise construction for domestic living rather than the commercial use that Le Corbusier had imagined for his soaring, urban cores. In many cases, especially in North American contexts, housing projects were built with little variety of use, poor transportation connections, and urban environments that isolated the poor and discouraged socio-economic mobility.[15]

Many of Hong Kong's largest housing estates resemble the gridded high-rise towers of Le Corbusier's urban centers. However, compared with the cities that Le Corbusier drafted in the early twentieth century, and other similar urban forms in Europe and North America built after Le Corbusier's model, Hong Kong's interpretation of the tower-in-the-park typology is far denser. The envisioned population density of Corbusier's Plan Voisin and his Radiant City was approximately 300 people per hectare in the residential areas.[16] Stuyvesant Town and Peter Cooper Village, built in Manhattan starting in the late 1940s, present a similar urban model to Le Corbusier's "Cities of Tomorrow" with high-rise towers scattered throughout an open park. Stuyvesant Town and Peter Cooper Village have a population density of approximately 940 people per hectare—three times as dense as Le Corbusier's model. Hong Kong's private estates present scale and density on an entirely different level. Figure 5 presents a side-by-side comparison of Le Corbusier's Plan Voisin (1925) with Hong Kong's Mei Foo Sun Chuen Estate (1965–78), with the two projects drawn at the same scale. With a site area of 16.5 hectares and an estimated population close to 40,000 inhabitants, Mei Foo Sun Chuen boasts a population density of about 2,370 people per hectare. That represents a population density of around eight times greater than what Le Corbusier had planned in his version of a "contemporary city." Two decades later, in another of Hong Kong's privately developed estates, Belvedere Gardens, the population density is about 3,733 people per hectare, more than twelve times the density of the Plan Voisin.

In terms of the proximity of buildings, Hong Kong's estates are also significantly denser than what Le Corbusier had imagined. While his Contemporary City had a central commercial core of 180-meter-tall high-rise towers, they were given generous space and arranged on a 400-meter grid. By comparison, Mei Foo Sun Chuen's 72-meter-tall tower blocks are closely packed on a 44-meter grid with minimal gaps between buildings.

The podium tower building type that makes up a significant number of Hong Kong's privately developed estates is one of the major departures from the tower-in-the-park urbanism of Le Corbusier. While some of Le Corbusier's proposals for commercial towers in his Contemporary City were linked to transportation hubs, they were generally located in verdant landscapes, set back from vehicular roads. Many of Hong Kong's housing estates are designed with towers that sit upon a low-rise podium that allows buildings to maintain a continuous edge along city streets.

This multilevel podium connects towers sometimes to each other and to the overall estate. Amenities such as transportation links, retail spaces, clubhouses, leisure spaces, recreational spaces, and landscaped communal spaces are housed in this podium zone. When combined with an extremely high population density in the residential towers above, a podium zone contributes to a more programmatically diverse urban environment.

> "This mix of functions provided some vibrancy at the building base and supported activities which inhabitants found useful and attractive ... The provision of the podium spread the development risk between housing and commercial property by permitting a substantial commercial component." (Shelton, Karakiewicz, and Kvan)[17]

The ten estates presented in the main section of this book make use of combination of freestanding towers and tower blocks with a podium base.

Despite the extreme density of Hong Kong's public and privately developed housing estates, the model of high-rise towers for dwelling has arguably been more culturally accepted in Hong Kong when compared with other models throughout the world. In contrast to the use of high-rise buildings within large estates to house low-income or marginalized populations in many Western states, high-rise dwelling in Hong Kong—and many cities in East Asia—is much more ubiquitous. Towers in Hong Kong make up much of the city's urban fabric and accommodate residents of every socio-economic group. Living in towers in Hong Kong, even within public housing estates, does not carry the stigma of high-rise public housing in the West. The acceptance, or even preference, for high-density urbanism in Hong Kong may be attributed to a more typical collective experience. Communities of towers are often clustered

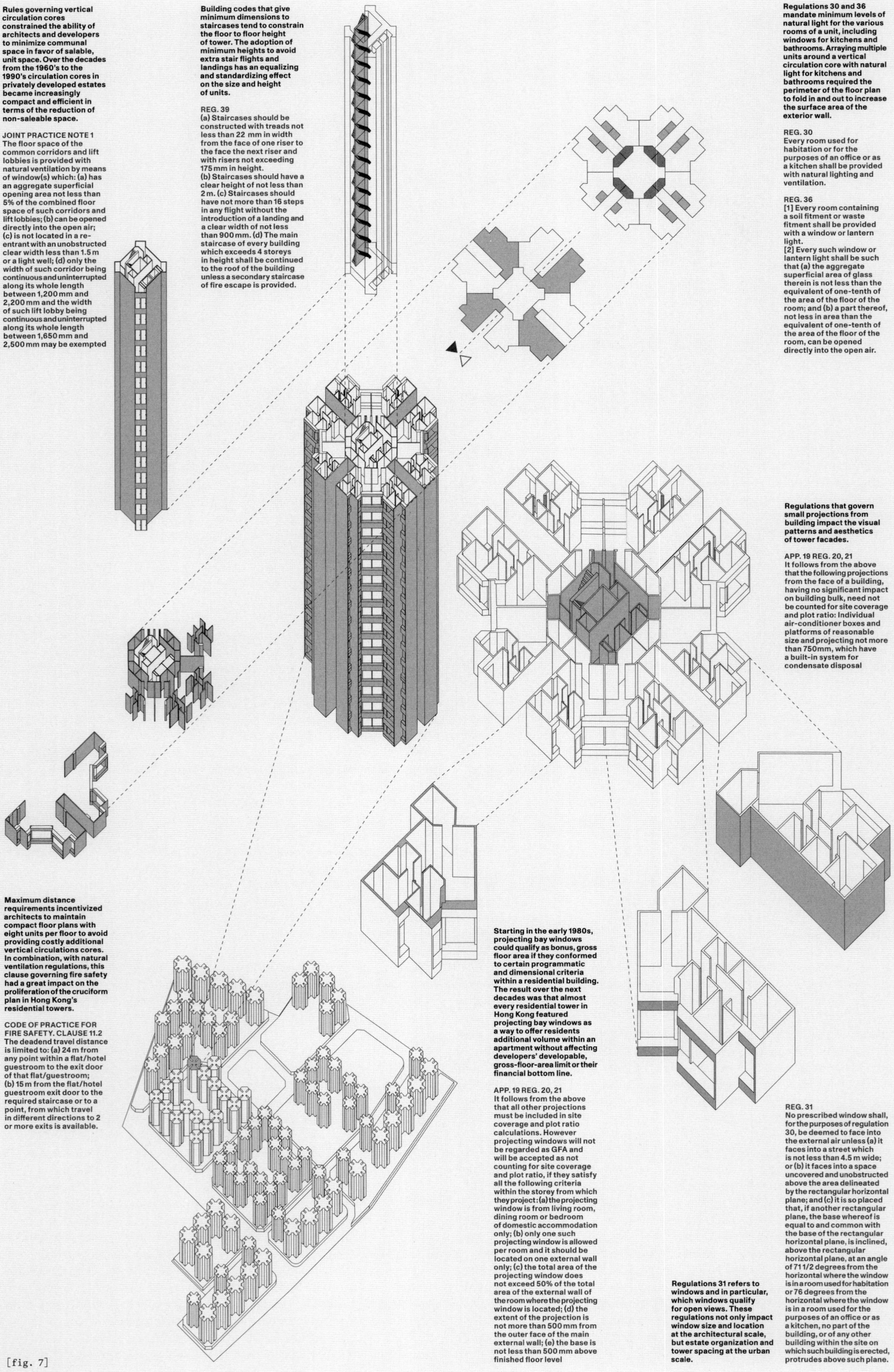

Rules governing vertical circulation cores constrained the ability of architects and developers to minimize communal space in favor of salable, unit space. Over the decades from the 1960's to the 1990's circulation cores in privately developed estates became increasingly compact and efficient in terms of the reduction of non-saleable space.

JOINT PRACTICE NOTE 1
The floor space of the common corridors and lift lobbies is provided with natural ventilation by means of window(s) which: (a) has an aggregate superficial opening area not less than 5% of the combined floor space of such corridors and lift lobbies; (b) can be opened directly into the open air; (c) is not located in a re-entrant with an unobstructed clear width less than 1.5 m or a light well; (d) only the width of such corridor being continuous and uninterrupted along its whole length between 1,200 mm and 2,200 mm and the width of such lift lobby being continuous and uninterrupted along its whole length between 1,650 mm and 2,500 mm may be exempted

Building codes that give minimum dimensions to staircases tend to constrain the floor to floor height of tower. The adoption of minimum heights to avoid extra stair flights and landings has an equalizing and standardizing effect on the size and height of units.

REG. 39
(a) Staircases should be constructed with treads not less than 22 mm in width from the face of one riser to the face the next riser and with risers not exceeding 175 mm in height.
(b) Staircases should have a clear height of not less than 2 m. (c) Staircases should have not more than 16 steps in any flight without the introduction of a landing and a clear width of not less than 900 mm. (d) The main staircase of every building which exceeds 4 storeys in height shall be continued to the roof of the building unless a secondary staircase of fire escape is provided.

Regulations 30 and 36 mandate minimum levels of natural light for the various rooms of a unit, including windows for kitchens and bathrooms. Arraying multiple units around a vertical circulation core with natural light for kitchens and bathrooms required the perimeter of the floor plan to fold in and out to increase the surface area of the exterior wall.

REG. 30
Every room used for habitation or for the purposes of an office or as a kitchen shall be provided with natural lighting and ventilation.

REG. 36
[1] Every room containing a soil fitment or waste fitment shall be provided with a window or lantern light.
[2] Every such window or lantern light shall be such that (a) the aggregate superficial area of glass therein is not less than the equivalent of one-tenth of the area of the floor of the room; and (b) a part thereof, not less in area than the equivalent of one-tenth of the area of the floor of the room, can be opened directly into the open air.

Regulations that govern small projections from building impact the visual patterns and aesthetics of tower facades.

APP. 19 REG. 20, 21
It follows from the above that the following projections from the face of a building, having no significant impact on building bulk, need not be counted for site coverage and plot ratio: Individual air-conditioner boxes and platforms of reasonable size and projecting not more than 750mm, which have a built-in system for condensate disposal

Maximum distance requirements incentivized architects to maintain compact floor plans with eight units per floor to avoid providing costly additional vertical circulations cores. In combination, with natural ventilation regulations, this clause governing fire safety had a great impact on the proliferation of the cruciform plan in Hong Kong's residential towers.

CODE OF PRACTICE FOR FIRE SAFETY. CLAUSE 11.2
The deadend travel distance is limited to: (a) 24 m from any point within a flat/hotel guestroom to the exit door of that flat/guestroom; (b) 15 m from the flat/hotel guestroom exit door to the required staircase or to a point, from which travel in different directions to 2 or more exits is available.

Starting in the early 1980s, projecting bay windows could qualify as bonus, gross floor area if they conformed to certain programmatic and dimensional criteria within a residential building. The result over the next decades was that almost every residential tower in Hong Kong featured projecting bay windows as a way to offer residents additional volume within an apartment without affecting developers' developable, gross-floor-area limit or their financial bottom line.

APP. 19 REG. 20, 21
It follows from the above that all other projections must be included in site coverage and plot ratio calculations. However projecting windows will not be regarded as GFA and will be accepted as not counting for site coverage and plot ratio, if they satisfy all the following criteria within the storey from which they project: (a) the projecting window is from living room, dining room or bedroom of domestic accommodation only; (b) only one such projecting window is allowed per room and it should be located on one external wall only; (c) the total area of the projecting window does not exceed 50% of the total area of the external wall of the room where the projecting window is located; (d) the extent of the projection is not more than 500 mm from the outer face of the main external wall; (e) the base is not less than 500 mm above finished floor level

Regulations 31 refers to windows and in particular, which windows qualify for open views. These regulations not only impact window size and location at the architectural scale, but estate organization and tower spacing at the urban scale.

REG. 31
No prescribed window shall, for the purposes of regulation 30, be deemed to face into the external air unless (a) it faces into a street which is not less than 4.5 m wide; or (b) it faces into a space uncovered and unobstructed above the area delineated by the rectangular horizontal plane; and (c) it is so placed that, if another rectangular plane, the base whereof is equal to and common with the base of the rectangular horizontal plane, is inclined, above the rectangular horizontal plane, at an angle of 71 1/2 degrees from the horizontal where the window is in a room used for habitation or 76 degrees from the horizontal where the window is in a room used for the purposes of an office or as a kitchen, no part of the building, or of any other building within the site on which such building is erected, protrudes above such plane.

[fig. 7]

around shopping centers, schools, parks, and public transportation. Urban compaction helps to improve standards of living for residents by conveniently connecting high-rise communities to urban amenities.[18]

Building Form and Building Code

As Hong Kong's public housing is not-profit-driven and is regulated by the Housing Authority and not by the Building Ordinance, many different attitudes toward site specificity, communal spaces, and variation in building type have developed in relation to privately developed buildings. Due to the very high value of land in Hong Kong, the organization and built form of most housing estates follows the planning code and building regulations very closely so that real estate developers can maximize the allowable, salable area in each estate. Since 1962, built area in Hong Kong is limited by plot ratio and gross floor area (GFA); therefore, building components and spaces that do not contribute to a sales profit are typically minimized.

> "From the perspective of developers, the gross floor area of a typical floor consists of two kinds of areas: the saleable area of the residential units and the non-saleable common area. It is because the saleable area and the common area both fall under the broadly defined measure of gross floor area that we see such constrained, real estate market-led architectural responses. The saleable area, the meat of the development, is always maximized, and the common area is always minimized. The effect of this zero-sum game on architecture is a lack of diversity and an inability to implement improved sustainability in residential design (Wong Wah Sang)".[19]

Areas of a typical tower block such as corridors, elevator lobbies, mechanical rooms, fire staircases, and community spaces are typically no bigger, wider, or larger than the minimum standards set out in the regulatory code. Distances between tower blocks are also regulated by minimum clearances allowing towers to be as close as 2.5 meters away from each other.

Building codes have had a significant effect on the typical shape and organization of floor plans. Regulation 30 of the Hong Kong Building Ordinance maintains that "every room used for habitation or for the purposes of an office or as a kitchen shall be provided with natural light and ventilation."[20] Regulation 36, from the same year, states that "every room containing a soil fitment or waste fitment shall be provided with a window or lantern light."[21] Arraying multiple units around a vertical circulation core with natural light for kitchens and bathrooms required the perimeter of the floor plan to fold in and out to increase the surface area of the exterior wall. Regulations that require a short distance between the windows of two adjacent units allowed bathrooms and kitchens to be tucked deep into vertical crevices or "re-entrants," closer to the building core, while bedrooms and living spaces were given privileged views.

Fire safety measures regulated by codes put limitations on the distance of habitable space from fire exits.

> "The dead-end travel distance is limited to: (i) 24 m from any point within the flat/hotel guestroom to the exit door of that flat/guestroom; (ii) 15 m from the flat/hotel guestroom exit door to the required staircase of a point, from which travel in different directions to 2 or more exits is available."[22]

In effect, these maximum distance requirements incentivized architects to maintain compact floor plans with eight units per floor to avoid providing costly additional vertical circulations cores. In combination, these fire safety and natural ventilation regulations had a great impact on the proliferation of the use of the cruciform plan in Hong Kong's residential towers.

Figure 7 is an exploded planometric drawing that shows the components of a typical Hong Kong residential housing estate at various scales. The drawing shows the organization of structural, circulation, and façade systems as well as the relationships of the towers to the individual apartment units. Excerpts from the Hong Kong Building Regulations are written next to relevant building components to illustrate the close connection between building code and building form. The strict relationship of building code to building form is apparent in that virtually all of the 529 towers in Hong Kong's ten largest private housing estates [fig. 1]share a nearly identical floor plan type despite being built by a number of different developers and architecture firms.

[fig. 7] Building codes have had a significant effect on the architecture of Hong Kong's privately developed estates. Areas of a typical tower block such as corridors, elevator lobbies, mechanical rooms, fire staircases, and community spaces are typically no bigger, wider, or larger than the minimum standards set out in the regulatory code. Distances between tower blocks are also regulated by minimum clearances allowing towers to be as close as 2.5 meters away from each other.

This exploded planometric drawing shows the components of a typical Hong Kong residential housing estate at various scales. The drawing presents Whampoa Garden and Block 13 as a typical example of an estate and tower to illustrate the organization of structural, circulation, and façade systems as well as the relationships of the towers to the individual apartment units. Excerpts from the Hong Kong Building Regulations are written next to relevant building components to demonstrate the close connection between building code and building form.

Beyond the evolution of Hong Kong's private housing into the predominant, cruciform building type in the latter half of the twentieth century, many building components were completely revolutionized to make the planning of a residential tower more efficient. The scissor stair is one example of a reinvented means of egress that was transformed for Hong Kong's buildings. This architectural device first appeared in Hong Kong housing in the 1960s and allows two staircases to be wound around each other in a double helix within the same vertical core of space.[23] This invention and subsequent approval and adoption into the regulatory code by the Hong Kong Department of Buildings meant that architects and developers could dramatically transform the efficiency and profitability of a floor plan in a tower by cutting the floor area required for two means of egress in half.

Other external, or projecting, building components such as bay windows have also been incentivized or discouraged over the years by building regulations. Starting in the early 1980s, projecting bay windows could qualify as bonus gross floor area if they conformed to certain programmatic and dimensional criteria within a residential building.[24] The result over the next decades was that almost every residential tower in Hong Kong featured projecting bay windows as a way to offer residents additional volume within an apartment without affecting developers' developable gross-floor-area limit or their financial bottom line. Through its unyielding repetition, the boxy protrusion of a bay window has a tremendous visual impact on the aesthetic fabric of an entire community when it is copied and pasted across an entire neighborhood of forty story towers.

Balconies are another architectural feature that have been influenced by building codes over the course of the last half century. The Mei Foo housing estate in Kowloon was the first large scale housing estate to be built in Hong Kong. During the time it was built in the 1960s, the size of the apartments it offered were relatively generous, and each apartment included an exterior balcony. Over subsequent decades as land values continued to rise, apartment sizes in large, privately developed housing estates continued to shrink, and balconies disappeared. Only in recent years have small balconies started to appear again on Hong Kong's residential towers because contemporary building regulations have incentivized small balconies as bonus space that does not take valuable GFA away from a floor plan.

Looking Forward

While the public housing produced in Hong Kong between 1950 and 2000 has introduced numerous models for mass housing that have evolved over time, privately developed estates from that era relied almost entirely upon a singular building type and similar planning strategies. Despite the Hong Kong government's support and subsidization of housing in the private sector, due to external financial pressures and priorities, the quality and quantity of space in a typical apartment flat in one of Hong Kong's privately developed estates has been significantly reduced through increased standardization. The profit-driven nature of real estate development in a city with a tight profit margin creates an architectural incentive to maximize leasable or salable floor area while minimizing costs. The analysis of estates in this book shows that Hong Kong's restrictive building code and high land value had all but erased distinctions between tower blocks within and across large-scale housing estates in the 1980s and 90s. Building policies and practices led to standardized, code-compliant floor plans which are mirrored, copied, extruded, and arrayed across sites to form thoroughly monotonous, overly repetitive urban environments.

The analysis attempts to shed light on how standardization, as a strategy in architecture and real estate development, has changed over recent decades. There is a marked difference between the estates developed in the 1960s and 70s, such as Mei Foo Sun Chuen, which provided communities with a collection of over two hundred and fifty unit types to select from, versus more recent estates from the 1990s like Laguna City with only four types for over 8,000 units. The study shows a direct relationship of higher instances of tower repetition in Hong Kong's large-scale, privately developed housing estates with increasing land prices in the city.

As authors, we are concerned with the consequences of an increasingly standardized world not just in terms of the architectural and aesthetic monotony of the built environment but also of the socio-economic impacts of conformity. The unyielding standardization of domestic space through repetitive floor plans limits the heterogeneity of communities and flexibility of social structures. Large-scale, mass housing projects with a broader selection of unit types, layouts, and sizes enable more diverse communities with less homogenous family structures to inhabit the same space. A wider spectrum of housing types provides flexibility and stability as personal or familial economic circumstances change. A community of varied housing types provides residents within a housing block options to adjust to a changing family structure over time, without having to leave the neighborhood in search of other alternatives. Overly standardized housing types have the potential to limit the quality of life, as residents must conform and adapt to spatial norms. As Hong Kong is often looked to as a model for high-density planning in rapidly developing mainland China and beyond, the wide-scale appropriation of architectural and urban standardization impacts the quality and habitability of cities for huge residential populations for decades to come.

While this book provides a study of some of the largest, most repetitive urban environments, the ten projects presented here represent a historical era of extreme standardization that occurred between 1953 and 1999. In recent years, regulators, architects, and developers in Hong Kong have shifted to new policies and new models of housing that have provided more domestic heterogeneity and variation within large scale developments. In the early 2000s, the Hong Kong Housing Authority shifted to site-specific housing strategies with a focus on providing more unit types with better views, solar orientation, natural ventilation, and improved public spaces on the ground level. Pre-fabrication and modular units were developed to keep costs low while providing a broader range of unit types. Engagement with residents through community planning workshops opened avenues for public housing estate residents to be heard and issues addressed. Overall, a focus on overall quality and a "socio-environmental" approach to housing design started to temper the "political-economic" demands for sheer numbers of housing units of the late twentieth century.[25]

Changes in the building code have also had an impact on the forms of private housing since the late 1990s. In the early 2000s, governing bodies in Hong Kong began to adopt building regulations to address growing concerns over environmental issues such as urban heat islands, natural ventilation at the scale of a neighborhood, and what is known as the "wall effect." The Hong Kong Buildings Department instituted a list of "Joint Practice Notes" that specified wider spaces in between buildings to encourage air circulation at the street level. Additional open spaces for building residents were required, encouraging designers to provide sky gardens within buildings, thus breaking up repetitive floor stacking and continuous building extrusions. New "Sustainable Building Design Guidelines," introduced in 2011, encouraged even more green spaces, leading to an increase in building separations and setbacks to further address urban heat island effects. New fire regulations adopted in 2011–12 allowed for open kitchens and windowless bathrooms in residential buildings, thus reducing the need for "re-entrants," or long perimeter walls with inward folds to provide natural ventilation for rooms closer to the central core.

While land prices in Hong Kong remain extremely high, buildings designed by the private housing sector are still, and will undoubtedly continue to be, directly influenced by building codes and regulations. However, new regulations in the last two decades have allowed building forms to more readily deviate from the ubiquitous, cruciform tower floor plans of the previous years. New building codes that no longer require bathrooms to be naturally ventilated provide much greater flexibility in layout planning. These codes combined with new regulations for setback configurations, greening ratios, and urban wind permeability have dramatically changed design approaches for large-scale developments.[26] Advances in design and construction technology as well as financial incentives for pre-fabrication of building components have made towers with a more heterogeneous range of units types more cost effective. Similar to recent qualitative changes in public housing design in Hong Kong, private developers and their architects must now take a more site-specific approach to housing design.

As researchers and educators, we hope that this work allows architects, planners, policy makers, developers, and the general public to better understand the benefits and drawbacks of serial planning and standardization of the built environment. While there are many positive aspects of the private estates in Hong Kong from the late twentieth century, including remarkable urban density and economic viability, building codes that allow for and encourage variation and heterogeneity for mass housing developments must continue to be developed and implemented if cities of the future are to provide humane and diverse modes of housing to sustain socially viable, heterogeneous communities.

[1] "The Historical Background of the Shek Kip Mei Estate," https://www.housingauthority.gov.hk/hdw/en/aboutus/events/community/heritage/about.html.

[2] "Exhibition Tells the Development of Public Housing," Press Release, June 1, 2004, Memories of Home—50 Years of Public Housing in Hong Kong, Hong Kong Heritage Museum, https://www.info.gov.hk/gia/general/200406/01/0601115.htm.

[3] https://hk.heritage.museum/documents/doc/en/downloads/materials/Public_Housing-E.pdf

[4] Otto Golger, "Resettling the Squatters", DEVELOPMENT DIGEST, Vol. 13, Issue 4 (October 1975): 84.

[5] Charlie Q. L. Xue, HONG KONG ARCHITECTURE 1945-2015: FROM COLONIAL TO GLOBAL (Singapore: Springer, 2016), 23.

[6] "Mark I & Mark II Blocks," Hong Kong Memory, https://www.hkmemory.hk/MHK/collections/public_housing/resettlement_estates/mark_I_and_mark_II_blocks/index.html.

[7] Paavo Monkkonen and Xiaohu Zhang, "Creating Mixed-Income Neighborhoods Unintentionally: Public Housing Revisualization and Socio-Economic Segregation in Hong Kong," in TRENDS AND ISSUES IN HOUSING IN ASIA: COMING OF AN AGE, eds. Urmi Sengupta and Annapurna Shaw (London: Routledge, 2018), 134.

[8] Hong Kong Housing Authority, Public Housing Development Timeline, January 2018, http://www.housingauthority.gov.hk/linear/en/about-us/public-housing-heritage/public-housing-development.

[9] Hong Kong Housing Authority Long Term Housing Strategy, A Policy Statement, April 1987, http://www.cityu.edu.hk/hkhousing/pdoc/LongTermHousingStrategyAPolicyStatement.htm.

[10] Hong Kong Housing Authority, Public Housing Development Timeline, January 2018, http://www.housingauthority.gov.hk/linear/en/about-us/public-housing-heritage/public-housing-development.

[11] Georg Kohlmaier and Barna von Sartory, HOUSES OF GLASS: A NINETEENTH-CENTURY BUILDING TYPE (Cambridge, MA: MIT Press, 1986), 4.

[12] Mary McLeod, "Architecture or Revolution: Taylorism, Technocracy, and Social Change," ART JOURNAL 43, no. 2 (Summer 1983): 135.

[13] Walter Gropius, THE NEW ARCHITECTURE AND THE BAUHAUS (Cambridge, MA: MIT Press, 1965), 33-34.

[14] Eric J. Jenkins, TO SCALE: ONE HUNDRED URBAN PLANS (Oxon: Routledge, 2008), 140.

[15] Baleshwar Thakur, George Pomeroy, Chris Cusack, and Sudhir K. Thakur, eds., CITY, SOCIETY AND PLANNING: ESSAYS IN HONOUR OF PROFESSOR A.K. DUTT, VOLUME ONE: CITY (New Delhi: Concept Publishing Company, 2007), 49.

[16] Le Corbusier, THE CITY OF TO-MORROW AND ITS PLANNING (Mineola: Dover, 1987), 172.

[17] Barrie Shelton, Justyna Karakiewicz, and Thomas Kvan, THE MAKING OF HONG KONG: FROM VERTICAL TO VOLUMETRIC (New York: Routledge, 2011), 118-19.

[18] Xing Quan Zhang, "High Rise and High-Density Compact Urban Form: The Development of Hong Kong," in COMPACT CITIES: SUSTAINABLE URBAN FORMS FOR DEVELOPING COUNTRIES, eds. Rod Burgess and Mike Jenks (London: E & FN Spon Press, 2000), 251.

[19] Wong Wah Sang, "Design Constraints on Residential Space, and a Possible Breakthrough" in TOWERS OF CHOICES: HONG KONG HOUSING BEYOND UNIFORMITY, eds., Winy Maas and Tihamér Hazarja et al. (Rotterdam: Marcel Witvoet and nai010 publishers, 2019), 120–21.

[20] https://www.elegislation.gov.hk/hk/cap123F!en@2015-02-06T00:00:00.

[21] Ibid.

[22] Hong Kong Code of Practice for Fire Safety, Clause 11.2.

[23] Shelton, Karakiewicz, and Kvan, THE MAKING OF HONG KONG, 137.

[24] Hong Kong Buildings Department, Practice Note for Authorized Persons, Registered Structural Engineers and Registered Geotechnical Engineers APP-19, Projections in relation to Site Coverage and Plot Ratio Building (Planning) Regulations 20 & 21, http://www.bd.gov.hk/english/documents/pnap/APP/APP019.pdf.

[25] John Ng, "Public Housing in Hong Kong in the Last 15 Years," in HOUSING ASIA 2020: PERSPECTIVES AND RESPONSES, eds. Christian J. Lange, Christiane Lange, and Eunice Seng (Hong Kong: HKU Architecture Press, 2017), 16.

[26] Carolin Fong, "Private Residential Housing in Hong Kong: A Brief History," in HOUSING ASIA 2020: PERSPECTIVES AND RESPONSES, eds. Christian J. Lange, Christiane Lange, and Eunice Seng (Hong Kong: HKU Architecture Press, 2017), 20.

100 Estates Mapping

The detailed maps in this section highlight the individual towers of each of the 100 largest private estates. They illustrate their organization within their respective urban or sub-urban contexts and the relationship of the towers to each other. The statistical information below each map provides the year of construction, population size, number of towers in each estate, and number of floors as well as flats in each tower. It also shows an indication of height variation and density.

The hundred estates are organized by population size. The estates identified on the maps comprise 1,719 towers with 352,221 flats and have a combined population of nearly one million residents—quite a significant population for a city of approximately seven million.

The maps reveal different attitudes toward urban forms. Many of the estates surveyed make efficient use of land by packing towers together into a grid. Some of these gridded estates follow an adjacent gridded street pattern, thus, providing a more integrated urban transition from estate to the surrounding city. In other estates, towers are more tenuously connected with the surrounding city due to barriers created by high speed roadways and thoroughfares. At the neighborhood scale, an internal, gridded organization sometimes helps to create a series of urban-scaled courtyards between the towers. The towers of some estates are organized on a staggered grid. The cores of the towers are aligned but with gaps between adjacent towers allowing for more open views between towers. Other estates are organized with towers laid out in sweeping, curvilinear arcs. These linear organizations of towers are not as densely packed as gridded estates and do not produce courtyards or enclosed communal spaces. Instead, they connect tower blocks directly to curving roadways and provide units with more open views of the surrounding city and landscape. These patterns of tower organization are often associated with more hilly topographies or estates set within a more rural setting.

These three organizational types—grids, staggered grids, and sweeping arcs—are discernable in the following maps. Urban organization is often constricted by the size and shape of the building plot, the topography of the site, and the views available.

At the end of this section, a comprehensive map presents the relative locations of the hundred largest privately developed housing estates across Hong Kong. The map reveals that most of the largest private estates are built outside the city center. Some of the largest of these estates are built on vast parcels previously used for industrial functions, such as dry docks or fuel storage depots, or on reclaimed land. Mapping the estates traces the residential, economic, and industrial transformation of Hong Kong over the second half of the twentieth century from a colonial outpost to a global city.

1. KINGSWOOD VILLAS [Yuen Long,1999]
Population: 39,361 Flats: 15,927
Floors min: 27 Floors max: 38
People/Flat: 2.47 Towers: 58

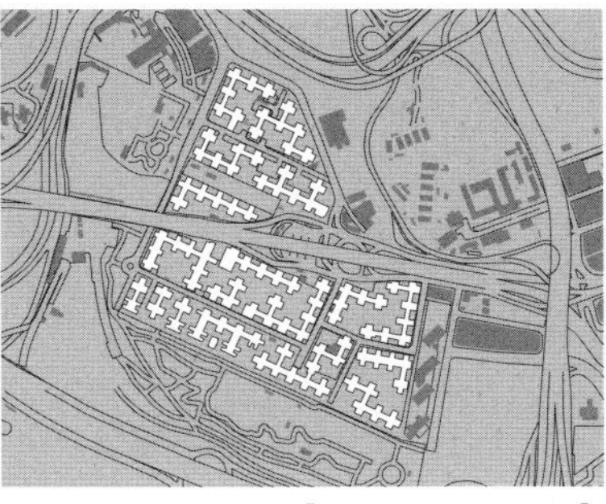

2. MEI FOO SUN CHUEN [Lai Chi Kok, 1978]
Population: 38,974 Flats: 13,149
Floors min: 17 Floors max: 21
People/Flat: 2.96 Towers: 99

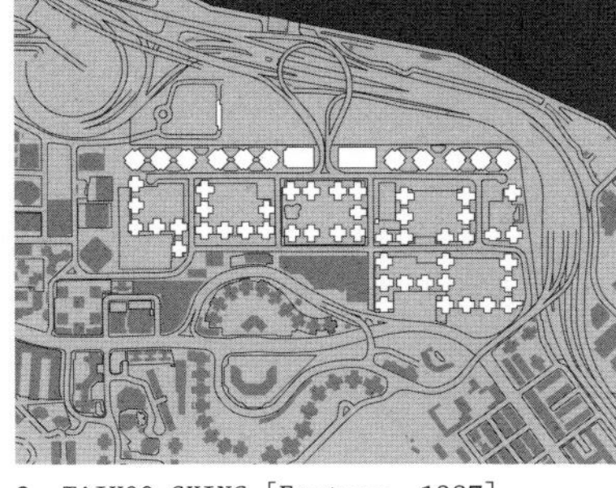

3. TAIKOO SHING [Eastern, 1987]
Population: 36,796 Flats: 12,696
Floors min: 24 Floors max: 30
People/Flat: 2.90 Towers: 61

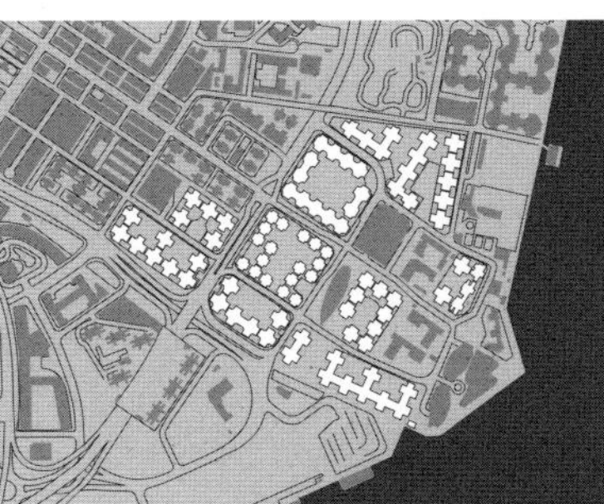

4. WHAMPOA GARDEN [Kowloon City, 1991]
Population: 31,613 Flats: 10,441
Floors min: 15 Floors max: 16
People/Flat: 3.03 Towers: 88

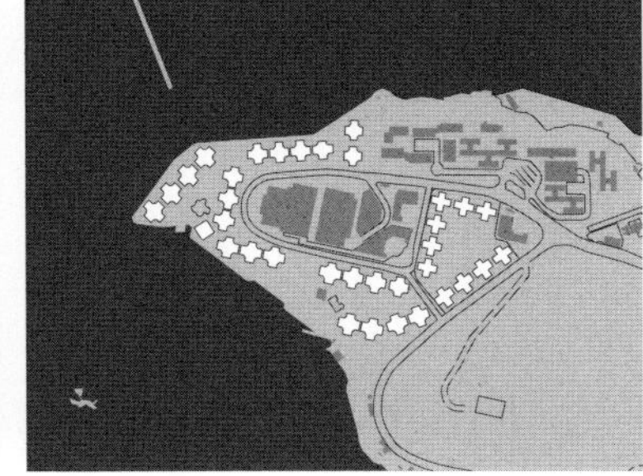

5. SOUTH HORIZONS [Southern, 1995]
Population: 31,496 Flats: 9,813
Floors min: 25 Floors max: 42
People/Flat: 3.21 Towers: 34

6. CITY ONE [Sha Tin, 1988]
Population: 24,758 Flats: 10,643
Floors min: 30 Floors max: 34
People/Flat: 2.33 Towers: 52

7. LAGUNA CITY [Kwun Tong, 1994]
Population: 23,354 Flats: 8,072
Floors min: 25 Floors max: 28
People/Flat: 2.89 Towers: 38

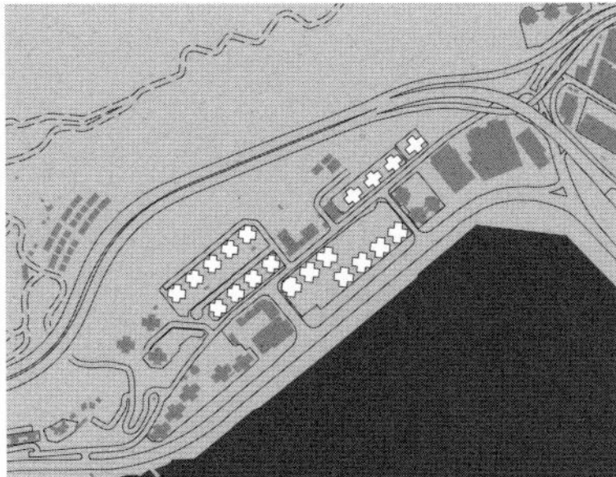

8. BELVEDERE GARDEN [Tsuen Wan, 1991]
Population: 20,423 Flats: 6,016
Floors min: 35 Floors max: 42
People/Flat: 3.39 Towers: 19

9. HENG FA CHUEN [Eastern, 1989]
Population: 18,921 Flats: 6,505
Floors min: 14 Floors max: 21
People/Flat: 2.91 Towers: 48

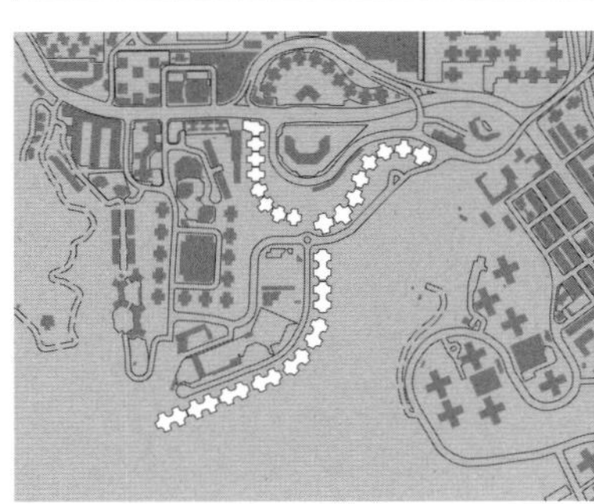

10. KORNHILL [Eastern, 1986]
Population: 18,450 Flats: 6,651
Floors min: 19 Floors max: 31
People/Flat: 2.77 Towers: 32

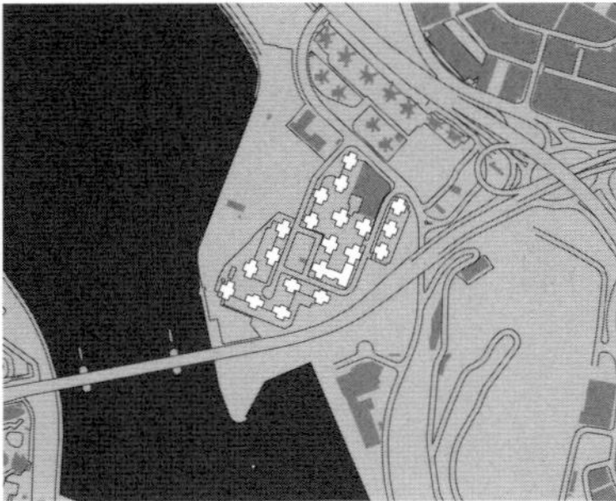

11. RIVIERA GARDENS [Tsuen Wan, 1989]
Population: 17,301 Flats: 5,692
Floors min: 30 Floors max: 41
People/Flat: 3.04 Towers: 20

12. OCEAN SHORES [Sai Kung, 2003]
Population: 15,007 Flats: 5,734
Floors min: 47 Floors max: 48
People/Flat: 2.62 Towers: 15

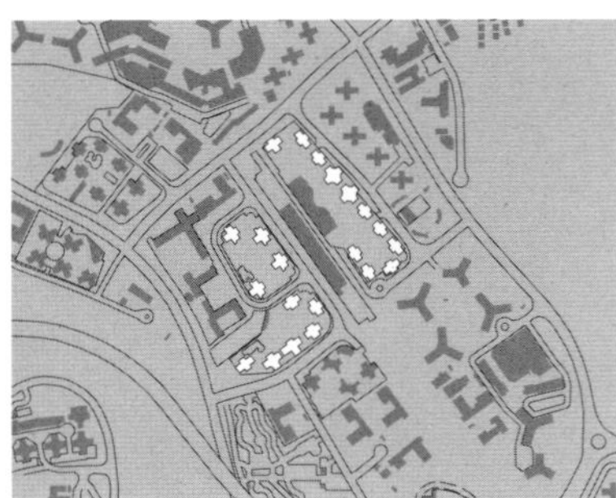

13. METRO CITY [Sai Kung, 2002]
Population: 14,234 Flats: 6,768
Floors min: 38 Floors max: 43
People/Flat: 2.10 Towers: 21

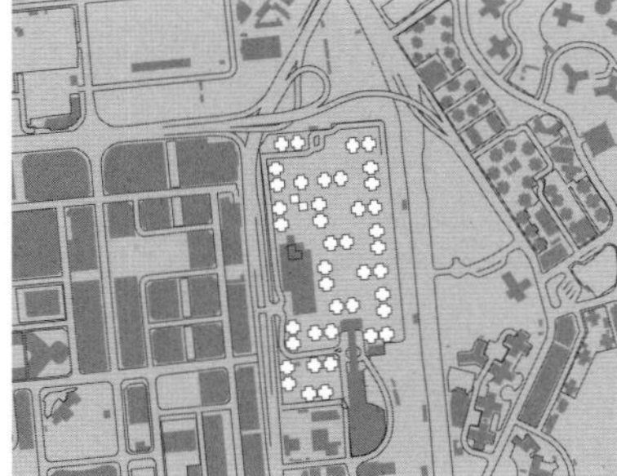

14. TELFORD GARDENS [Kwun Tong, 1982]
Population: 13,700 Flats: 4,991
Floors min: 11 Floors max: 26
People/Flat: 2.74 Towers: 21

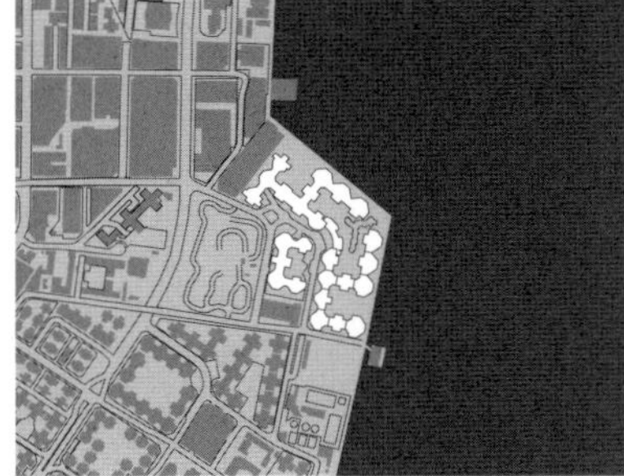

15. LAGUNA VERDE [Kowloon City, 2001]
Population: 12,569 Flats: 4,757
Floors min: 19 Floors max: 32
People/Flat: 2.64 Towers: 25

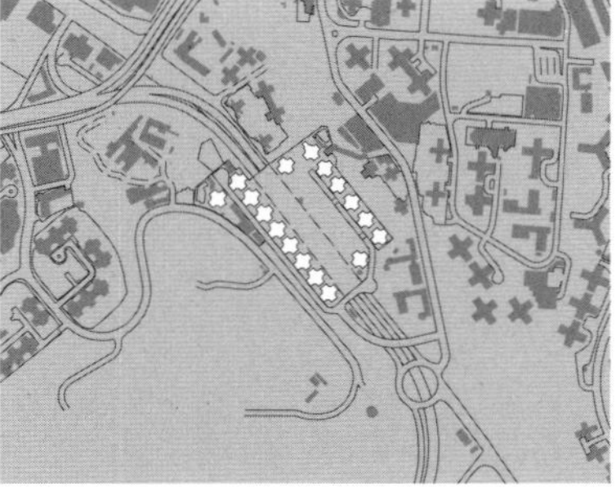

16. SCENEWAY GARDEN [Kwun Tong, 1992]
Population: 12,490 Flats: 4,114
Floors min: 28 Floors max: 34
People/Flat: 3.04 Towers: 17

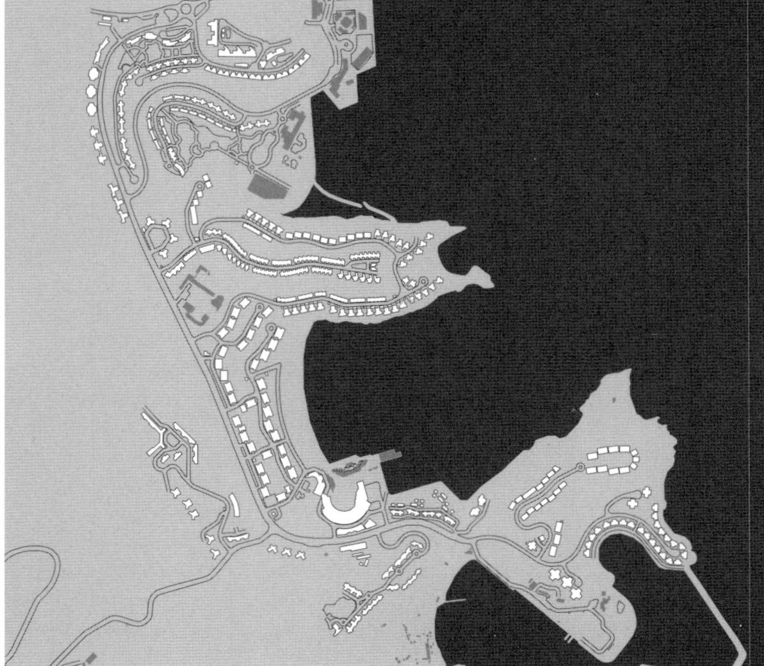
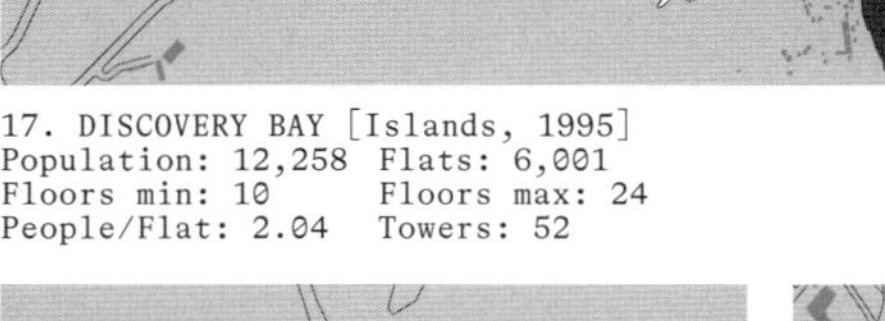

17. DISCOVERY BAY [Islands, 1995]
Population: 12,258 Flats: 6,001
Floors min: 10 Floors max: 24
People/Flat: 2.04 Towers: 52

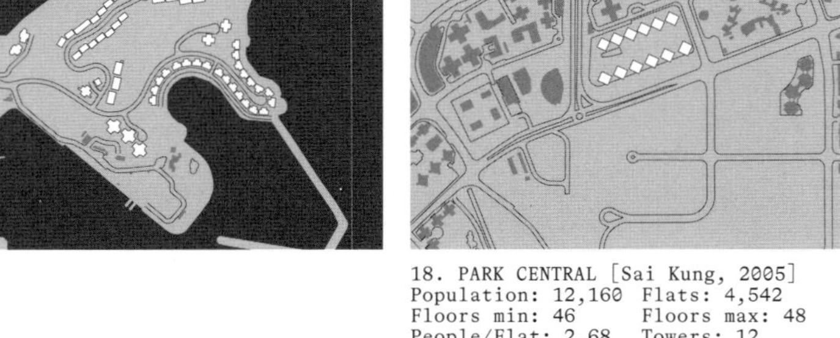

18. PARK CENTRAL [Sai Kung, 2005]
Population: 12,160 Flats: 4,542
Floors min: 46 Floors max: 48
People/Flat: 2.68 Towers: 12

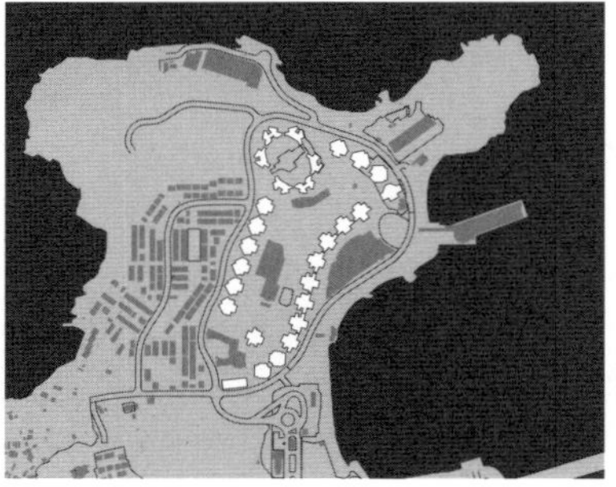

19. PARK ISLAND [Tsuen Wan, 2009]
Population: 11,529 Flats: 5,300
Floors min: 7 Floors max: 27
People/Flat: 2.18 Towers: 31

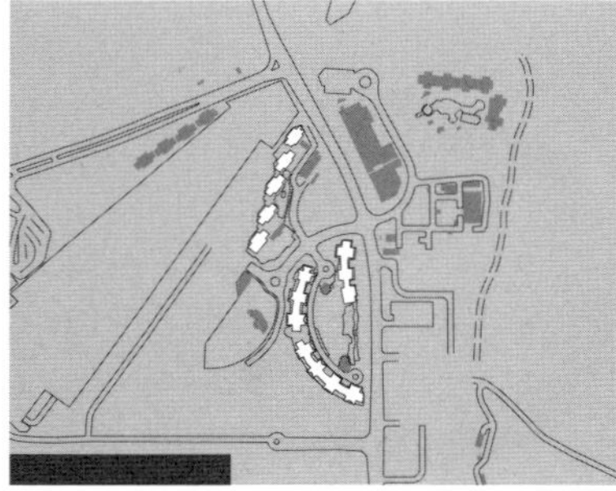

20. LOHAS PARK [Sai Kung, 2012]
Population: 11,501 Flats: 6,369
Floors min: 49 Floors max: 59
People/Flat: 1.81 Towers: 30

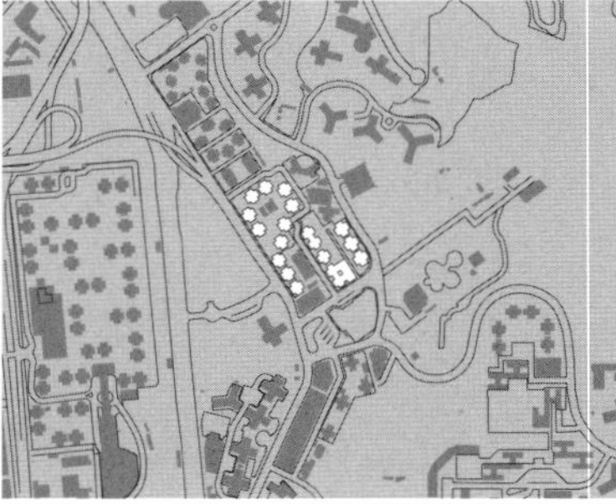

21. AMOY GARDENS [Kwun Tong, 1987]
Population: 11,373 Flats: 4,896
Floors min: 26 Floors max: 35
People/Flat: 2.32 Towers: 19

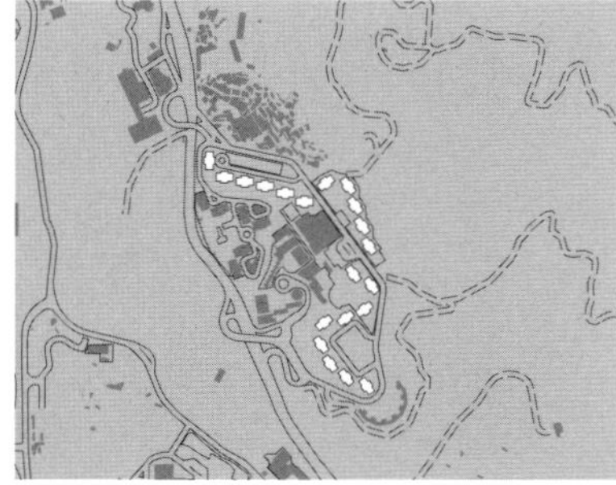

22. CHI FU FA YUEN [Southern, 1981]
Population: 11,203 Flats: 4,333
Floors min: 27 Floors max: 28
People/Flat: 2.59 Towers: 20

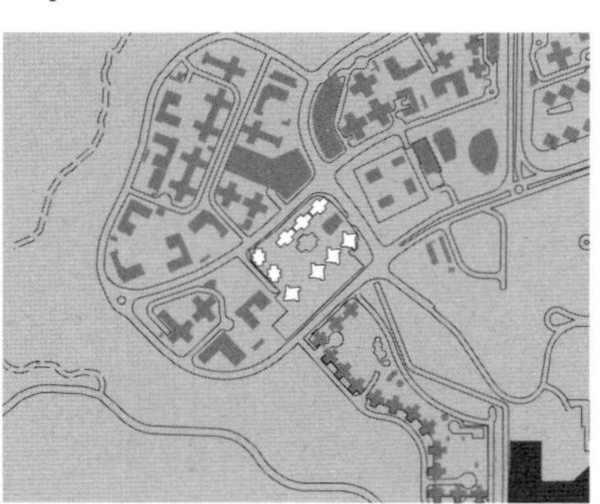

23. METRO TOWN/LE POINT [Sai Kung, 2007]
Population: 10,967 Flats: 3,778
Floors min: 50 Floors max: 53
People/Flat: 2.90 Towers: 9

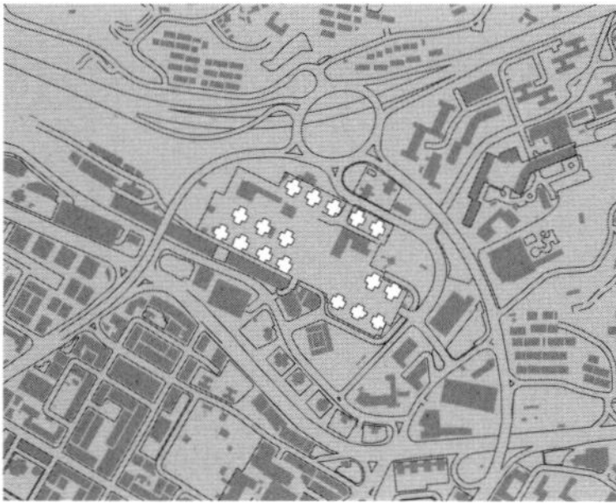

24. LUK YEUNG SUN CHUEN [Tsuen Wan, 1984]
Population: 10,953 Flats: 4,056
Floors min: 28 Floors max: 30
People/Flat: 2.70 Towers: 17

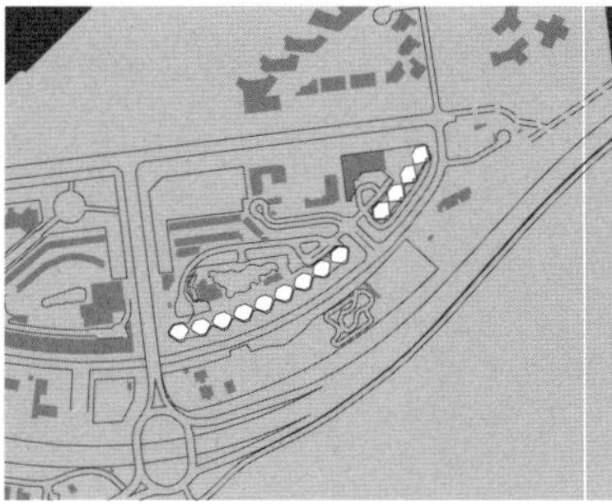

25. CARIBBEAN COAST [Islands, 2011]
Population: 10,621 Flats: 5,291
Floors min: 47 Floors max: 52
People/Flat: 2.01 Towers: 13

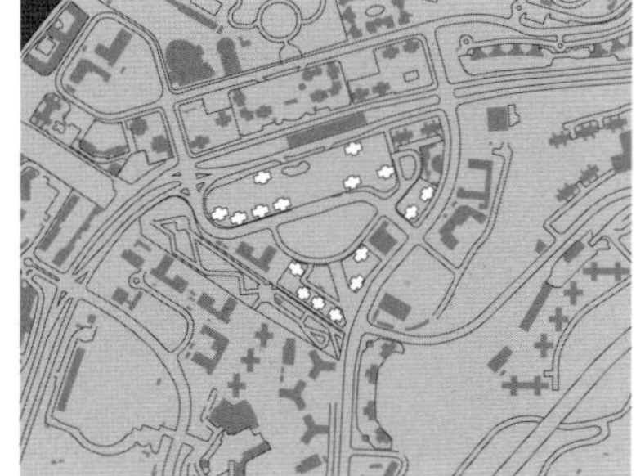

26. SUNSHINE CITY [Sha Tin, 1996]
Population: 10,543 Flats: 4,761
Floors min: 28 Floors max: 39
People/Flat: 2.21 Towers: 20

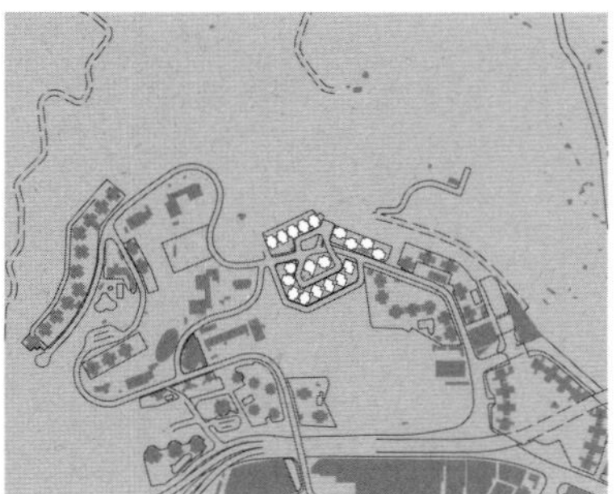

27. TSUEN WAN CENTRE [Tsuen Wan, 1982]
Population: 10,303 Flats: 4,516
Floors min: 26 Floors max: 35
People/Flat: 2.28 Towers: 19

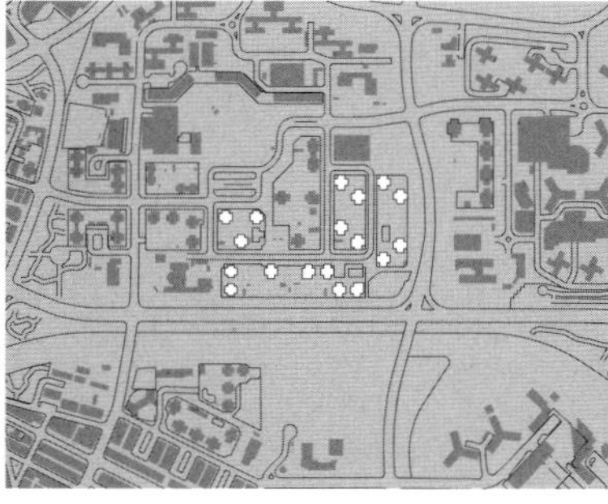

28. TAI PO CENTRE [Tai Po, 1987]
Population: 10,122 Flats: 4,080
Floors min: 20 Floors max: 30
People/Flat: 2.48 Towers: 22

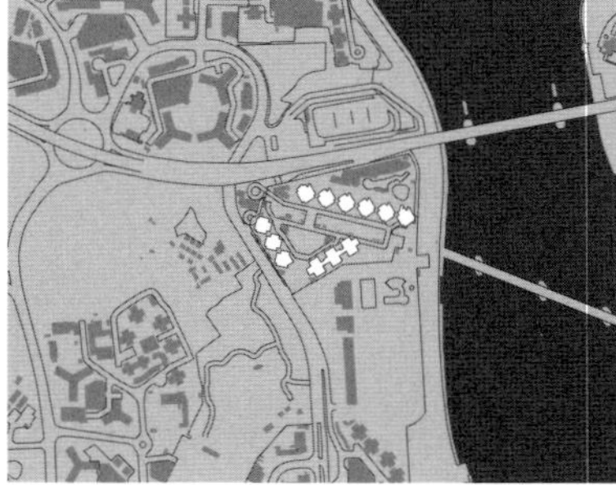

29. TIERRA VERDE [Kwai Tsing, 1999]
Population: 9,981 Flats: 3,460
Floors min: 34 Floors max: 40
People/Flat: 2.88 Towers: 12

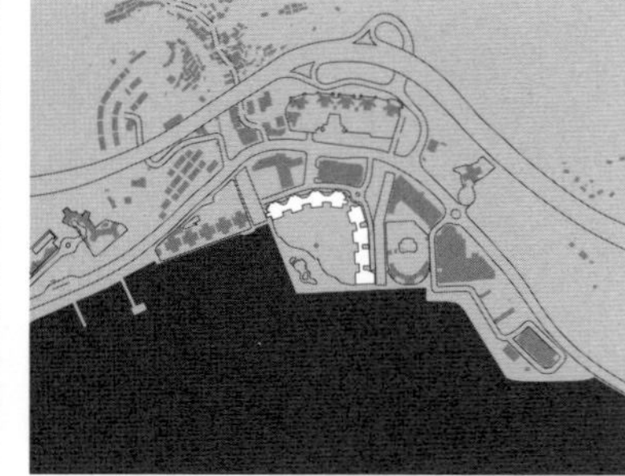

30. BELLAGIO [Tsuen Wan, 2006]
Population: 9,565 Flats: 3,346
Floors min: 54 Floors max: 54
People/Flat: 2.86 Towers: 8

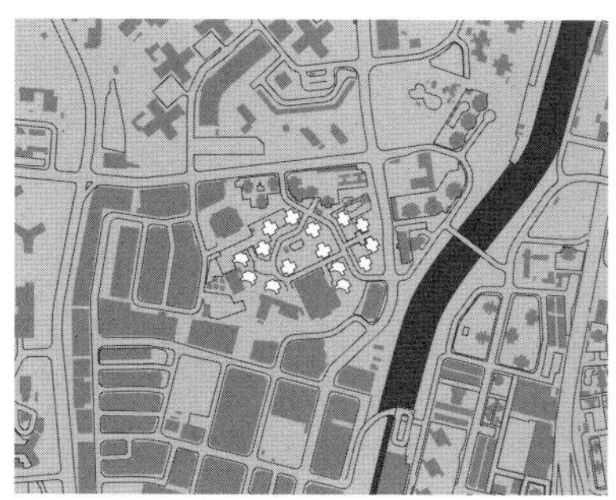

31. TAI HING GARDENS [Yuen Long, 1994]
Population: 9,513 Flats: 3,647
Floors min: 30 Floors max: 35
People/Flat: 2.61 Towers: 15

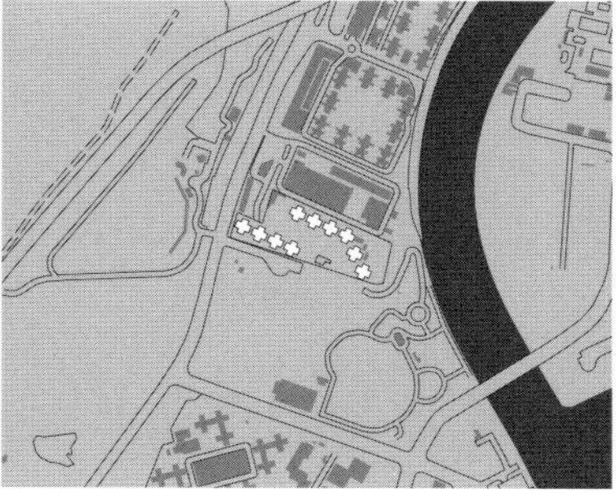

32. SUN TUEN MUN CENTRE [Tuen Mun, 1990]
Population: 9,486 Flats: 3,500
Floors min: 44 Floors max: 44
People/Flat: 2.71 Towers: 10

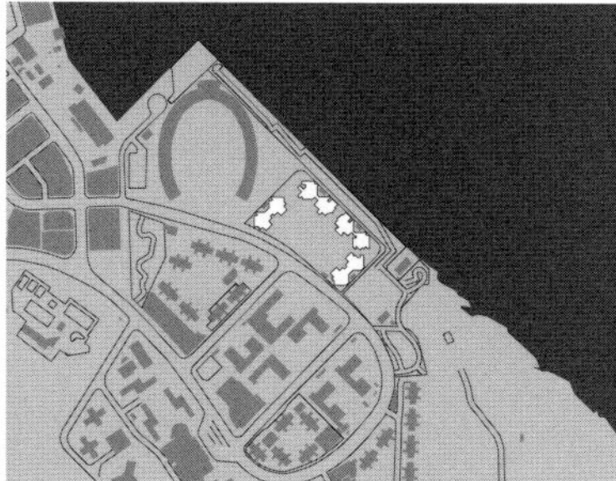

33. ISLAND RESORT [Eastern, 2001]
Population: 9,466 Flats: 3,111
Floors min: 50 Floors max: 51
People/Flat: 3.04 Towers: 8

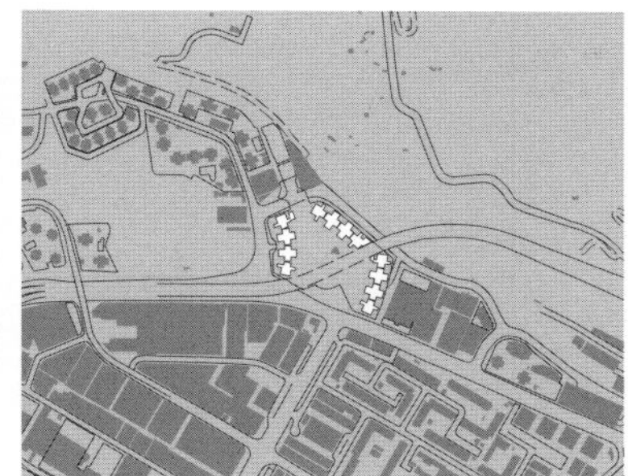

34. DISCOVERY PARK [Tsuen Wan, 1998]
Population: 9,457 Flats: 3,360
Floors min: 40 Floors max: 40
People/Flat: 2.81 Towers: 12

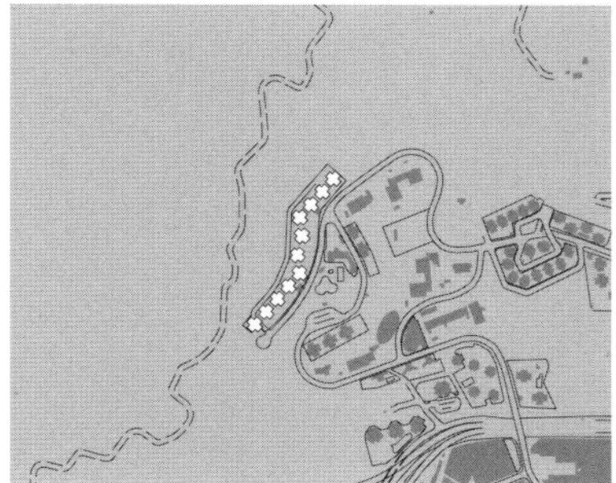

35. ALLWAY GARDENS [Tsuen Wan, 1981]
Population: 9,180 Flats: 3,424
Floors min: 22 Floors max: 29
People/Flat: 2.68 Towers: 16

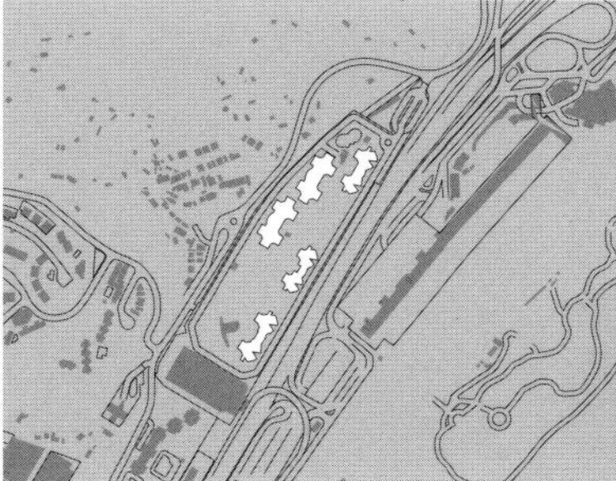

36. ROYAL ASCOT [Sha Tin, 1996]
Population: 8,999 Flats: 2,504
Floors min: 32 Floors max: 40
People/Flat: 3.59 Towers: 10

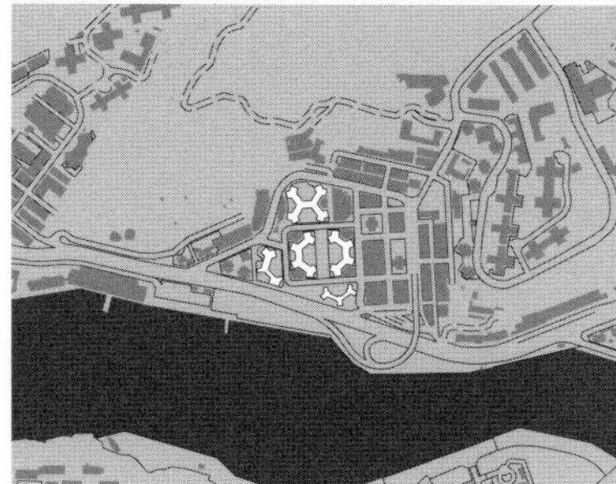

37. ABERDEEN CENTRE [Southern, 1982]
Population: 8,722 Flats: 2,804
Floors min: 26 Floors max: 27
People/Flat: 3.11 Towers: 20

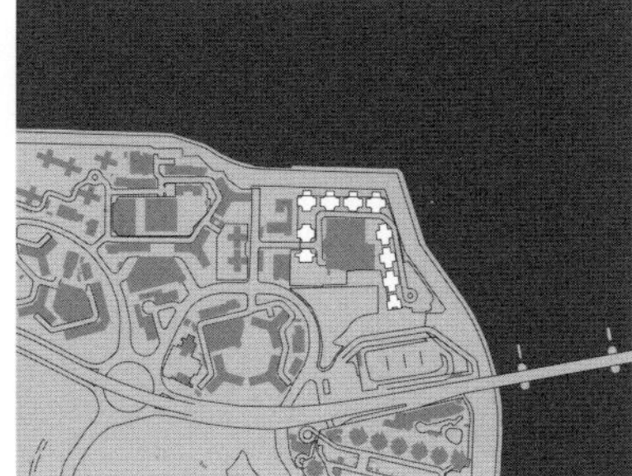

38. VILLA ESPLANADA [Kwai Tsing, 2000]
Population: 8,689 Flats: 2,824
Floors min: 35 Floors max: 40
People/Flat: 3.08 Towers: 10

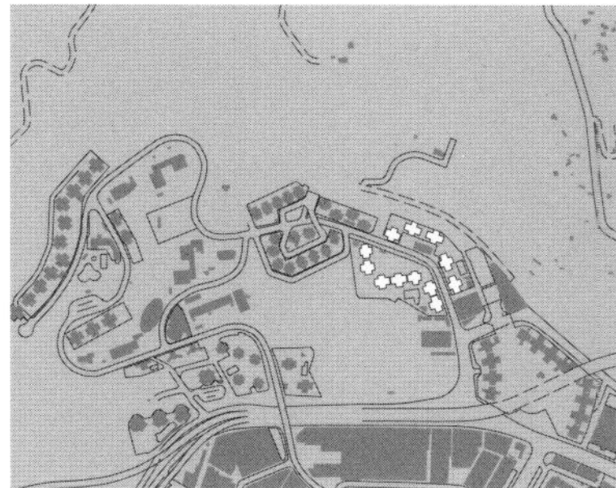

39. TSUEN KING GARDEN [Tsuen Wan, 1988]
Population: 8,600 Flats: 3,024
Floors min: 29 Floors max: 33
People/Flat: 2.84 Towers: 12

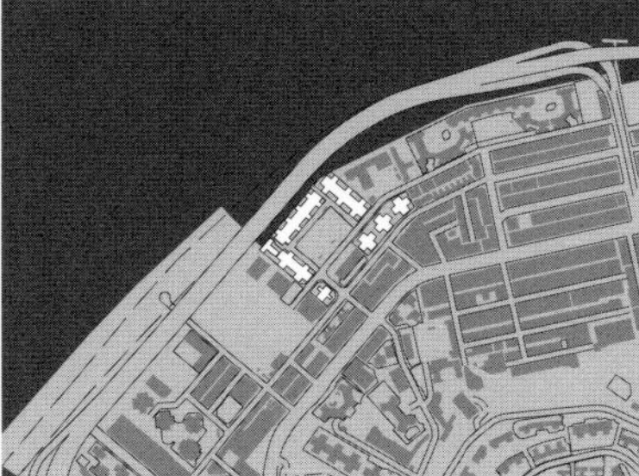

40. CITY GARDEN [Eastern, 1986]
Population: 8,158 Flats: 2,406
Floors min: 27 Floors max: 27
People/Flat: 3.39 Towers: 14

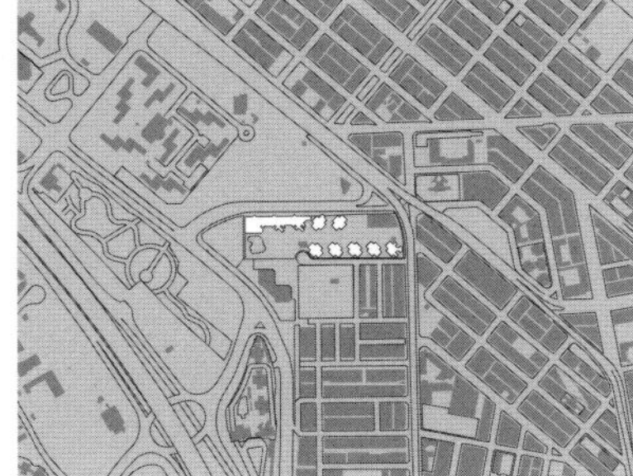

41. METRO HARBOUR VIEW [Yau Tsim Mong, 2003]
Population: 8,145 Flats: 3,520
Floors min: 44 Floors max: 44
People/Flat: 2.31 Towers: 10

42. TSEUNG KWAN O PLAZA [Sai Kung, 2003]
Population: 8,079 Flats: 2,882
Floors min: 45 Floors max: 45
People/Flat: 2.80 Towers: 8

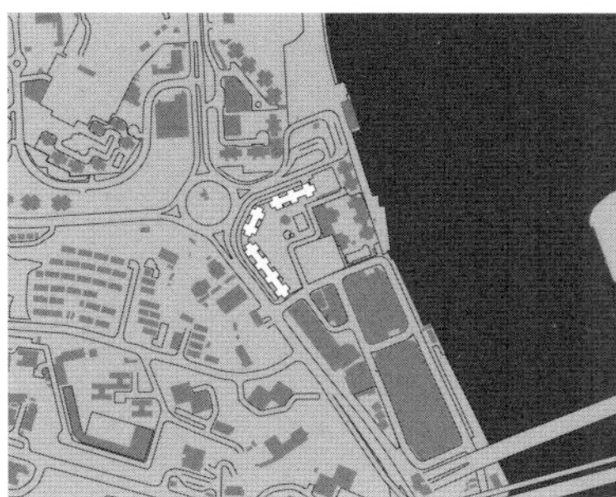

43. GREENFIELD GARDEN [Kwai Tsing, 1990]
Population: 7,871 Flats: 3,216
Floors min: 34 Floors max: 37
People/Flat: 2.45 Towers: 11

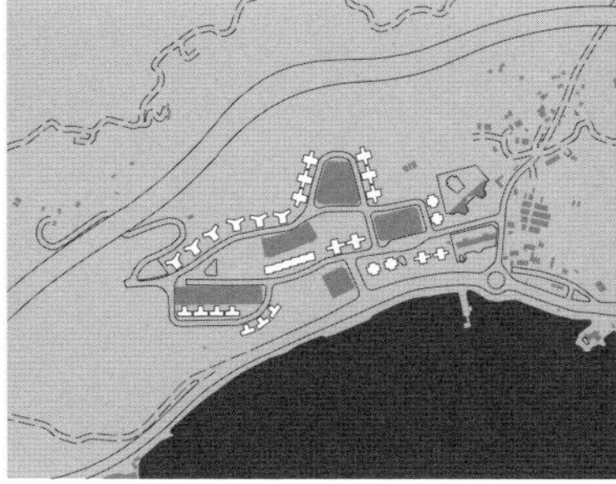

44. HONG KONG GARDEN [Tsuen Wan, 2002]
Population: 7,582 Flats: 2,830
Floors min: 12 Floors max: 29
People/Flat: 2.68 Towers: 28

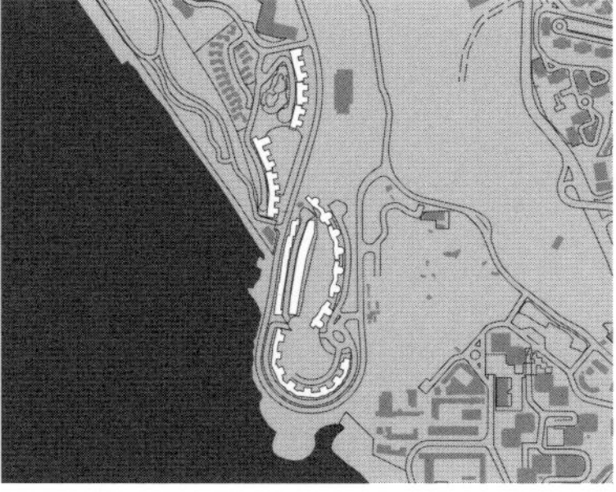

45. BEL-AIR [Southern, 2008]
Population: 7,372 Flats: 2,857
Floors min: 26 Floors max: 47
People/Flat: 2.58 Towers: 30

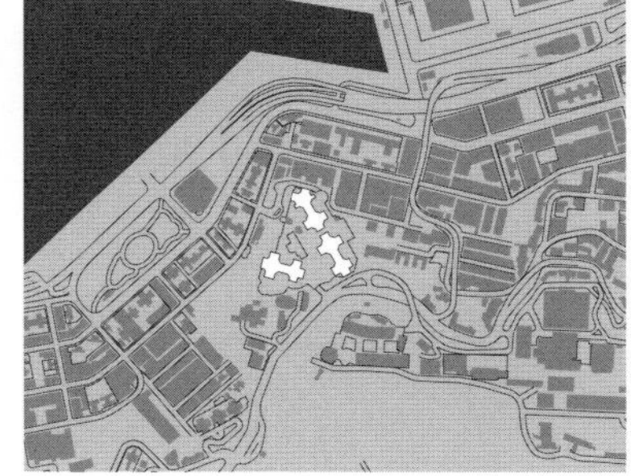

46. THE BELCHER'S [Central/Western, 2001]
Population: 7,295 Flats: 2,214
Floors min: 44 Floors max: 48
People/Flat: 3.29 Towers: 6

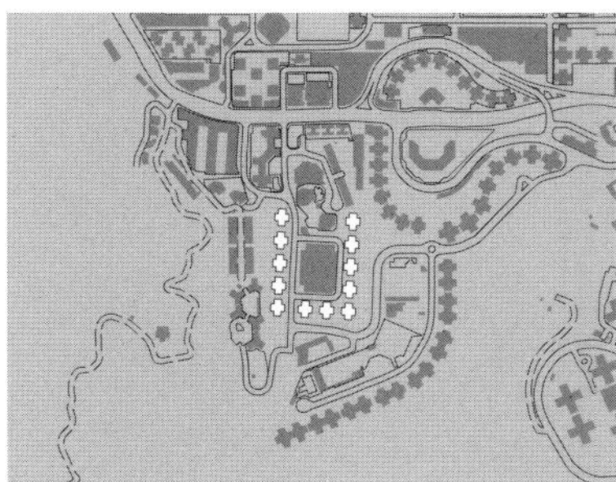

47. NAN FUNG SUN CHUEN [Eastern, 1978]
Population: 7,243 Flats: 2,826
Floors min: 28 Floors max: 33
People/Flat: 2.56 Towers: 12

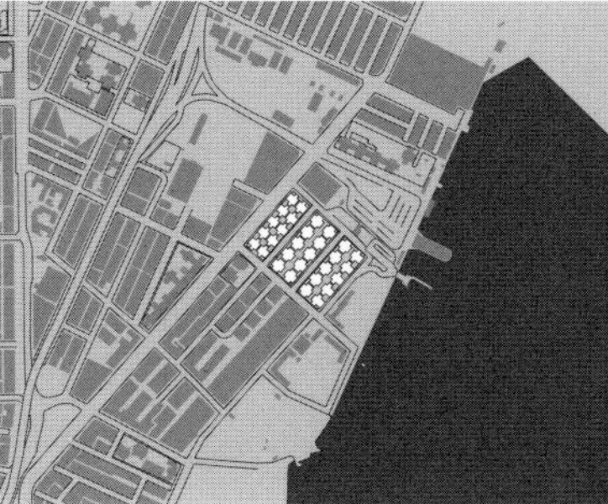

48. WYLER GARDENS [Kowloon City, 1979]
Population: 7,140 Flats: 2,730
Floors min: 13 Floors max: 13
People/Flat: 2.62 Towers: 30

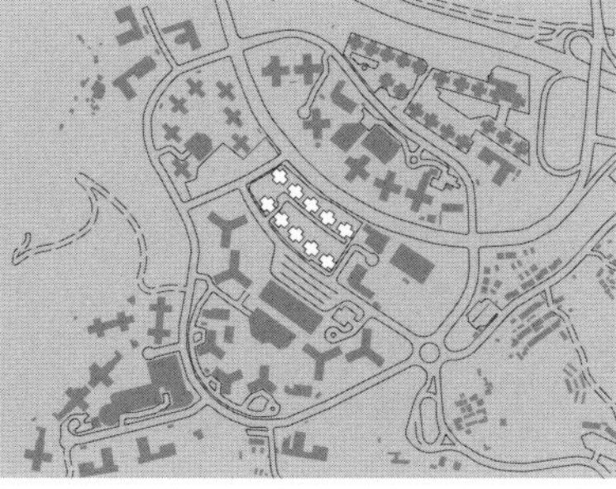

49. FLORA PLAZA [North, 1995]
Population: 7,082 Flats: 2,710
Floors min: 34 Floors max: 34
People/Flat: 2.61 Towers: 10

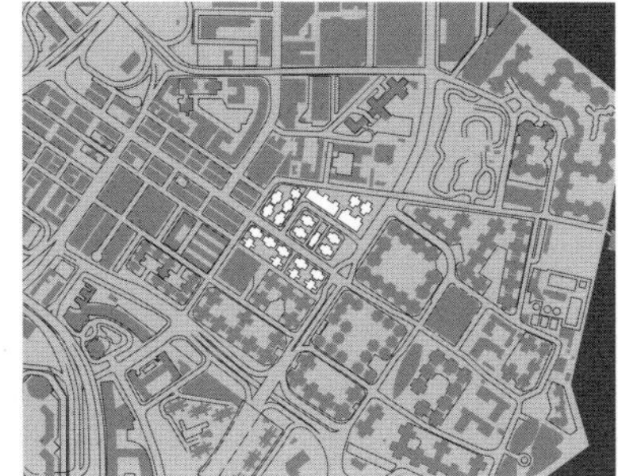

50. WHAMPOA ESTATE [Kowloon City, 1978]
Population: 7,048 Flats: 2,742
Floors min: 13 Floors max: 15
People/Flat: 2.57 Towers: 25

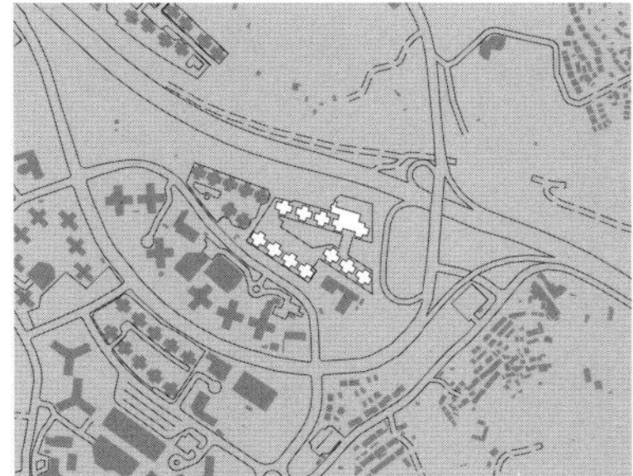

51. DAWNING VIEWS [North 1999]
Population: 7,010 Flats: 2,688
Floors min: 29 Floors max: 29
People/Flat: 2.61 Towers: 12

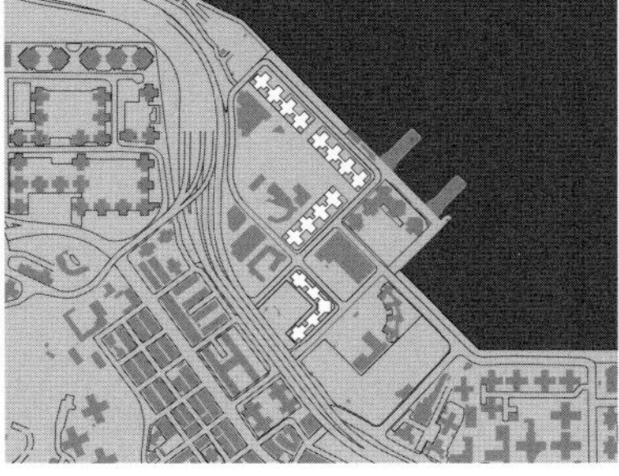

52. LEI KING WAN [Eastern, 1989]
Population: 6,999 Flats: 2,295
Floors min: 17 Floors max: 19
People/Flat: 3.05 Towers: 17

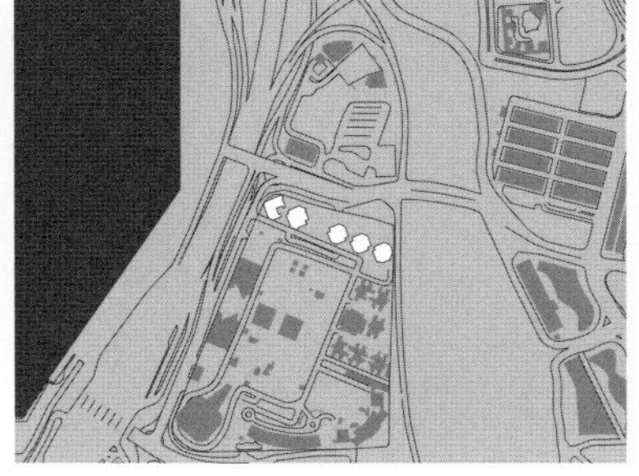

53. SORRENTO [Yau Tsim Mong, 2003]
Population: 6,675 Flats: 2,127
Floors min: 51 Floors max: 65
People/Flat: 3.14 Towers: 5

54. VISTA PARADISO [Sha Tin, 1999]
Population: 6,574 Flats: 2,032
Floors min: 23 Floors max: 30
People/Flat: 3.24 Towers: 11

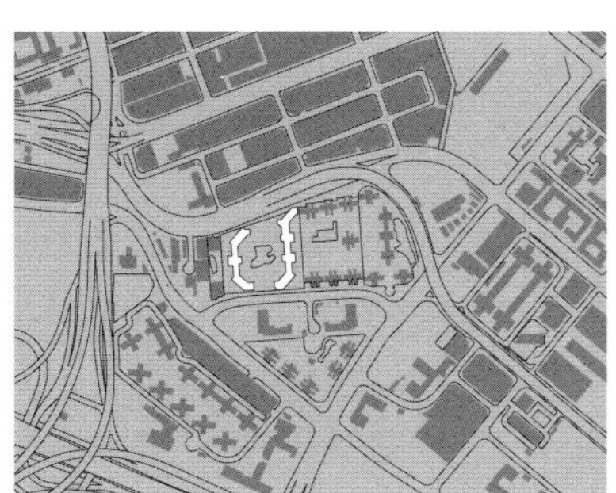

55. BANYAN GARDEN [Sham Shui Po, 2004]
Population: 6,442 Flats: 2,528
Floors min: 40 Floors max: 49
People/Flat: 2.55 Towers: 7

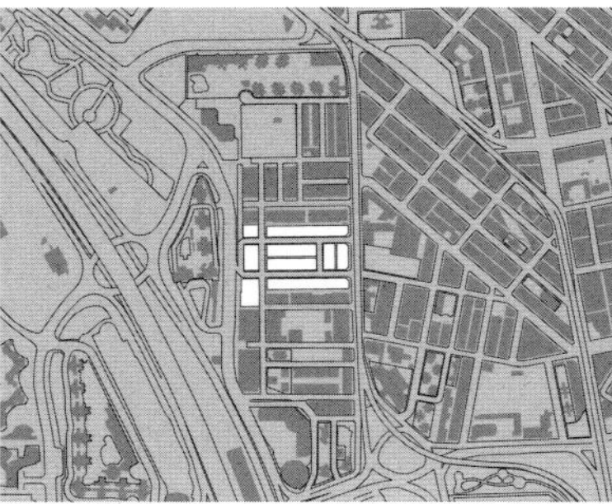

56. COSMOPOLITAN EST. [Yau Tsim Mong, 1975]
Population: 6,401 Flats: 2,125
Floors min: 12 Floors max: 12
People/Flat: 3.01 Towers: 15

57. IS. HARBOURVIEW [Yau Tsim Mong, 2000]
Population: 6,293 Flats: 2,316
Floors min: 35 Floors max: 39
People/Flat: 2.72 Towers: 9

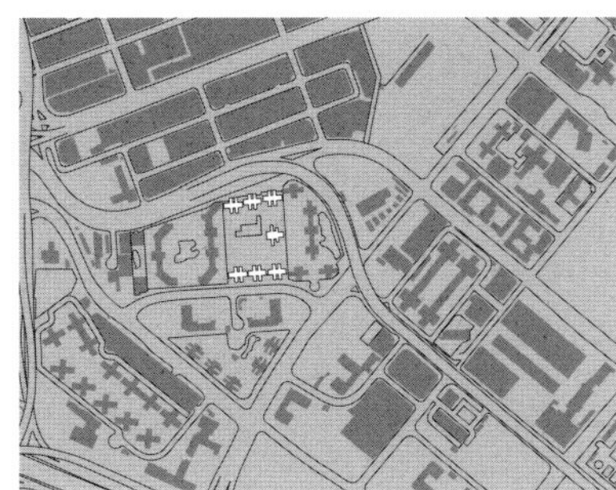

58. LIBERTE [Sham Shui Po, 2003]
Population: 6,256 Flats: 2,436
Floors min: 37 Floors max: 46
People/Flat: 2.57 Towers: 7

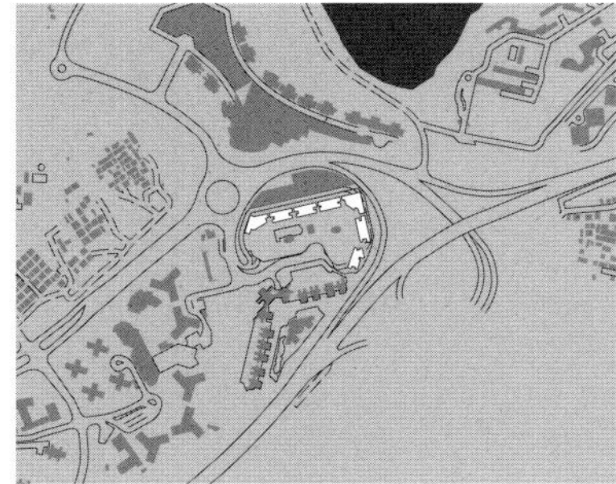

59. LAKE SILVER [Sha Tin, 2009]
Population: 6,149 Flats: 2,169
Floors min: 38 Floors max: 46
People/Flat: 2.83 Towers: 7

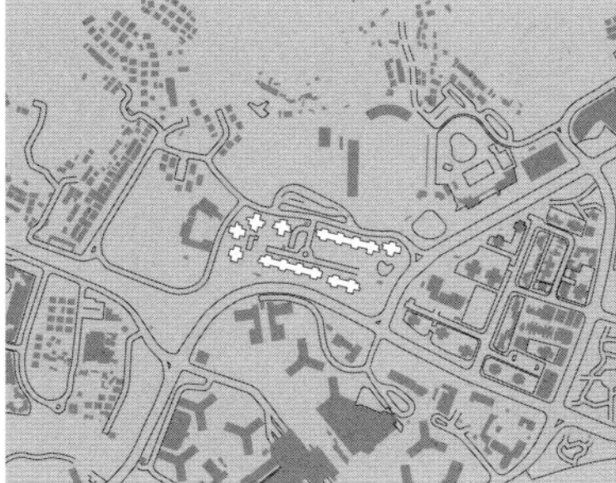

60. SERENITY PARK [Tai Po, 1994]
Population: 6,146 Flats: 2,476
Floors min: 21 Floors max: 22
People/Flat: 2.48 Towers: 15

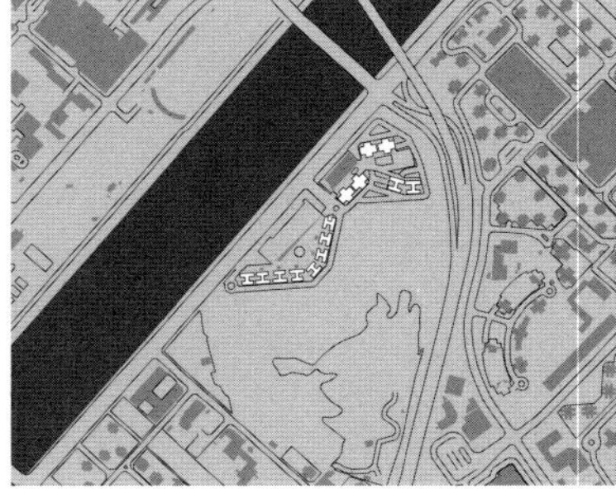

61. BELAIR GARDENS [Sha Tin, 1987]
Population: 6,119 Flats: 1,944
Floors min: 25 Floors max: 28
People/Flat: 3.15 Towers: 14

62. HARBOUR PLACE [Kowloon City, 2002]
Population: 6,105 Flats: 2,470
Floors min: 35 Floors max: 36
People/Flat: 2.47 Towers: 7

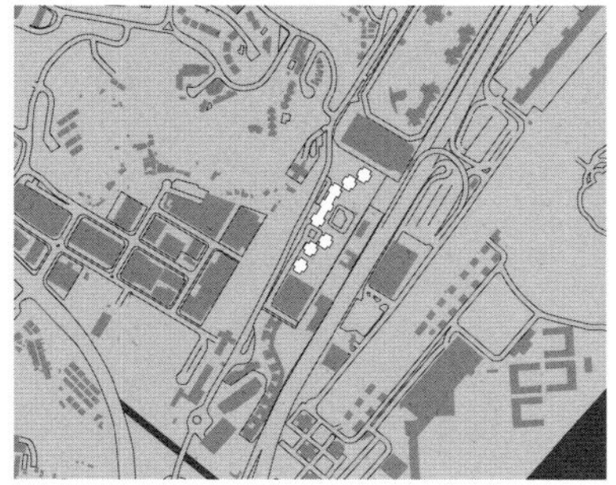

63. JUBILEE GARDEN [Sha Tin, 1986]
Population: 5,880 Flats: 2,261
Floors min: 6 Floors max: 34
People/Flat: 2.60 Towers: 9

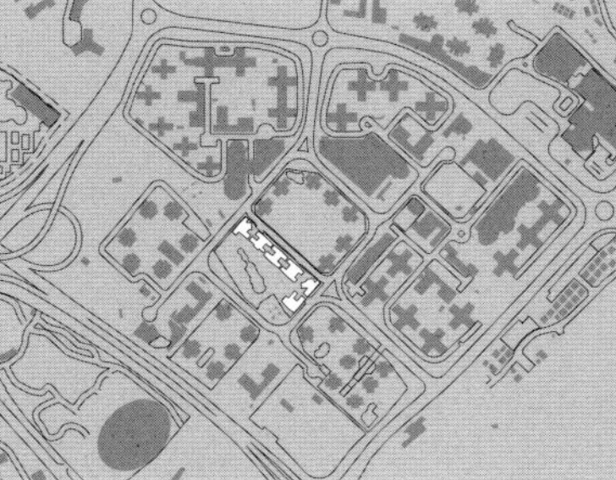

64. RESIDENCE OASIS [Sai Kung, 2004]
Population: 5,807 Flats: 2,135
Floors min: 48 Floors max: 49
People/Flat: 2.72 Towers: 6

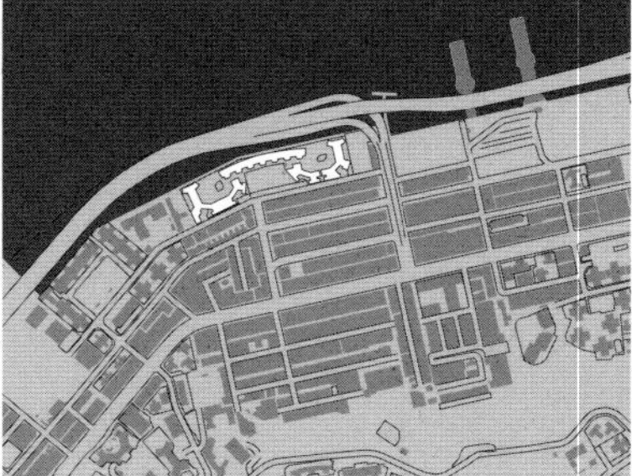

65. PROVIDENT CENTRE [Eastern, 1984]
Population: 5,750 Flats: 1,450
Floors min: 25 Floors max: 25
People/Flat: 3.97 Towers: 17

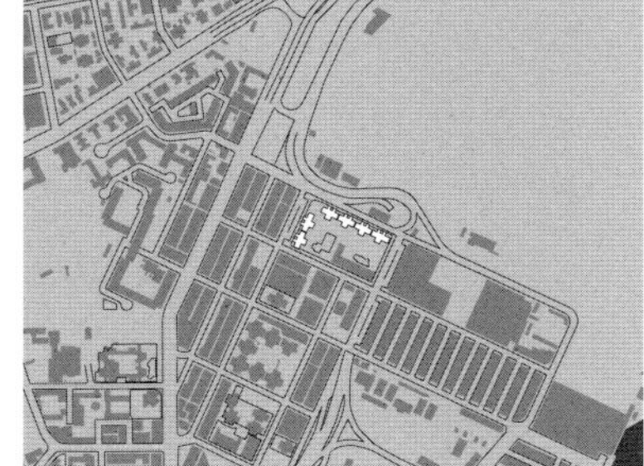

66. SKY TOWER [Kowloon City, 2004]
Population: 5,660 Flats: 2,209
Floors min: 47 Floors max: 47
People/Flat: 2.56 Towers: 6

67. CASTELLO [Sha Tin, 1999]
Population: 5,600 Flats: 1,744
Floors min: 31 Floors max: 32
People/Flat: 3.21 Towers: 7

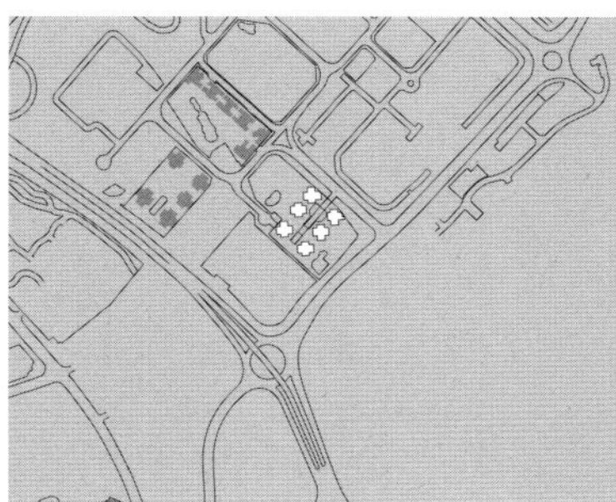

68. LA CITE NOBLE [Sai Kung, 1999]
Population: 5,562 Flats: 2,184
Floors min: 44 Floors max: 47
People/Flat: 2.55 Towers: 6

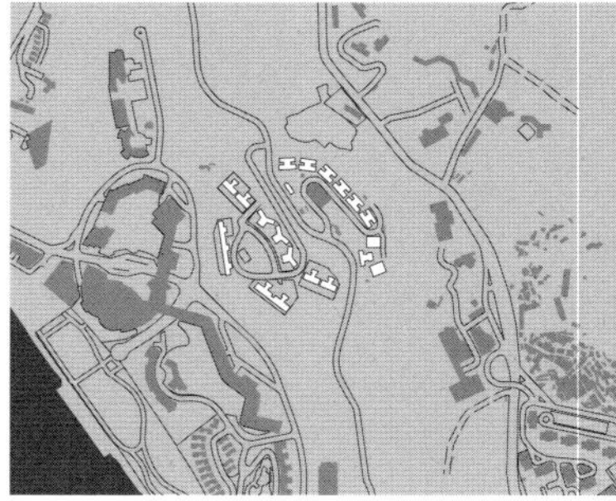

69. BAGUIO VILLA [Southern, 1979]
Population: 5,475 Flats: 1,543
Floors min: 10 Floors max: 30
People/Flat: 3.55 Towers: 33

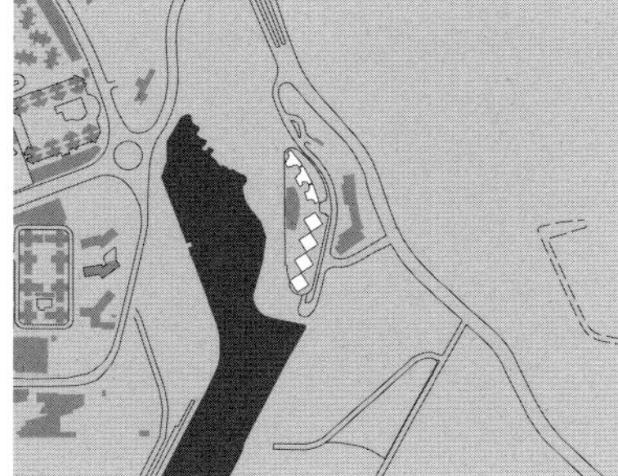

70. OSCAR BY THE SEA [Sai Kung, 2002]
Population: 5,458 Flats: 1,959
Floors min: 32 Floors max: 49
People/Flat: 2.79 Towers: 7

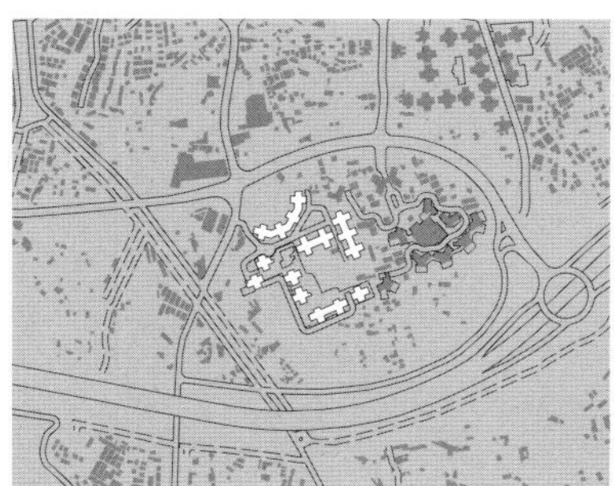

71. SERENO VERDE [Yuen Long, 2002]
Population: 5,331 Flats: 1,674
Floors min: 13 Floors max: 14
People/Flat: 3.18 Towers: 16

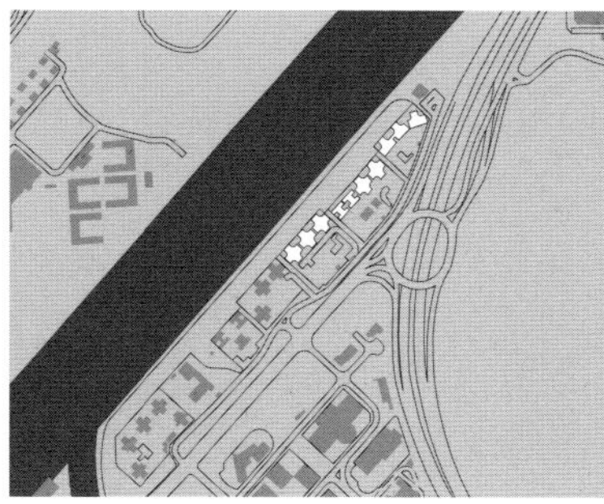

72. PICTORIAL GARDEN [Sha Tin, 1993]
Population: 5,291 Flats: 1,734
Floors min: 23 Floors max: 29
People/Flat: 3.05 Towers: 10

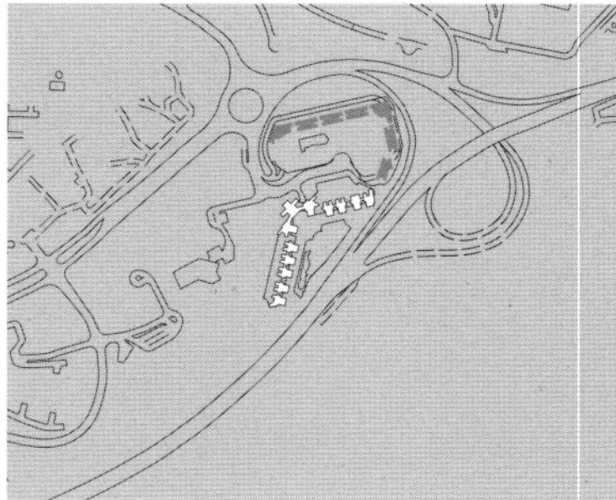

73. MONTE VISTA [Sha Tin, 2000]
Population: 5,286 Flats: 1,606
Floors min: 26 Floors max: 30
People/Flat: 3.29 Towers: 12

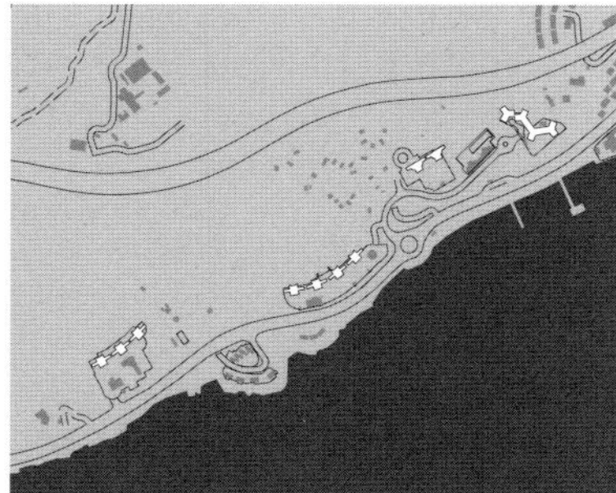

74. SEA CREST VILLA [Tsuen Wan, 1995]
Population: 5,167 Flats: 2,224
Floors min: 24 Floors max: 37
People/Flat: 2.32 Towers: 13

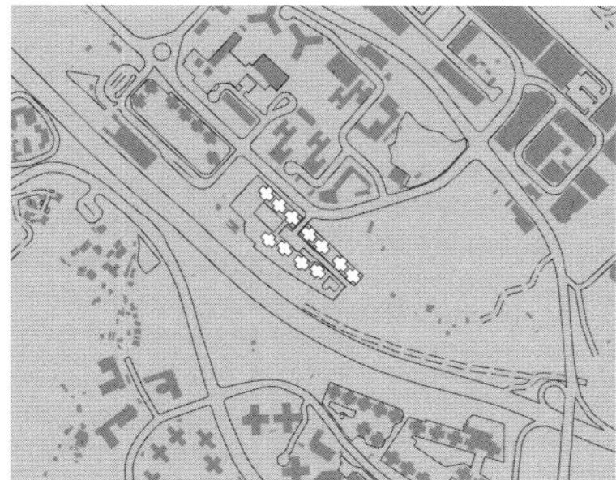

75. FANLING CENTRE [North, 1991]
Population: 5,089 Flats: 2,200
Floors min: 25 Floors max: 25
People/Flat: 2.31 Towers: 11

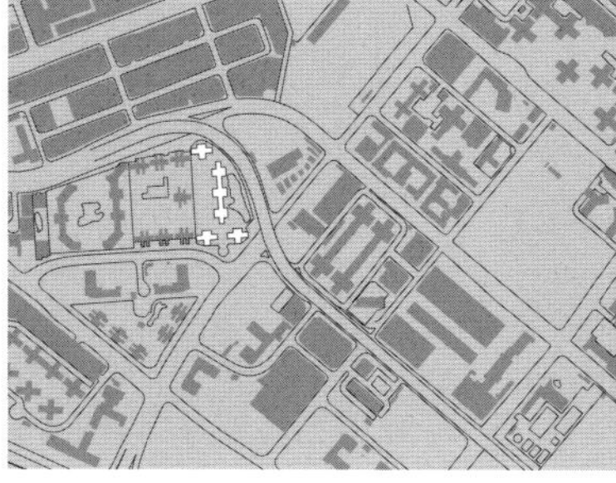

76. THE PACIFICA [Sham Shui Po, 2005]
Population: 5,084 Flats: 2,308
Floors min: 45 Floors max: 51
People/Flat: 2.20 Towers: 6

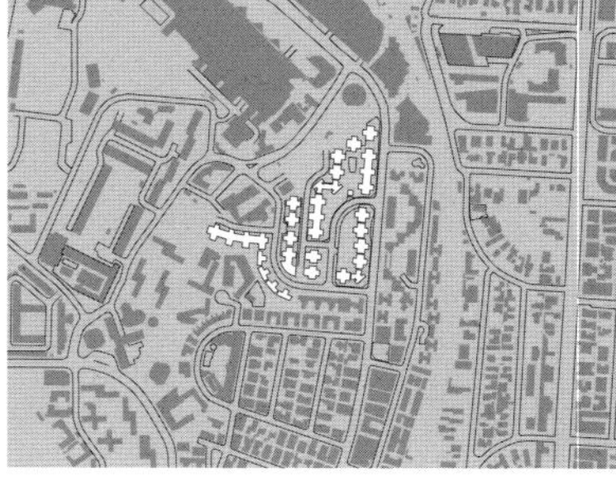

77. PARC OASIS [Sham Shui Po, 1995]
Population: 4,989 Flats: 1,823
Floors min: 8 Floors max: 10
People/Flat: 2.74 Towers: 32

78. NAN FUNG PLAZA [Sai Kung, 1999]
Population: 4,988 Flats: 1,614
Floors min: 39 Floors max: 43
People/Flat: 3.09 Towers: 5

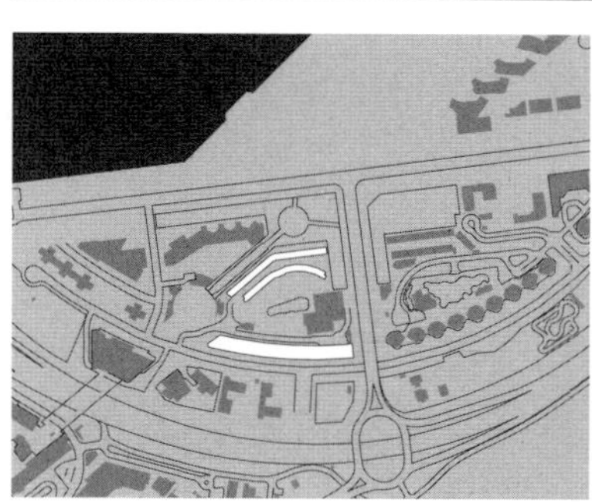

79. COASTAL SKYLINE [Islands, 2007]
Population: 4,959 Flats: 3,350
Floors min: 15 Floors max: 50
People/Flat: 1.48 Towers: 13

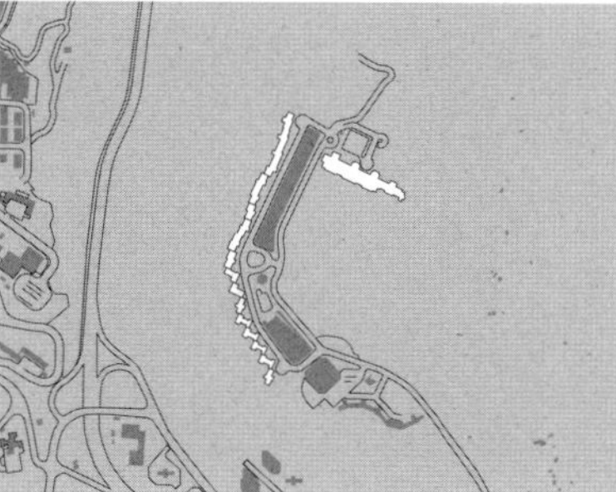

80. WONDERLAND VILLAS [Kwai Tsing, 1987]
Population: 4,927 Flats: 1,503
Floors min: 14 Floors max: 35
People/Flat: 3.28 Towers: 22

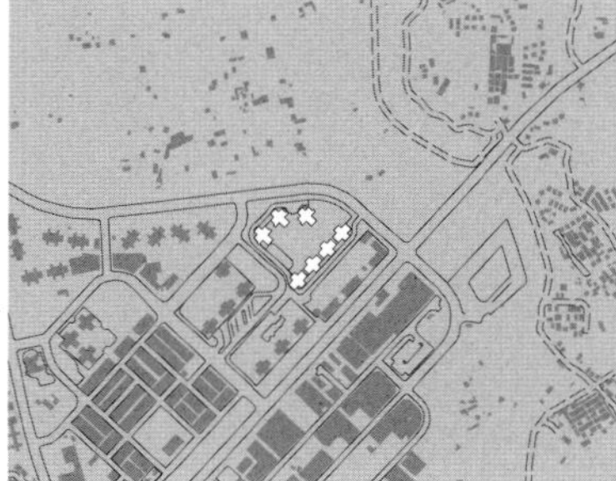

81. BELAIR MONTE [North, 1998]
Population: 4,908 Flats: 1,680
Floors min: 30 Floors max: 30
People/Flat: 2.92 Towers: 7

82. MAYFAIR GARDENS [Kwai Tsing, 1984]
Population: 4,848 Flats: 1,912
Floors min: 28 Floors max: 33
People/Flat: 2.54 Towers: 8

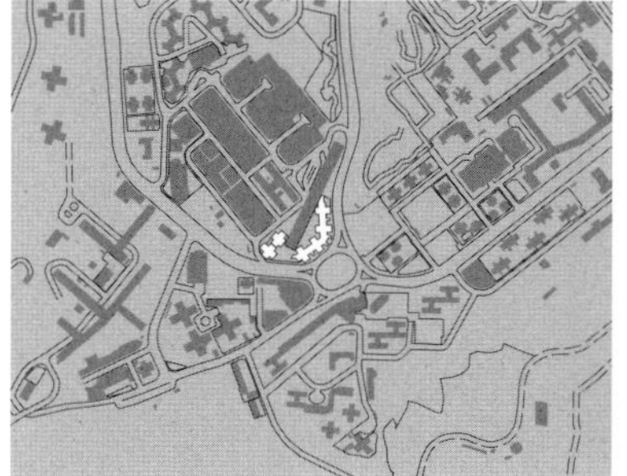

83. NEW JADE GARDEN [Eastern, 1988]
Population: 4,771 Flats: 1,488
Floors min: 31 Floors max: 31
People/Flat: 3.21 Towers: 6

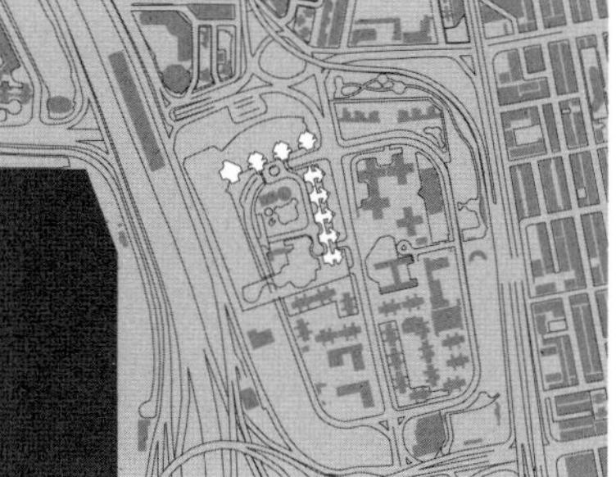

84. PARK AVENUE [Yau Tsim Mong, 2001]
Population: 4,741 Flats: 2,948
Floors min: 35 Floors max: 45
People/Flat: 1.61 Towers: 9

85. YOHO TOWN [Yuen Long, 2004]
Population: 4,587 Flats: 2,200
Floors min: 32 Floors max: 37
People/Flat: 2.09 Towers: 8

86. CHELSEA HEIGHTS [Tuen Mun, 2000]
Population: 4,469 Flats: 1595
Floors min: 33 Floors max: 34
People/Flat: 2.80 Towers: 6

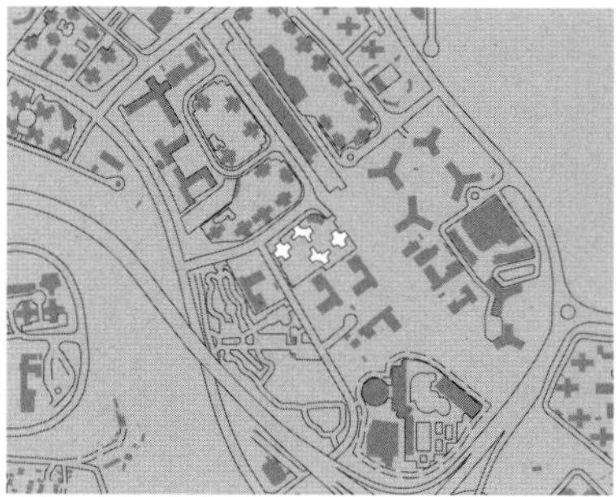

87. THE PINNACLE [Sai Kung, 1998]
Population: 4,432 Flats: 1,424
Floors min: 44 Floors max: 45
People/Flat: 3.11 Towers: 4

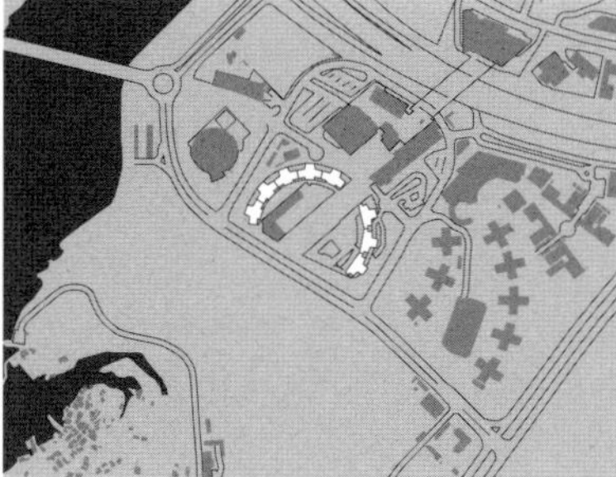

88. TUNG CHUNG CRESCENT [Islands, 1999]
Population: 4,403 Flats: 2,158
Floors min: 28 Floors max: 42
People/Flat: 2.04 Towers: 8

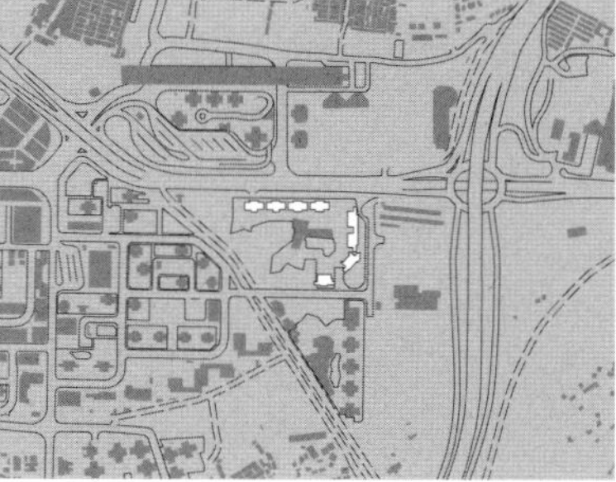

89. YOHO MIDTOWN [Yuen Long, 2010]
Population: 4,360 Flats: 1,891
Floors min: 33 Floors max: 40
People/Flat: 2.31 Towers: 8

90. VISION CITY [Tsuen Wan, 2007]
Population: 4,340 Flats: 1,469
Floors min: 42 Floors max: 44
People/Flat: 2.95 Towers: 5

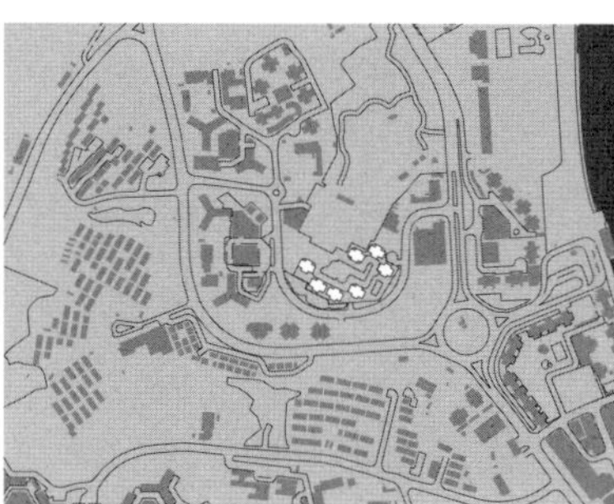

91. TSING YI GARDEN [Kwai Tsing, 1986]
Population: 4,323 Flats: 1,520
Floors min: 25 Floors max: 30
People/Flat: 2.84 Towers: 7

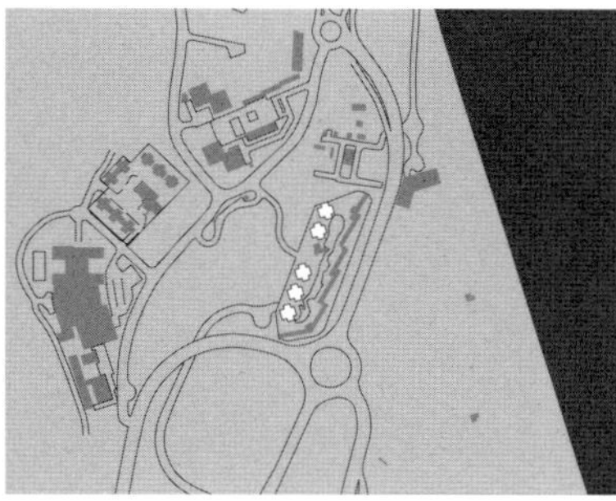

92. RAMBLER CREST [Kwai Tsing, 2003]
Population: 4,228 Flats: 1,587
Floors min: 40 Floors max: 40
People/Flat: 2.66 Towers: 5

93. ROYAL PENINSULA [Kowloon City, 2000]
Population: 4,206 Flats: 1,669
Floors min: 35 Floors max: 42
People/Flat: 2.52 Towers: 5

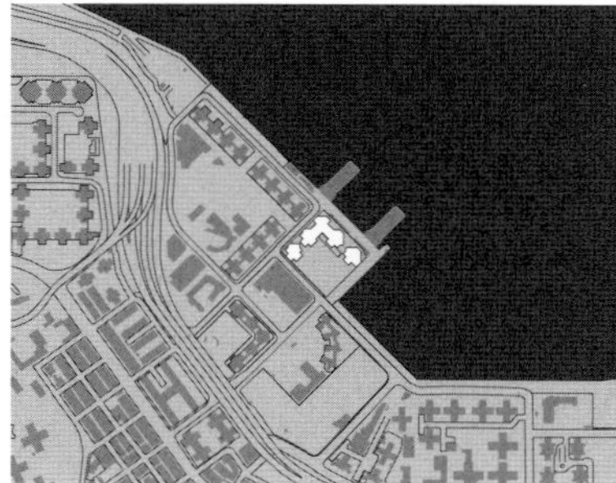

94. GRAND PROMENADE [Eastern, 2005]
Population: 4,137 Flats: 2,021
Floors min: 55 Floors max: 58
People/Flat: 2.05 Towers: 5

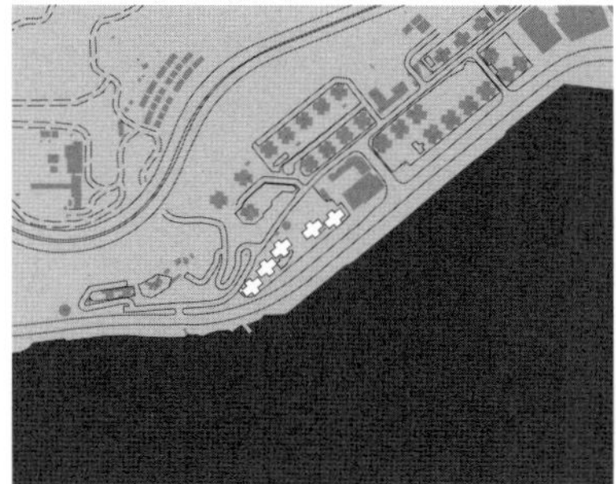

95. BAYVIEW GARDEN [Tsuen Wan, 1992]
Population: 4,118 Flats: 1,200
Floors min: 30 Floors max: 30
People/Flat: 3.43 Towers: 5

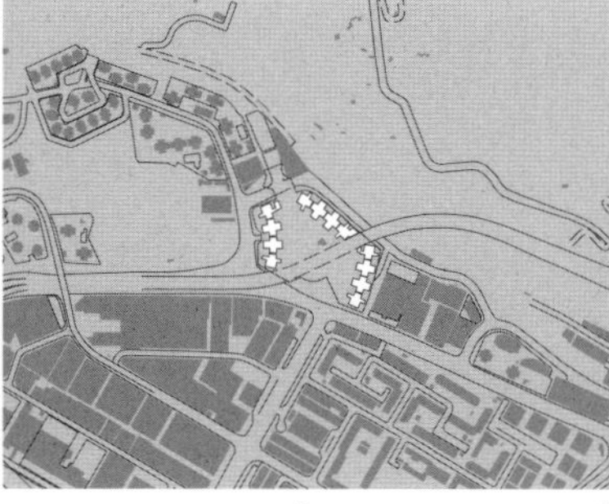

96. THE WATERFRONT [Yau Tsim Mong, 2000]
Population: 4,094 Flats: 1,290
Floors min: 34 Floors max: 37
People/Flat: 3.17 Towers: 6

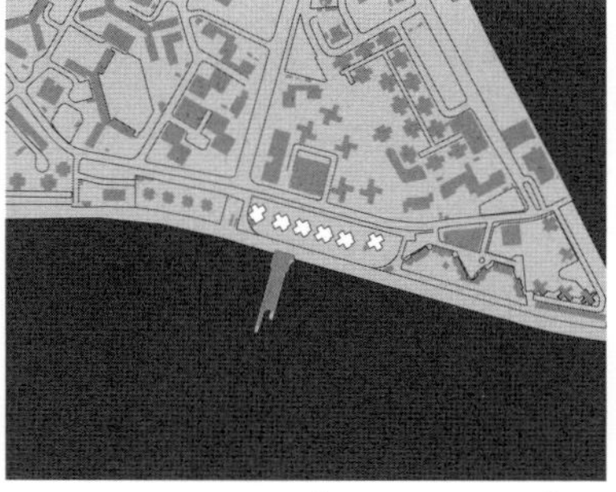

97. PIERHEAD GARDEN [Tuen Mun, 1988]
Population: 4,051 Flats: 1,432
Floors min: 29 Floors max: 30
People/Flat: 2.83 Towers: 6

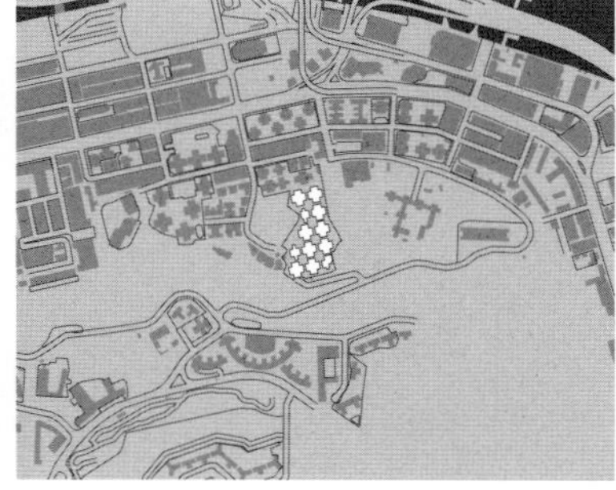

98. BEDFORD GARDENS [Eastern, 1981]
Population: 4,037 Flats: 1,421
Floors min: 18 Floors max: 19
People/Flat: 2.84 Towers: 12

99. TUEN MUN TOWN PLAZA [Tuen Mun, 1992]
Population: 4,036 Flats: 1,968
Floors min: 27 Floors max: 29
People/Flat: 2.05 Towers: 8

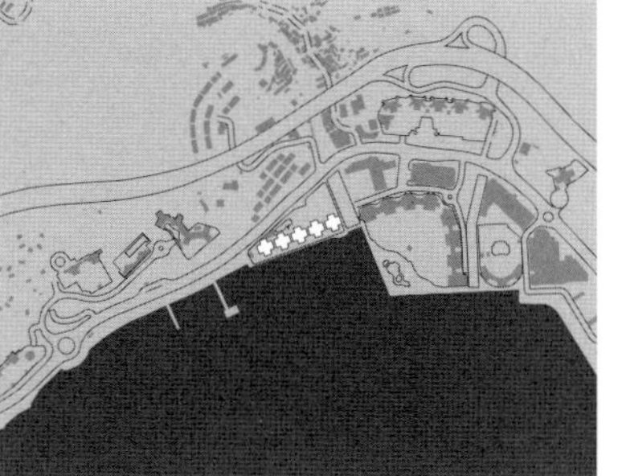

100. LIDO GARDEN [Tsuen Wan, 1989]
Population: 4,016 Flats: 1,392
Floors min: 35 Floors max: 35
People/Flat: 2.89 Towers: 5

100 Largest Estates

1. KINGSWOOD VILLAS
2. MEI FOO SUN CHUEN
3. TAIKOO SHING
4. WHAMPOA GARDEN
5. SOUTH HORIZONS
6. CITY ONE
7. LAGUNA CITY
8. BELVEDERE GARDEN
9. HENG FA CHUEN
10. KORNHILL
11. RIVIERA GARDENS
12. OCEAN SHORES
13. METRO CITY
14. TELFORD GARDENS
15. LAGUNA VERDE
16. SCENEWAY GARDEN
17. DISCOVERY BAY
18. PARK CENTRAL
19. PARK ISLAND
20. LOHAS PARK
21. AMOY GARDENS
22. CHI FU FA YUEN
23. METRO TOWN/LE POINT
24. LUK YEUNG SUN CHUEN
25. CARIBBEAN COAST
26. SUNSHINE CITY
27. TSUEN WAN CENTRE
28. TAI PO CENTRE
29. TIERRA VERDE
30. BELLAGIO
31. TAI HING GARDENS
32. SUN TUEN MUN CENTRE
33. ISLAND RESORT
34. DISCOVERY PARK
35. ALLWAY GARDENS
36. ROYAL ASCOT
37. ABERDEEN CENTRE
38. VILLA ESPLANADA
39. TSUEN KING GARDEN
40. CITY GARDEN
41. METRO HARBOUR VIEW
42. TSEUNG KWAN O PLAZA
43. GREENFIELD GARDEN
44. HONG KONG GARDEN
45. BEL AIR
46. THE BELCHERS
47. NAN FUNG SUN CHUEN
48. WYLER GARDENS
49. FLORA PLAZA
50. WHAMPOA ESTATE
51. DAWNING VIEWS
52. LEI KING WAN
53. SORRENTO
54. VISTA PARADISO
55. BANYAN GARDEN
56. COSMOPOLITAN ESTATE
57. ISLAND HABOURVIEW
58. LIBERTE
59. LAKE SILVER
60. SERENITY PARK
61. BELAIR GARDENS
62. HARBOUR PLACE
63. JUBILEE GARDEN
64. RESIDENCE OASIS
65. PROVIDENT CENTRE
66. SKY TOWER
67. CASTELLO
68. LA CITE NOBLE
69. BAGUIO VILLA
70. OSCAR BY THE SEA
71. SERENO VERDE
72. PICTORIAL GARDEN
73. MONTE VISTA
74. SEA CREST VILLA
75. FANLING CENTRE
76. THE PACIFICA
77. PARC OASIS
78. NAN FUNG PLAZA
79. COASTAL SKYLINE
80. WONDERLAND VILLAS
81. BELAIR MONTE
82. MAYFAIR GARDEN
83. NEW JADE GARDEN
84. PARK AVENUE
85. YOHO TOWN
86. CHELSEA HEIGHTS
87. THE PINNACLE
88. TUNG CHUNG CRESCENT
89. YOHO MIDTOWN
90. VISION CITY
91. TSING YI GARDEN
92. RAMBLER CREST
93. ROYAL PENINSULA
94. GRAND PROMENADE
95. BAYVIEW GARDEN
96. THE WATERFRONT
97. PIERHEAD GARDEN
98. BEDFORD GARDENS
99. TUEN MUN TOWN PLAZA
100. LIDO GARDEN

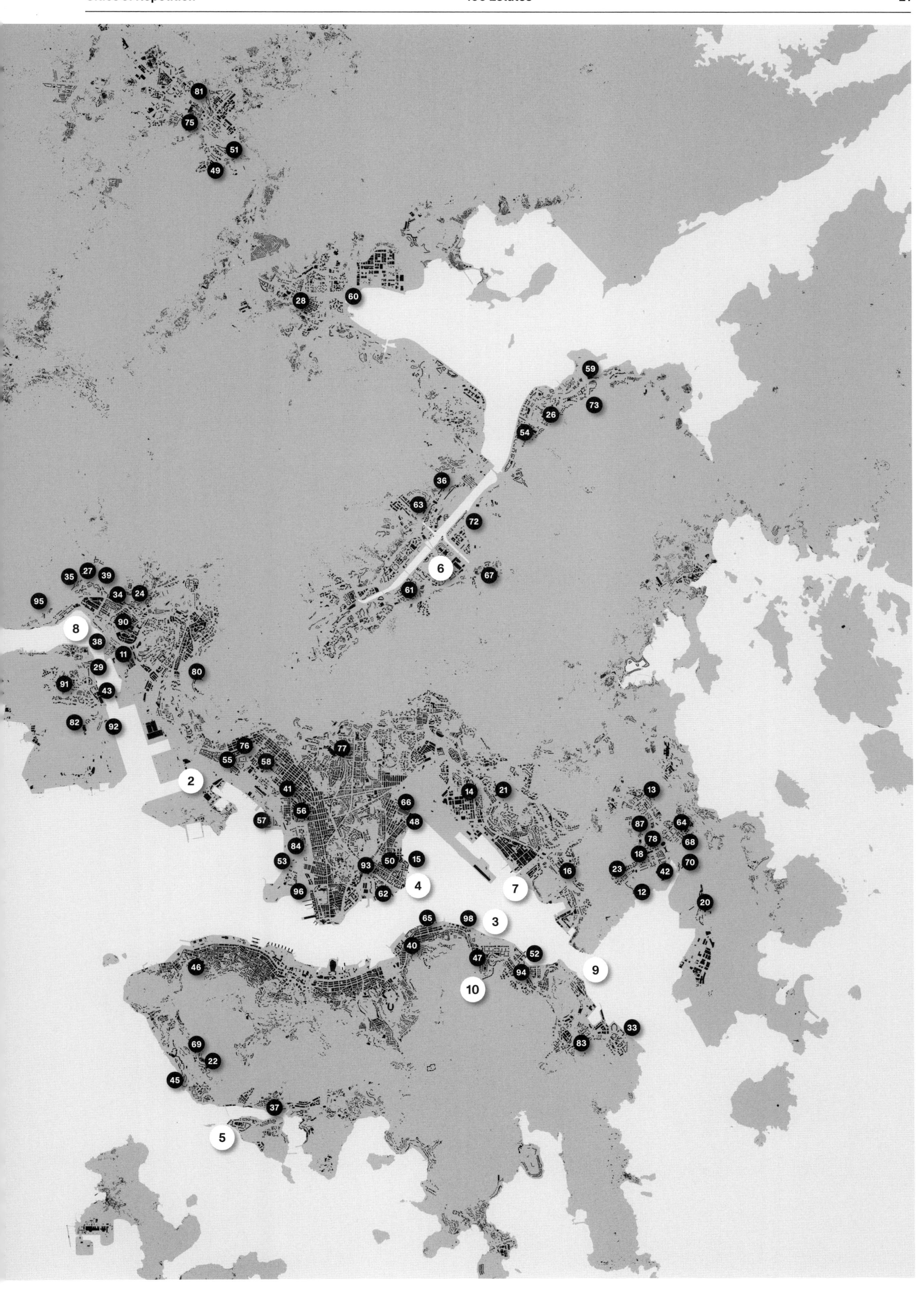
81
75
51
49
28
60
59
73
26
54
36
63
72
6
67
61
35
27
39
95
34
24
90
8
38
11
29
80
91
43
82
92
76
77
55
58
2
41
14
21
13
66
56
57
48
87
64
78
68
84
18
53
15
50
93
70
16
23
42
4
7
12
96
62
20
65
98
3
40
52
47
46
94
9
10
33
69
83
22
45
37
5

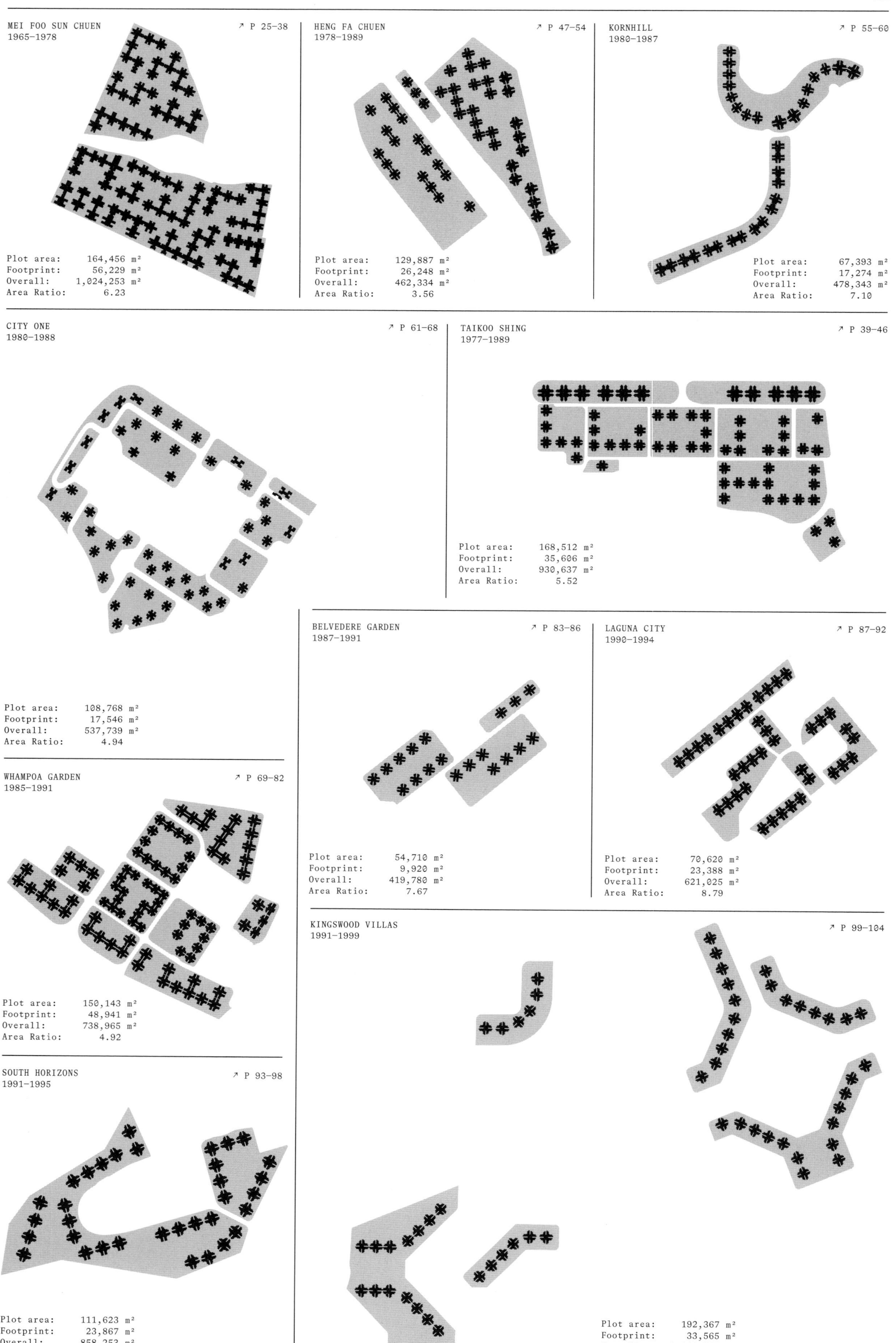
MEI FOO SUN CHUEN
1965–1978
↗ P 25–38
Plot area: 164,456 m²
Footprint: 56,229 m²
Overall: 1,024,253 m²
Area Ratio: 6.23
HENG FA CHUEN
1978–1989
↗ P 47–54
Plot area: 129,887 m²
Footprint: 26,248 m²
Overall: 462,334 m²
Area Ratio: 3.56
KORNHILL
1980–1987
↗ P 55–60
Plot area: 67,393 m²
Footprint: 17,274 m²
Overall: 478,343 m²
Area Ratio: 7.10
CITY ONE
1980–1988
↗ P 61–68
Plot area: 108,768 m²
Footprint: 17,546 m²
Overall: 537,739 m²
Area Ratio: 4.94
TAIKOO SHING
1977–1989
↗ P 39–46
Plot area: 168,512 m²
Footprint: 35,606 m²
Overall: 930,637 m²
Area Ratio: 5.52
WHAMPOA GARDEN
1985–1991
↗ P 69–82
Plot area: 150,143 m²
Footprint: 48,941 m²
Overall: 738,965 m²
Area Ratio: 4.92
BELVEDERE GARDEN
1987–1991
↗ P 83–86
Plot area: 54,710 m²
Footprint: 9,920 m²
Overall: 419,780 m²
Area Ratio: 7.67
LAGUNA CITY
1990–1994
↗ P 87–92
Plot area: 70,620 m²
Footprint: 23,388 m²
Overall: 621,025 m²
Area Ratio: 8.79
SOUTH HORIZONS
1991–1995
↗ P 93–98
Plot area: 111,623 m²
Footprint: 23,867 m²
Overall: 858,253 m²
Area Ratio: 7.69
KINGSWOOD VILLAS
1991–1999
↗ P 99–104
Plot area: 192,367 m²
Footprint: 33,565 m²
Overall: 1,148,732 m²
Area Ratio: 5.97

10 Estates Documenting Repetition

Drawings of tower plans comprise an important section of the book. The following section presents comparative floor-plan drawings of towers and units from each of the ten largest privately developed housing estates in Hong Kong. The dozens of pages dedicated to documenting these tower types is in and of itself a relentless exercise in repetition. These ten estates comprise 529 towers with 99,913 flats and roughly 290,000 residents—representing 4% of Hong Kong's population. The estates are presented chronologically by their development dates in order to reveal how degrees of repetition and standardization have changed over time.

Each estate section in the book is introduced with an axonometric drawing of the entire estate from above—to provide the reader with an understanding of the overall urban form. The context for each estate is simplified; pared down drawings reveal estate proximity to adjacent water frontages, vehicular roadways, recreational landscapes, and significant or interconnected commercial buildings or complexes. Although the ten estates are represented with their full urban contexts in the map section of the book, the deliberate removal of other elements in the urban context in the axonometric drawings allows the estates to be seen as miniature cities onto themselves. These three-dimensional drawings also reveal organizational urban design strategies and ideas about the importance of collective space and place-making for each of these skyscraper communities.

The tower and unit drawings that form the majority of this section reveal the similarities and subtle differences between the various design and layout strategies in each of the estates. Stylistic cues and color schemes for drawings were appropriated from the graphic language of sales catalogs developed by real estate agencies. Many Hong Kong-based real estate agents have redrawn the estates into catalogs and maps for potential renters and buyers. In trying to sell or lease apartments in each of the estates, elaborate maps with color-coded floor plans allow prospective buyers and renters to distinguish between slightly differing tower phases, apartment types, numbers of rooms, and net floor areas. Each agency has developed its own color palette in an attempt to clarify between two and three bedroom units or flats that face southeast versus northwest. Exploring each of the housing estates through real estate office literature and maps yields a virtual rainbow of color schemes, each designed to help real estate agents point out differences.

The graphics in this book use color in a manner designed to show consistency and difference across tower blocks within one estate and across multiple estates. Due to the limitations of offset printing, there are too many unique unit types in all of the ten estates documented to provide each type with its own discernably different color. To overcome this limitation in project documentation and analysis, a broad color spectrum, that relates to unit sizes and helps to illustrate slight variations between apartment types and sizes, was adopted. From smallest to largest unit floor area, the colors represent an area range from 30 to 150 square meters. The gradient code starts with blue and moves through green, yellow, orange, pink, and violet. The same color gradient is used across the ten estates and allows readers to see how diverse or how repetitive an estate is. More colorful estate plan layouts, therefore, signify a broader variety of unit sizes and more architectural heterogeneity; estates with a limited number of colors signify the opposite. Similar colors in adjacent units or across opposite sides of the elevator core reveal architectural strategies for mirroring, or flipping, units to achieve greater standardization.

Base plans for the residential towers and units were carefully redrawn from blueprints obtained from Hong Kong Buildings Department record drawings to ensure accuracy according to the original plans. The tower plan section in each estate documentation shows the unique towers at scale. The estates range between 99 towers in Mei Foo and 19 towers in Belvedere Garden; however, only the distinctive towers in each estate are shown. The number of duplicates of unique towers is indicated in each tower plan's upper, right-hand corner. In general, the fewer individual tower plans, the more repetition.

The logic of the towers is essentially based on a standard core that is fitted with a range of unit types. In retrospect, a "plug and play" methodology has been utilized by the estate architects to design different tower configurations. The layout of vertical circulation cores therefore provides important criteria for the categorization of tower types and the analysis of repetition. Though many of the estates were built over years in various phases, towers are presented here by similarities in their vertical circulation system, or core type. Core types were often used repetitively and not always applied chronologically by developers. Therefore, categorizations of tower blocks by core type include towers built in different phases and with different unit configurations.

Almost all the tower blocks documented in these ten estates—with some exceptions in Mei Foo, City One and Whampoa Garden—consist of a central core with eight units, identified by clockwise lettering around the core from A to H. Some deviation between actual tower block numbers or unit numbers and those presented in the book must be noted. Though every effort was made to match tower or block numbers in the book with tower numbers in the actual estates, in some cases where no block numbers existed, or numbers were skipped, tower numbering has been adjusted for consistency in this comparative study. Tower numbering in the book follows a hierarchical system that begins numbering with the first tower in the first building phase, then accounts for vertical circulation core types, and finally considers the numbering of the tower block in the actual estate. Similarly, tower orientation and therefore unit numbering/lettering in the book is based not on cardinal directions, but on a consistent orientation of the elevator core across estates. Numerical and alphabetical notes next to each floor plan list the tower variations with respect to the core type, tower copies, and block and phase numbers for each copy.

The unit type section of the book is conceived of conceptually as a horizontal cross section through all estate towers. It follows the same hierarchical numbering order as the tower section. However, duplicates highlighted through the outlined numbers are represented and reveal the repetitiveness of an estate via the color-coding method. Units are unwrapped from the core in a clockwise manner and laid out horizontally from A to H—with a few exceptions in the Mei Foo and City One estates. The unique unit types in each estate are highlighted with color-coded backgrounds and the square-meter indication. Symbols for 90- or 180-degree rotation, or mirroring indicators, shown in the corner of unit drawings notate when floor plans are virtual copies of another unit with a different orientation.

The challenge of the *Cities of Repetition* project was the task of documenting a highly repetitious set of information into drawings and diagrams that make subtle degrees of standardization legible. Through layering, cataloging, superimposition, color-coding, and analytical diagramming, we have sought to reveal difference within a context of radical sameness.

145.0 → 149.9
140.0 → 144.9
135.0 → 139.9
130.0 → 134.9
125.0 → 129.9
120.0 → 124.9
115.0 → 119.9
110.0 → 114.9
105.0 → 109.9
100.0 → 104.9
95.0 → 99.9
93.0 → 94.9
91.0 → 92.9
89.0 → 90.9
87.0 → 88.9
85.0 → 86.9
83.0 → 84.9
81.0 → 82.9
79.0 → 80.9
77.0 → 79.9
75.0 → 76.9
74.0 → 74.9
73.0 → 73.9
72.0 → 72.9
71.0 → 71.9
70.0 → 70.9
69.0 → 69.9
68.0 → 68.9
67.0 → 67.9
66.0 → 66.9
65.0 → 65.9
64.5 → 64.9
64.0 → 64.4
63.5 → 63.9
63.0 → 63.4
62.5 → 62.9
62.0 → 62.4
61.5 → 61.9
61.0 → 61.4
60.5 → 60.9
60.0 → 60.4
59.0 → 59.9
58.0 → 58.9
57.0 → 57.9
56.0 → 56.9
55.0 → 55.9
54.0 → 54.9
53.0 → 53.9
52.0 → 52.9
51.0 → 51.9
50.0 → 50.9
48.0 → 49.9
46.0 → 47.9
44.0 → 45.9
42.0 → 43.9
40.0 → 41.9
38.0 → 39.9
36.0 → 37.9
34.0 → 35.9
32.0 → 33.9
30.0 → 31.9

Number ranges on the margin of this page provide a legend for unit sizes (measured in square meters). Colored tabs on the pages of the following section alilgn with these area calculations and reveal the range of unit sizes and types in each estate.

Mei Foo Sun Chuen 1965–78

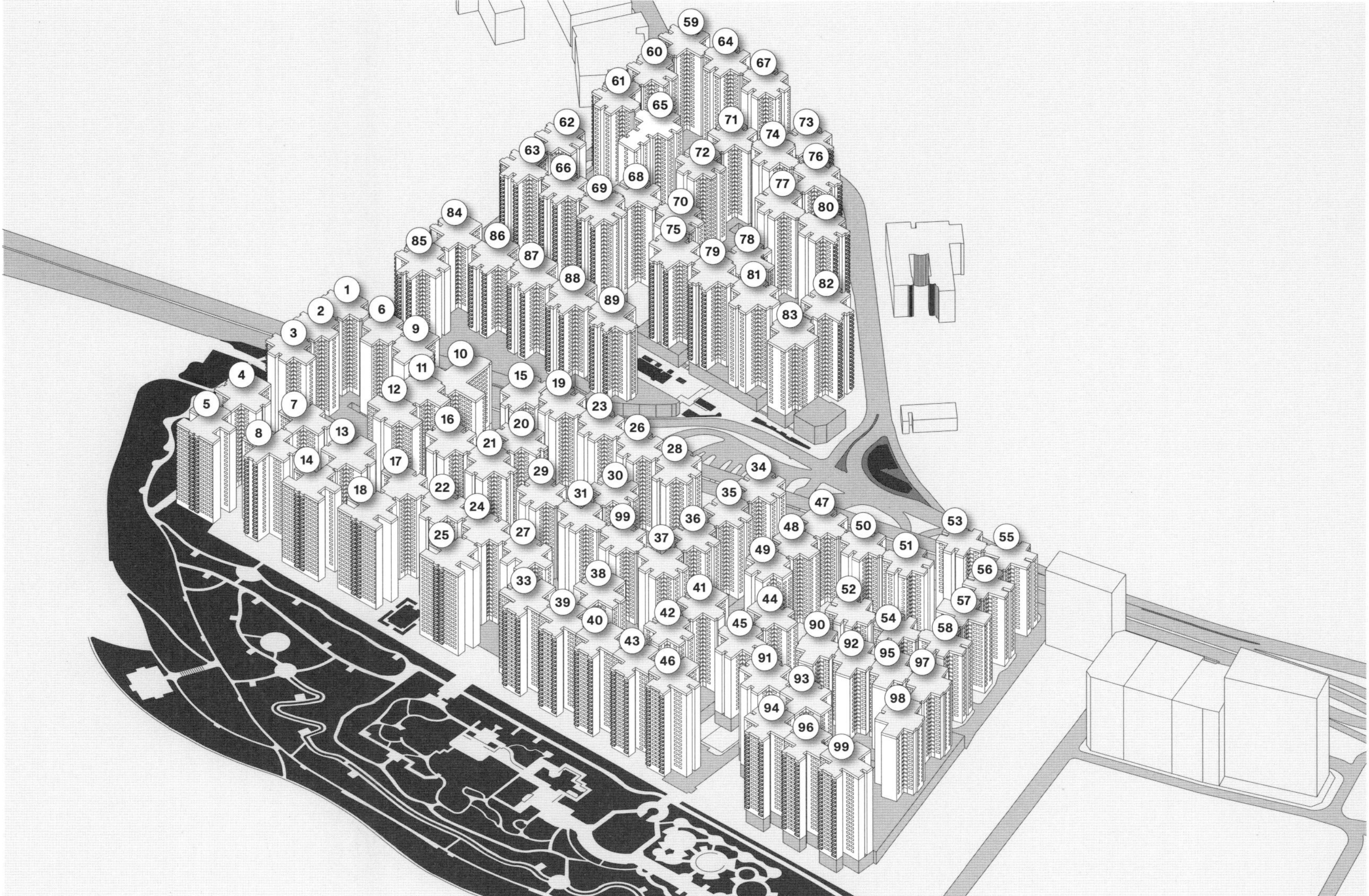

DEVELOPER:	Mei Foo Investments Ltd.		
ARCHITECT:	Wong Tung & Partners Ltd.		
LOCATION:	Lai Chi Kok		
POPULATION:	38,974		
TOWERS:	99	UNIQUE:	68
APARTMENTS:	13,149	UNIT TYPES:	223
CORES:	2	PHASES:	8

Mei Foo Sun Chuen, or simply Mei Foo, was the first large scale, privately developed housing estate built in Hong Kong. The estate sits on a forty-acre site and was built in eight phases over thirteen years between 1965 and 1978. Upon the completion of the estate's 99, twenty-story tower blocks in 1978, the complex of over 13,000 apartments was considered the largest private housing development in the world. The estate was built in reclaimed land that was previously used for fuel storage tanks by the Mobil Oil Company; "Mei Foo" is Mobil's trading name in Chinese.

The towers within the estate are built in a loose gridded pattern, clustered and organized by construction phase. An elevated multi-lane, vehicular thoroughfare, Kwai Chung Road, cuts through the estate, effectively dividing it into two parts. Towers rise up from a series of stacked horizontal levels that separate pedestrian areas above from vehicular traffic below. This thick, multi-layered podium space houses the public amenities of the estate and provide large courtyards between towers.

With a current residential community around 40,000 inhabitants (once estimated at 70,000–80,000), the complex was designed with a diverse range of amenities and programs. Effectively, it was and remains to this day, a city within a city. The Mei Foo Estate includes schools, playgrounds, supermarkets, retail stores, medical clinics, a police station, a post office, restaurants, and local food markets. Since its original development, a large bus terminus has connected the community to the larger urban context of Hong Kong and Kowloon. Mei Foo was later connected to Hong Kong's MTR (Mass Transit Railway) Tsuen Wan Line in 1982 and the West Rail Line in 2003. By Hong Kong standards the apartment sizes are generous. Every apartment in the complex offers at least two rooms, one bathroom, and a balcony. For such a large estate, Mei Foo exhibits the most diverse collection of tower block types and apartment types within the ten estates surveyed in this research project.

Type A

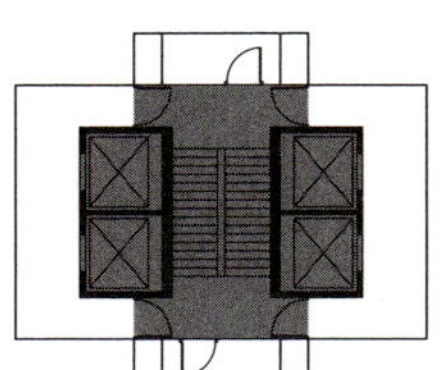

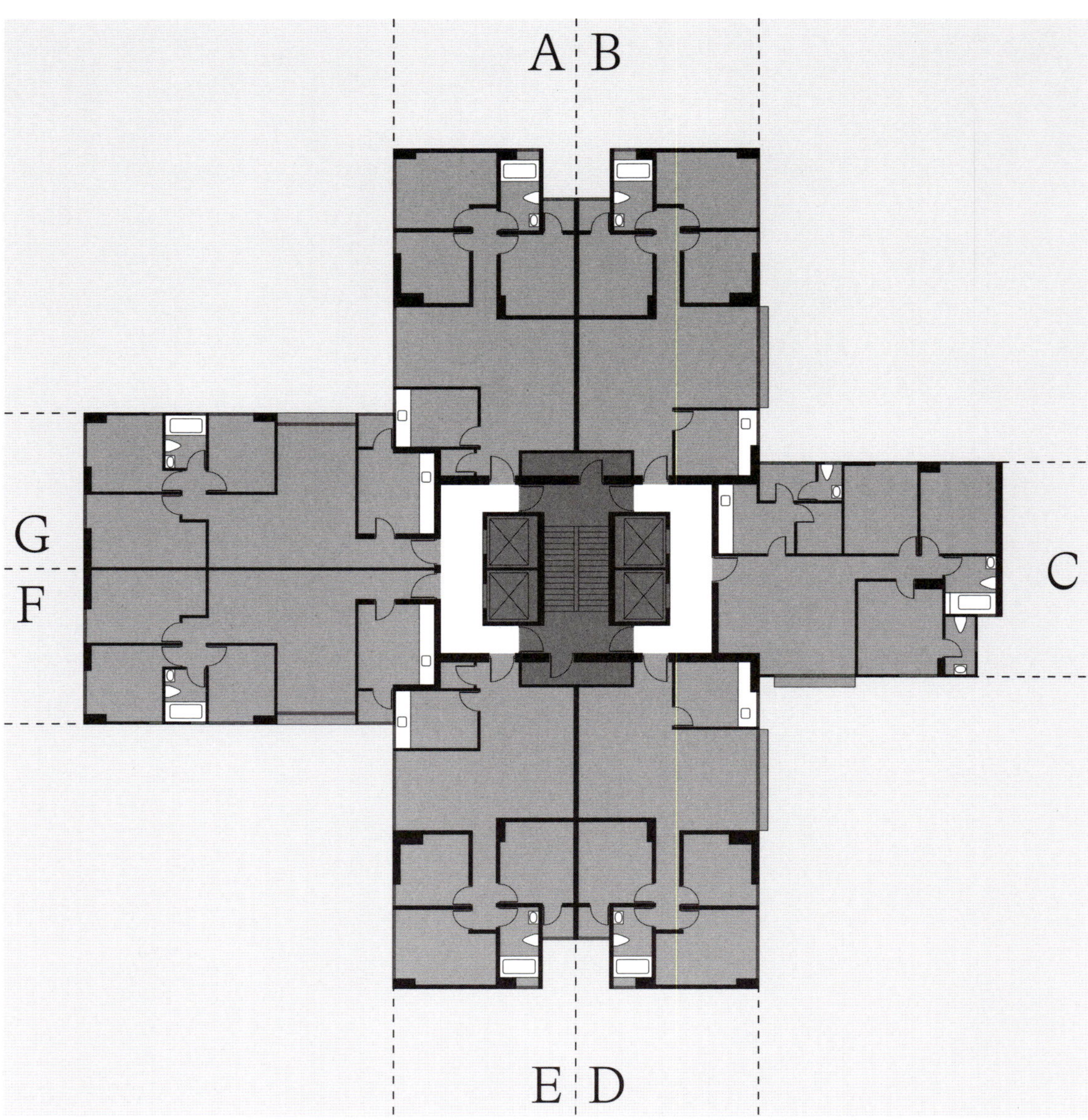

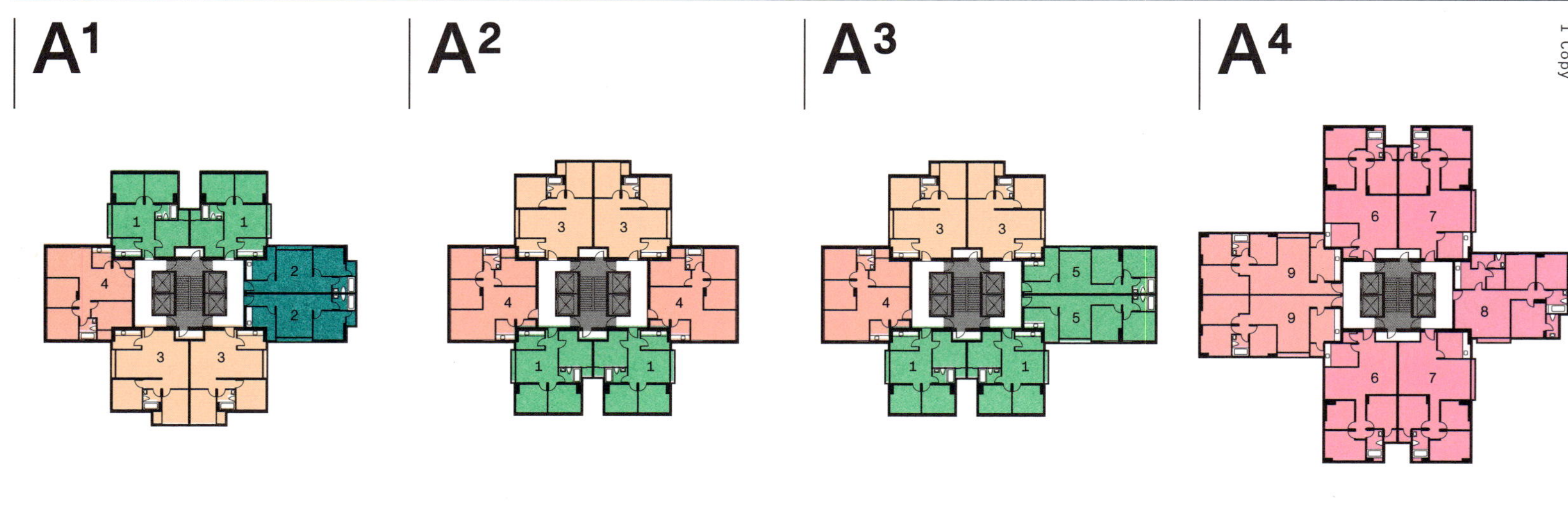

Phase 1 →/ **Block 1**

Phase 1 →/ **Block 2**

Phase 1 →/ **Block 3**

Phase 1 →/ **Block 4, 13**

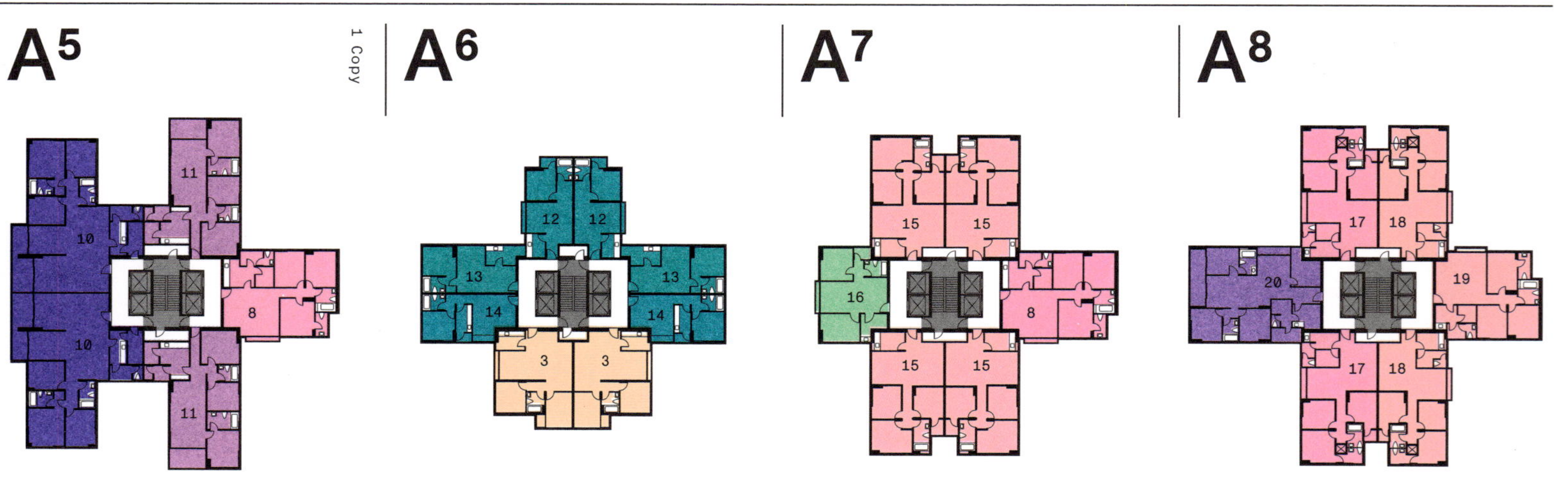

Phase 1 →/ **Block 5, 14**

Phase 1 →/ **Block 6**

Phase 1 →/ **Block 7**

Phase 1 →/ **Block 8**

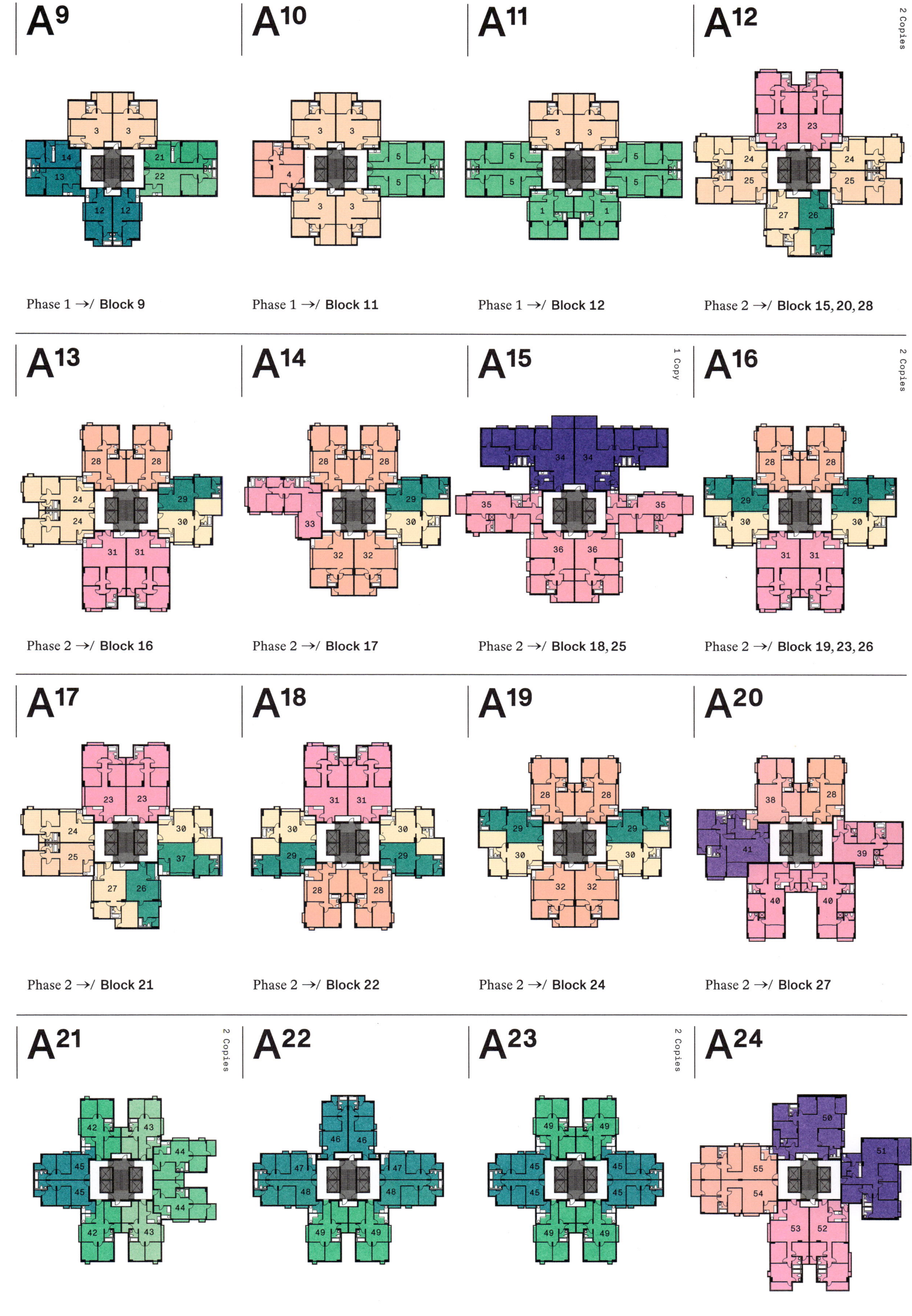
A9
3
3
14
13
21
22
12
12
Phase 1 →/ Block 9
A10
3
3
4
5
5
3
3
Phase 1 →/ Block 11
A11
3
3
5
5
5
5
1
1
Phase 1 →/ Block 12
A12
2 Copies
23
23
24
25
24
25
27
26
Phase 2 →/ Block 15, 20, 28
A13
28
28
24
24
29
30
31
31
Phase 2 →/ Block 16
A14
28
28
33
29
30
32
32
Phase 2 →/ Block 17
A15
1 Copy
34
34
35
35
36
36
Phase 2 →/ Block 18, 25
A16
2 Copies
28
28
29
30
29
30
31
31
Phase 2 →/ Block 19, 23, 26
A17
23
23
24
25
30
37
27
26
Phase 2 →/ Block 21
A18
31
31
30
29
30
29
28
28
Phase 2 →/ Block 22
A19
28
28
29
30
29
30
32
32
Phase 2 →/ Block 24
A20
38
28
41
39
40
40
Phase 2 →/ Block 27
A21
2 Copies
42
43
44
45
45
44
42
43
Phase 3 →/ Block 29, 30, 34
A22
46
46
47
48
47
48
49
49
Phase 3 →/ Block 31
A23
2 Copies
49
49
45
45
45
45
49
49
Phase 3 →/ Block 32, 35, 36
A24
50
51
55
54
53
52
Phase 3 →/ Block 33

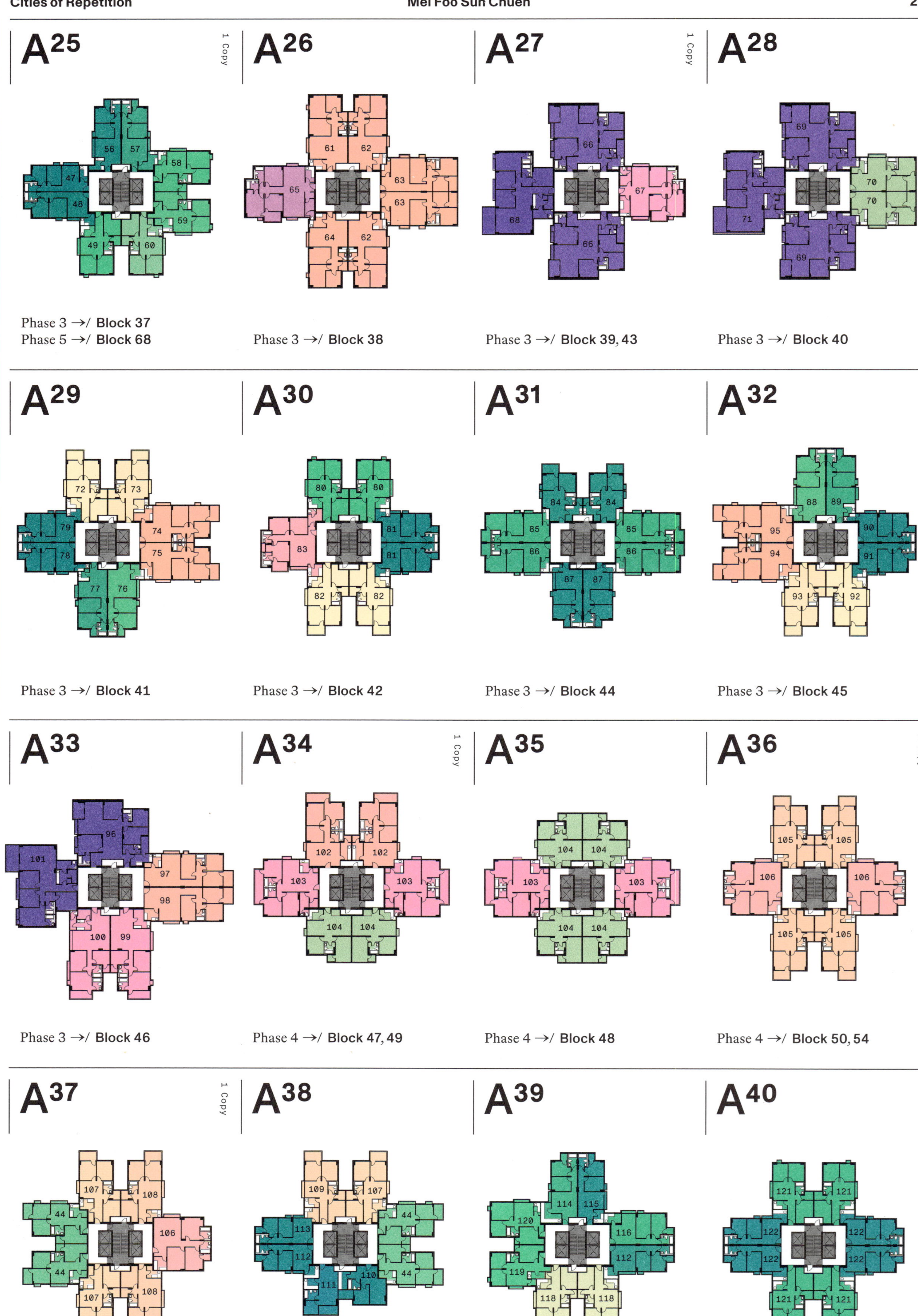

Phase 3 →/ Block 37
Phase 5 →/ Block 68

Phase 3 →/ Block 38

Phase 3 →/ Block 39, 43

Phase 3 →/ Block 40

Phase 3 →/ Block 41

Phase 3 →/ Block 42

Phase 3 →/ Block 44

Phase 3 →/ Block 45

Phase 3 →/ Block 46

Phase 4 →/ Block 47, 49

Phase 4 →/ Block 48

Phase 4 →/ Block 50, 54

Phase 4 →/ Block 51, 52

Phase 4 →/ Block 53

Phase 4 →/ Block 55

Phase 4 →/ Block 56

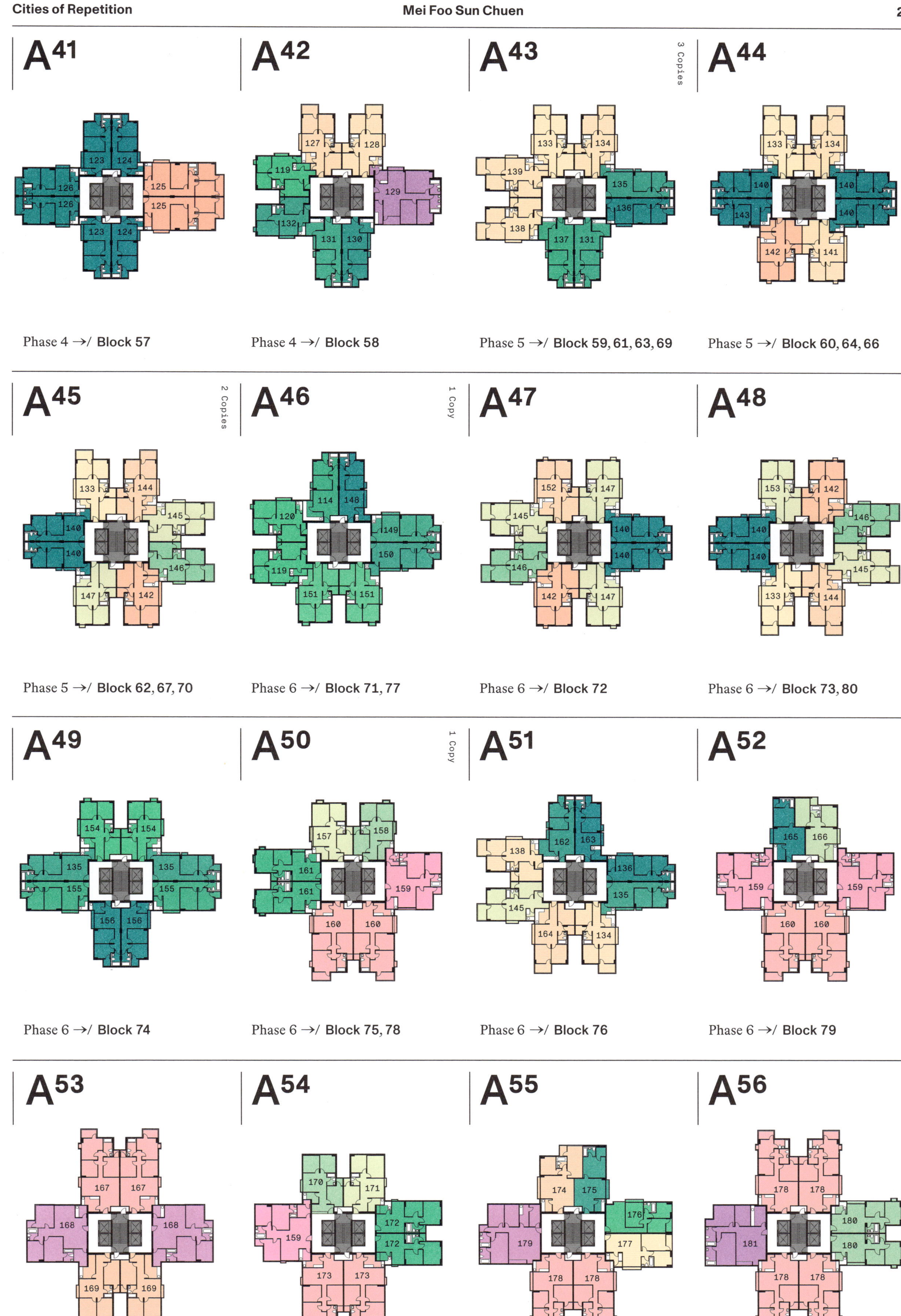
A41
Phase 4 →/ Block 57
A42
Phase 4 →/ Block 58
A43
3 Copies
Phase 5 →/ Block 59, 61, 63, 69
A44
2 Copies
Phase 5 →/ Block 60, 64, 66
A45
2 Copies
Phase 5 →/ Block 62, 67, 70
A46
1 Copy
Phase 6 →/ Block 71, 77
A47
Phase 6 →/ Block 72
A48
1 Copy
Phase 6 →/ Block 73, 80
A49
Phase 6 →/ Block 74
A50
1 Copy
Phase 6 →/ Block 75, 78
A51
Phase 6 →/ Block 76
A52
Phase 6 →/ Block 79
A53
Phase 6 →/ Block 81
A54
Phase 6 →/ Block 82
A55
Phase 7 →/ Block 84
A56
Phase 7 →/ Block 85

A57

2 Copies

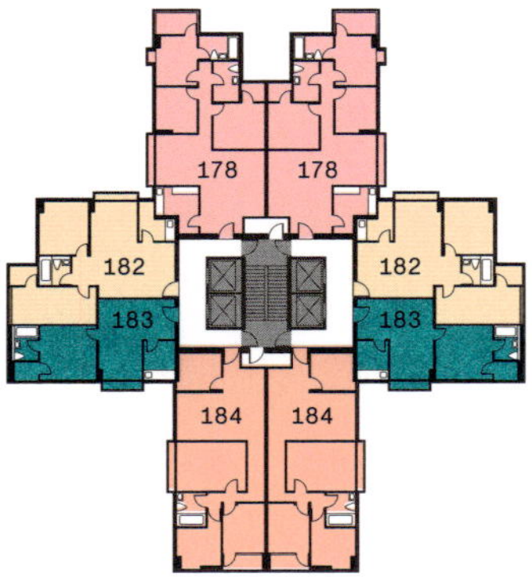

Phase 7 →/ **Block 86, 87, 88**

A58

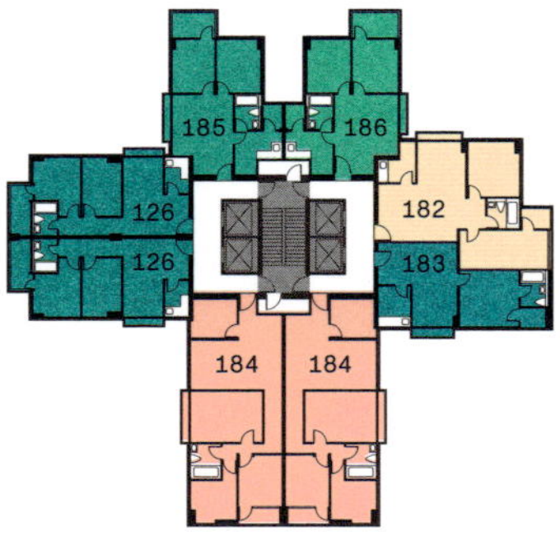

Phase 7 →/ **Block 89**

A59

3 Copies

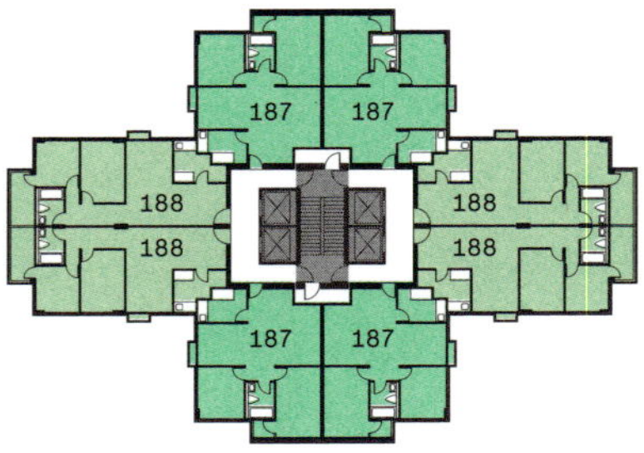

Phase 8 →/ **Block 90, 92, 95, 97**

A60

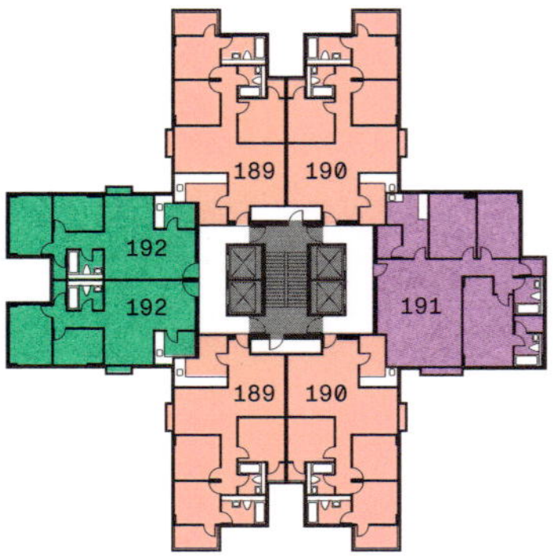

Phase 8 →/ **Block 91**

A61

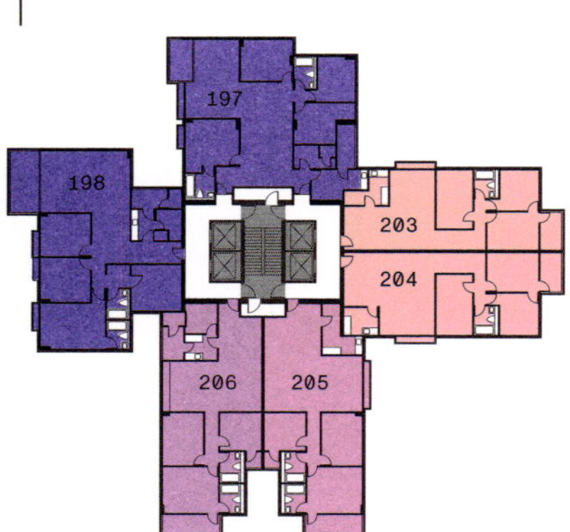

Phase 8 →/ **Block 93**

A62

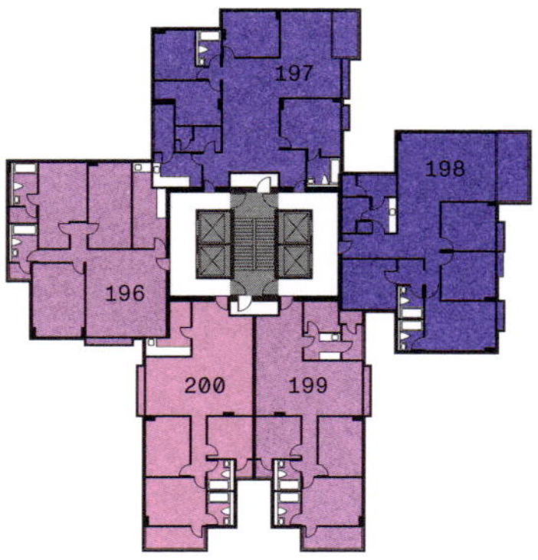

Phase 8 →/ **Block 94**

A63

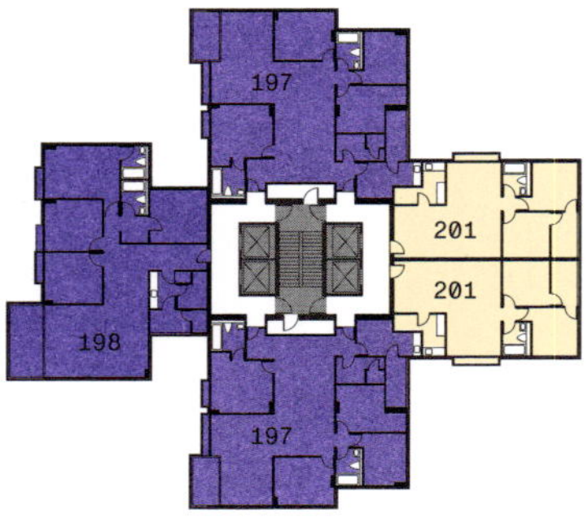

Phase 8 →/ **Block 96**

A64

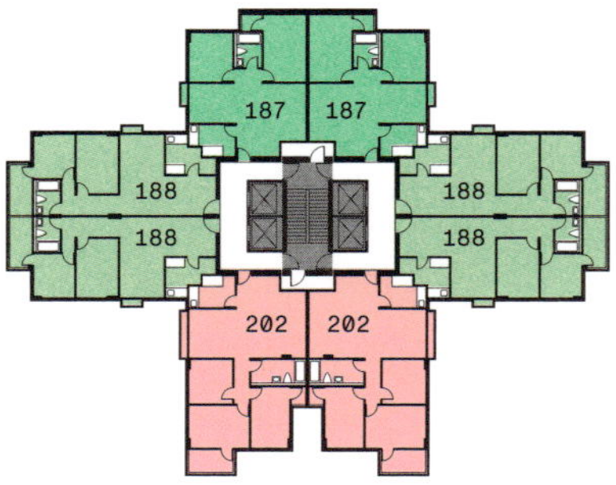

Phase 8 →/ **Block 98**

A65

Phase 8 →/ **Block 99**

Type B

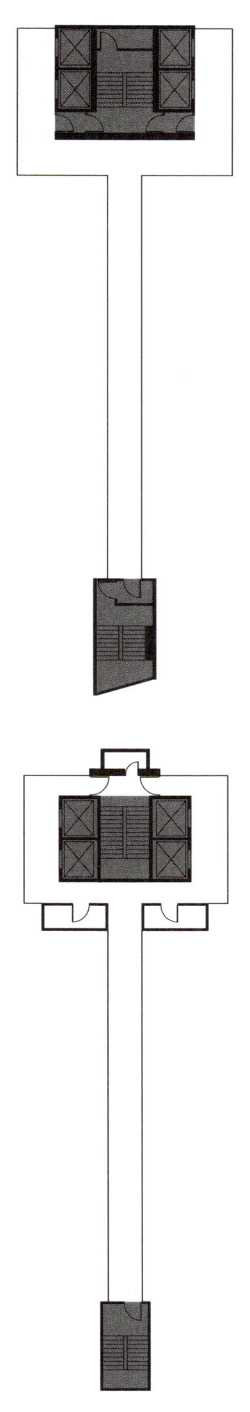

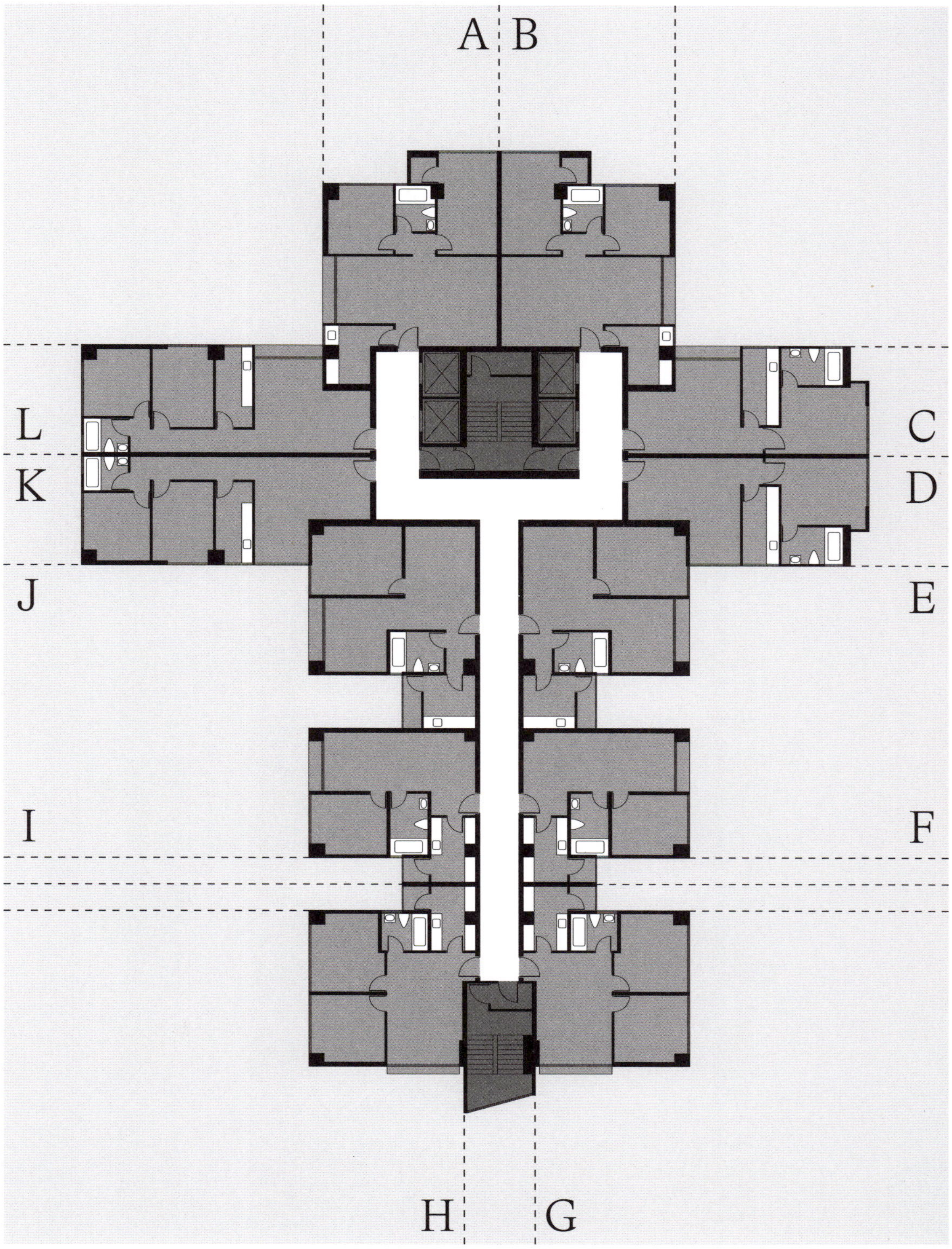

A B

L C

K D

J E

I F

H G

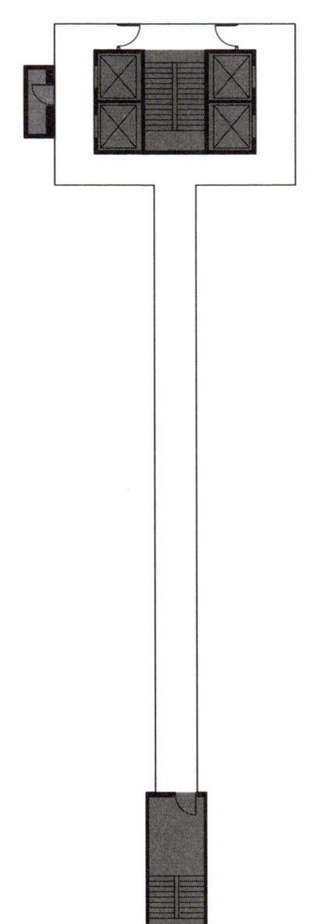

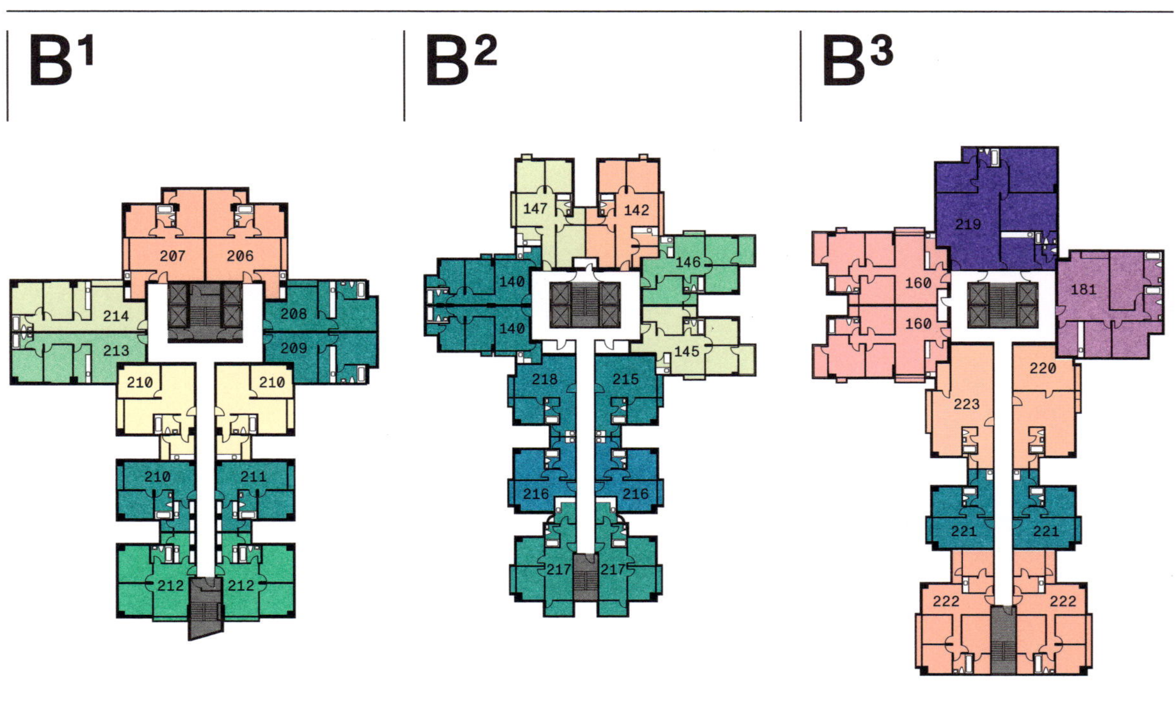

Phase 1 →/ **Block 10**

Phase 5 →/ **Block 65**

Phase 6 →/ **Block 83**

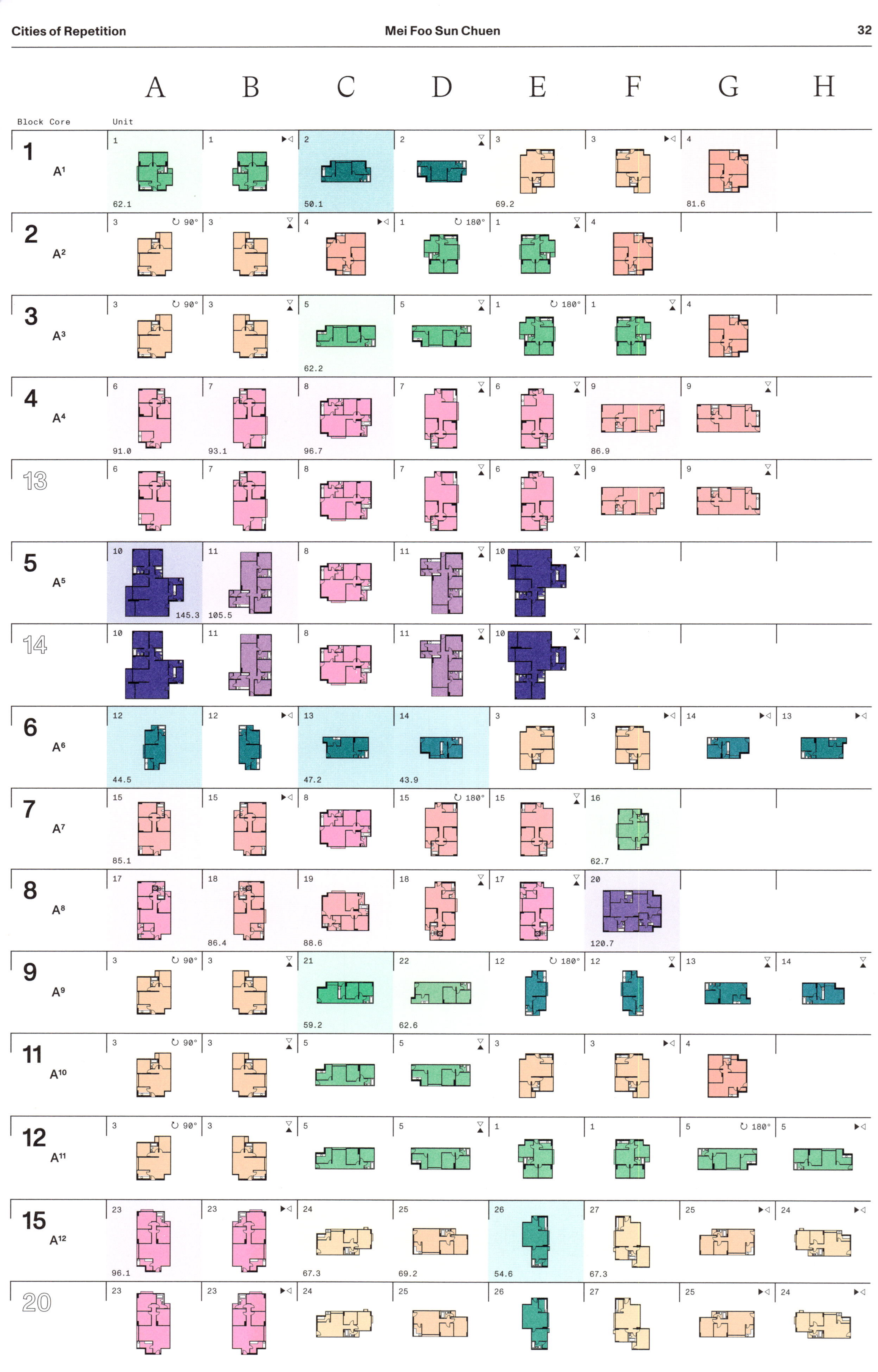
A B C D E F G H
Block Core
Unit
1 A1
1 62.1
1
2 50.1
2
3 69.2
3
4 81.6
2 A2
3 90°
3
4
1 180°
1
4
3 A3
3 90°
3
5 62.2
5
1 180°
1
4
4 A4
6 91.0
7 93.1
8 96.7
7
6
9 86.9
9
13
6
7
8
7
6
9
9
5 A5
10 145.3
11 105.5
8
11
10
14
10
11
8
11
10
6 A6
12 44.5
12
13 47.2
14 43.9
3
3
14
13
7 A7
15 85.1
15
8
15 180°
15
16 62.7
8 A8
17
18 86.4
19 88.6
18
17
20 120.7
9 A9
3 90°
3
21 59.2
22 62.6
12 180°
12
13
14
11 A10
3 90°
3
5
5
3
3
4
12 A11
3 90°
3
5
5
1
1
5 180°
5
15 A12
23 96.1
23
24 67.3
25 69.2
26 54.6
27 67.3
25
24
20
23
23
24
25
26
27
25
24

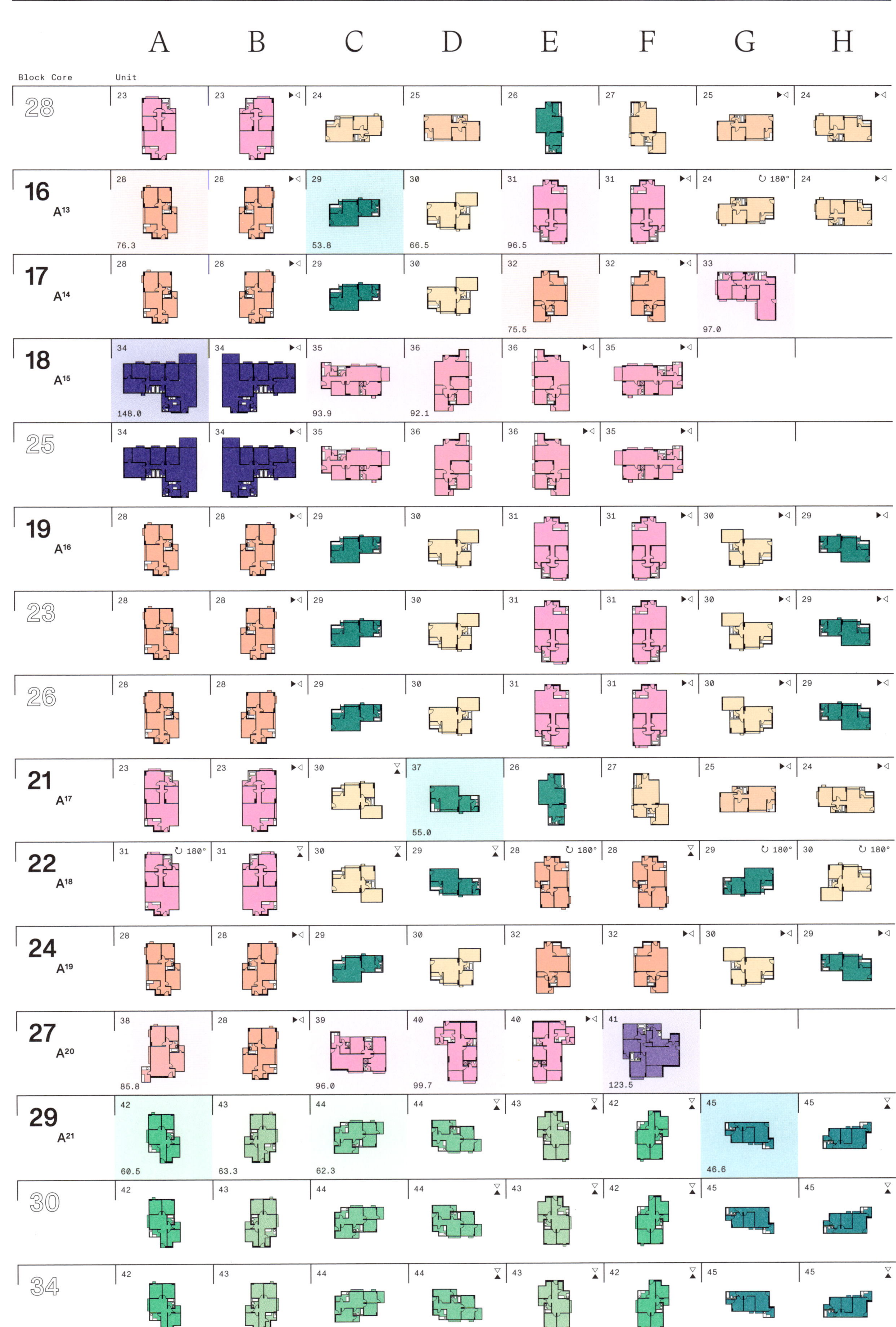
A B C D E F G H
Block Core
Unit
28
23 23 24 25 26 27 25 24
16 A13
28 76.3 28 29 53.8 30 66.5 31 96.5 31 24 180° 24
17 A14
28 28 29 30 32 75.5 32 33 97.0
18 A15
34 148.0 34 35 93.9 36 92.1 36 35
25
34 34 35 36 36 35
19 A16
28 28 29 30 31 31 30 29
23
28 28 29 30 31 31 30 29
26
28 28 29 30 31 31 30 29
21 A17
23 23 30 37 55.0 26 27 25 24
22 A18
31 180° 31 30 29 28 180° 28 29 180° 30 180°
24 A19
28 28 29 30 32 32 30 29
27 A20
38 85.8 28 39 96.0 40 99.7 40 41 123.5
29 A21
42 60.5 43 63.3 44 62.3 44 43 42 45 46.6 45
30
42 43 44 44 43 42 45 45
34
42 43 44 44 43 42 45 45

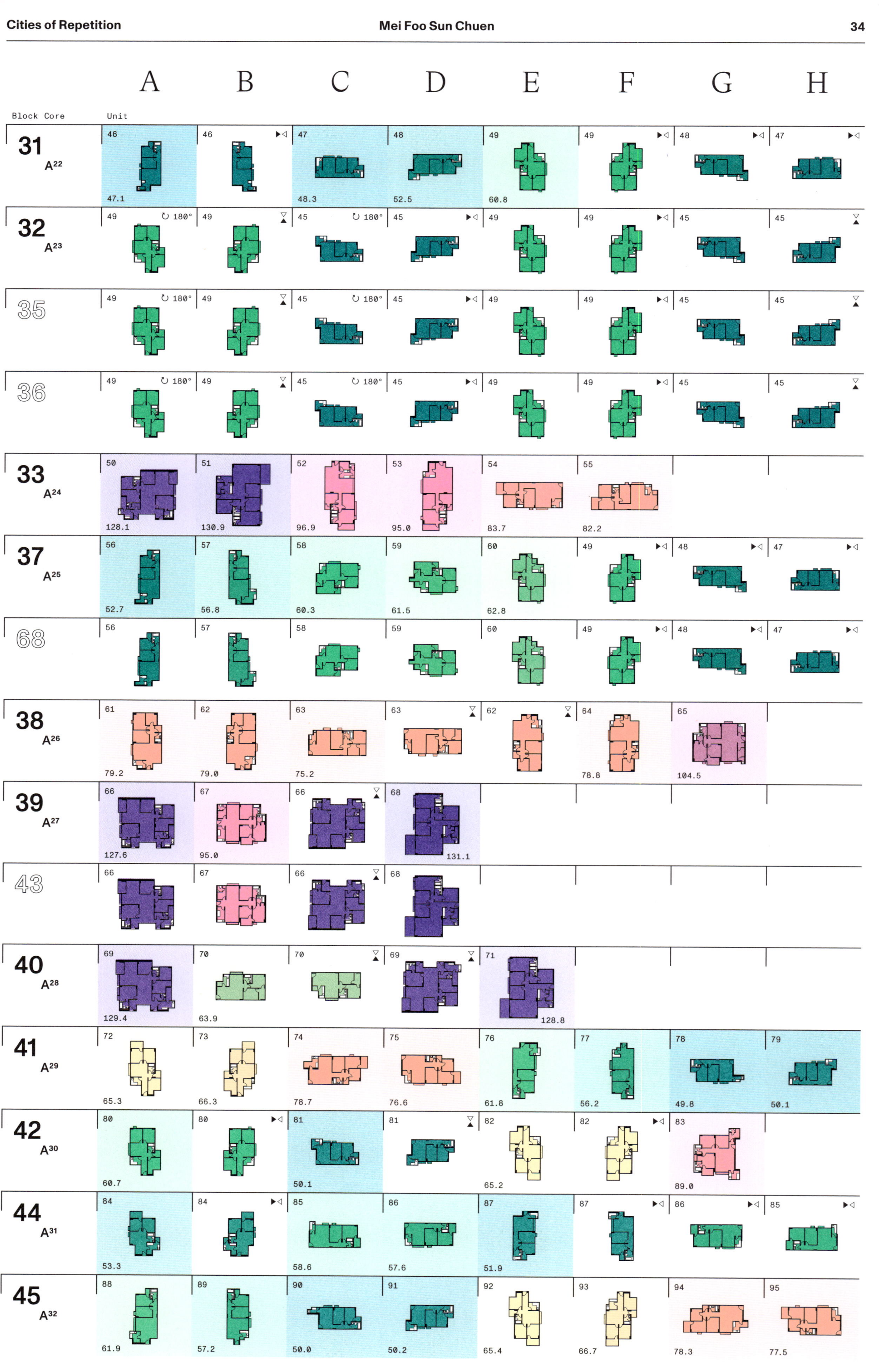
A B C D E F G H
Block Core
Unit
31 A22
46 47.1
46
47 48.3
48 52.5
49 60.8
49
48
47
32 A23
49 180°
49
45 180°
45
49
49
45
45
35
49 180°
49
45 180°
45
49
49
45
45
36
49 180°
49
45 180°
45
49
49
45
45
33 A24
50 128.1
51 130.9
52 96.9
53 95.0
54 83.7
55 82.2
37 A25
56 52.7
57 56.8
58 60.3
59 61.5
60 62.8
49
48
47
68
56
57
58
59
60
49
48
47
38 A26
61 79.2
62 79.0
63 75.2
63
62
64 78.8
65 104.5
39 A27
66 127.6
67 95.0
66
68 131.1
43
66
67
66
68
40 A28
69 129.4
70 63.9
70
69
71 128.8
41 A29
72 65.3
73 66.3
74 78.7
75 76.6
76 61.8
77 56.2
78 49.8
79 50.1
42 A30
80 60.7
80
81 50.1
81
82 65.2
82
83 89.0
44 A31
84 53.3
84
85 58.6
86 57.6
87 51.9
87
86
85
45 A32
88 61.9
89 57.2
90 50.0
91 50.2
92 65.4
93 66.7
94 78.3
95 77.5

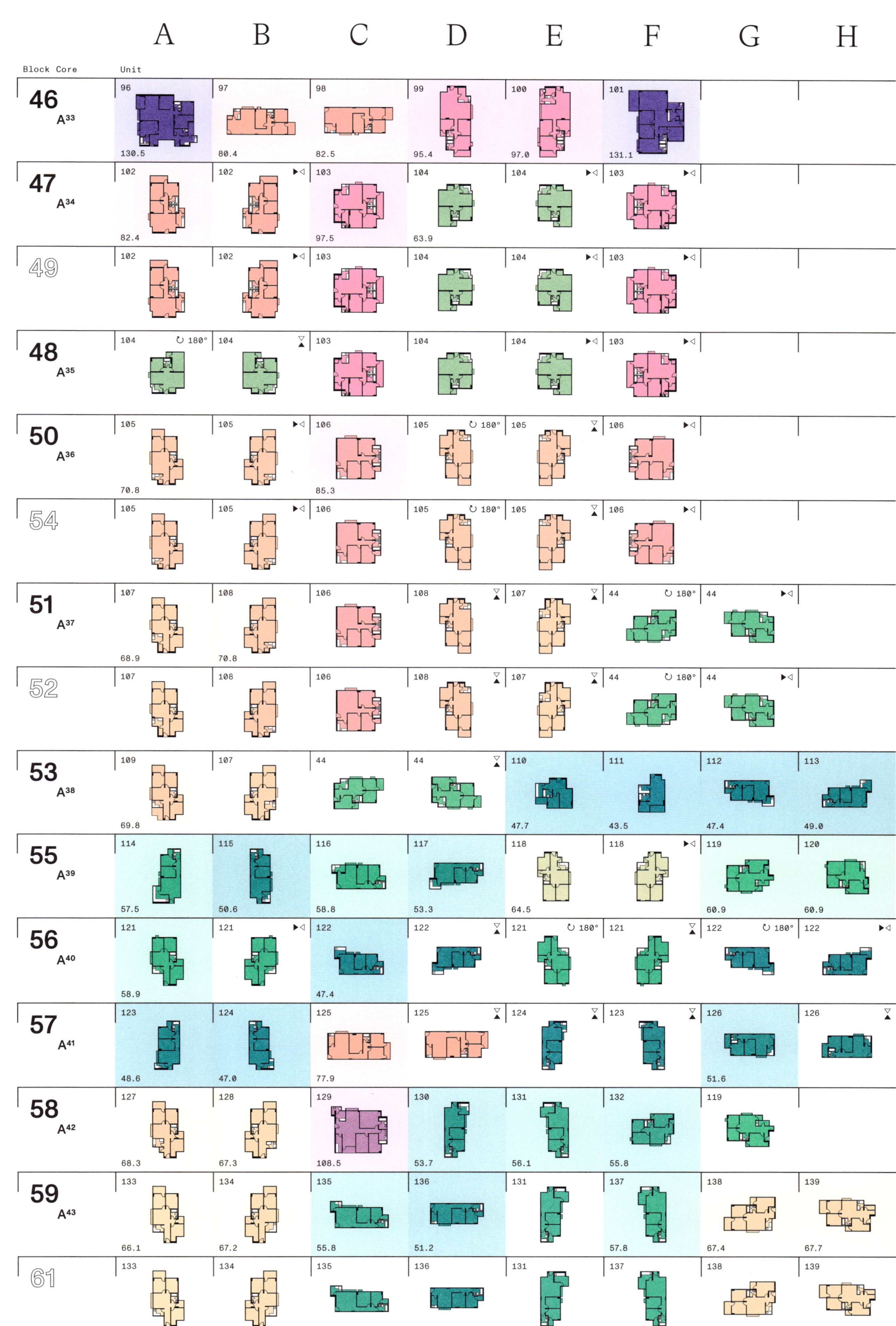
A B C D E F G H
Block Core
Unit
46 A33
96 130.5
97 80.4
98 82.5
99 95.4
100 97.0
101 131.1
47 A34
102 82.4
102
103 97.5
104 63.9
104
103
49
102
102
103
104
104
103
48 A35
104 180°
104
103
104
104
103
50 A36
105 70.8
105
106 85.3
105 180°
105
106
54
105
105
106
105 180°
105
106
51 A37
107 68.9
108 70.8
106
108
107
44 180°
44
52
107
108
106
108
107
44 180°
44
53 A38
109 69.8
107
44
44
110 47.7
111 43.5
112 47.4
113 49.0
55 A39
114 57.5
115 50.6
116 58.8
117 53.3
118 64.5
118
119 60.9
120 60.9
56 A40
121 58.9
121
122 47.4
122
121 180°
121
122 180°
122
57 A41
123 48.6
124 47.0
125 77.9
125
124
123
126 51.6
126
58 A42
127 68.3
128 67.3
129 108.5
130 53.7
131 56.1
132 55.8
119
59 A43
133 66.1
134 67.2
135 55.8
136 51.2
131
137 57.8
138 67.4
139 67.7
61
133
134
135
136
131
137
138
139

Block Core	A	B	C	D	E	F	G	H
	Unit							
63	133	134	135	136	131	137	138	139
69	133	134	135	136	131	137	138	139
60 A44	133	134	140 49.9	140 ▽▲	141 67.3	142 72.7	143 46.8	140 ▶◁
64	133	134	140	140 ▽▲	141	142	143	140 ▶◁
66	133	134	140	140 ▽▲	141	142	143	140 ▶◁
62 A45	133	144 69.8	145 64.9	146 62.8	142 ▶◁	147 64.5	140 ↻ 180°	140 ▶◁
67	133	144	145	146	142 ▶◁	147	140 ↻ 180°	140 ▶◁
70	133	144	145	146	142 ▶◁	147	140 ↻ 180°	140 ▶◁
71 A46	114	148 50.9	149 56.5	150 55.9	151 61.9	151 ▶◁	119	120
77	114	148	149	150	151	151 ▶◁	119	120
72 A47	152 69.9	147 ↻ 180°	140	140 ▽▲	147 ▶◁	142	146 ▶◁	145 ▶◁
73 A48	153 64.7	142 ↻ 180°	146 ▽▲	145 ▽▲	144 ▽▲	133 ▽▲	140 ↻ 180°	140 ▶◁
80	153	142 ↻ 180°	146 ▽▲	145 ▽▲	144 ▽▲	133 ▽▲	140 ↻ 180°	140 ▶◁
74 A49	154 60.8	154 ▶◁	135	155 57.5	156 50.6	156 ▶◁	155 ▶◁	135 ▶◁
75 A50	157 64.8	158 63.0	159 93.8	160 86.8	160 ▶◁	161 58.9	161 ▽▲	

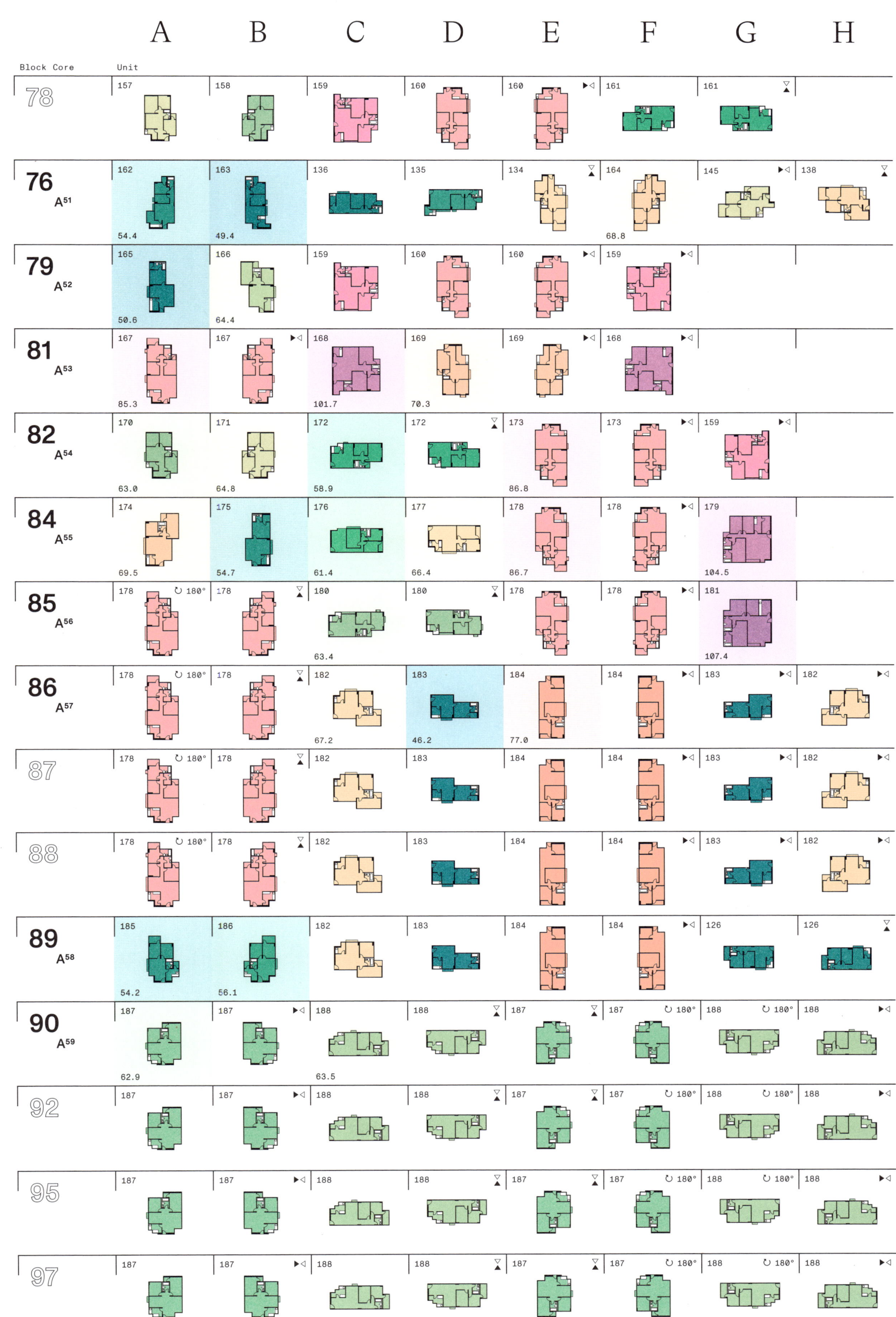

A B C D E F G H
Block Core
Unit
78
157
158
159
160
160
161
161
76
A51
162 54.4
163 49.4
136
135
134
164 68.8
145
138
79
A52
165 50.6
166 64.4
159
160
160
159
81
A53
167 85.3
167
168 101.7
169 70.3
169
168
82
A54
170 63.0
171 64.8
172 58.9
172
173 86.8
173
159
84
A55
174 69.5
175 54.7
176 61.4
177 66.4
178 86.7
178
179 104.5
85
A56
178 180°
178
180 63.4
180
178
178
181 107.4
86
A57
178 180°
178
182 67.2
183 46.2
184 77.0
184
183
182
87
178 180°
178
182
183
184
184
183
182
88
178 180°
178
182
183
184
184
183
182
89
A58
185 54.2
186 56.1
182
183
184
184
126
126
90
A59
187 62.9
187
188 63.5
188
187
187 180°
188 180°
188
92
187
187
188
188
187
187 180°
188 180°
188
95
187
187
188
188
187
187 180°
188 180°
188
97
187
187
188
188
187
187 180°
188 180°
188

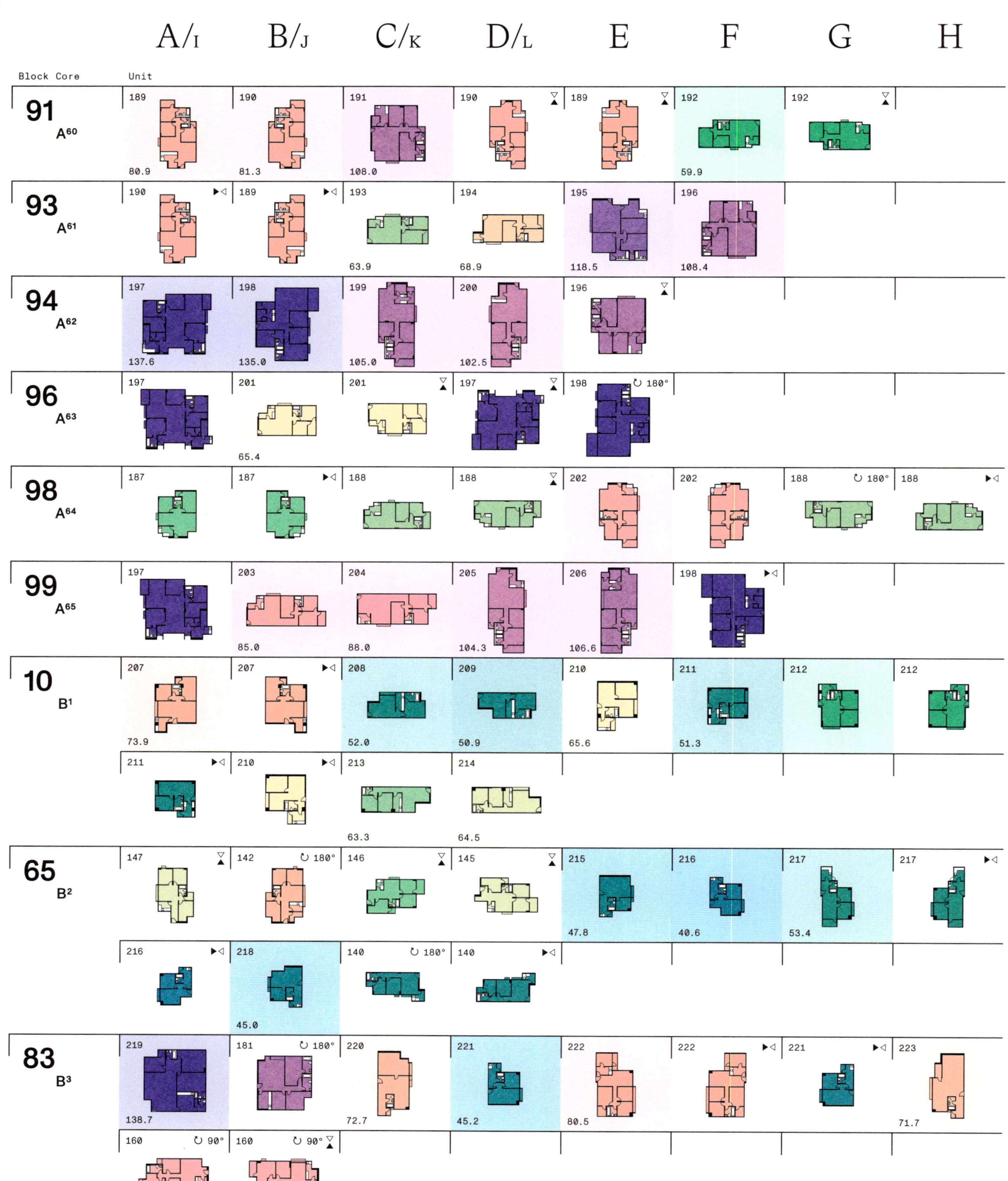
A/I
B/J
C/K
D/L
E
F
G
H
Block Core
Unit
91
A60
189
80.9
190
81.3
191
108.0
190
189
192
59.9
192
93
A61
190
189
193
63.9
194
68.9
195
118.5
196
108.4
94
A62
197
137.6
198
135.0
199
105.0
200
102.5
196
96
A63
197
201
65.4
201
197
198
180°
98
A64
187
187
188
188
202
202
188
180°
188
99
A65
197
203
85.0
204
88.0
205
104.3
206
106.6
198
10
B1
207
73.9
207
208
52.0
209
50.9
210
65.6
211
51.3
212
212
211
210
213
63.3
214
64.5
65
B2
147
142
180°
146
145
215
47.8
216
40.6
217
53.4
217
216
218
45.0
140
180°
140
83
B3
219
138.7
181
180°
220
72.7
221
45.2
222
80.5
222
221
223
71.7
160
90°
160
90°

Taikoo Shing 1977–89

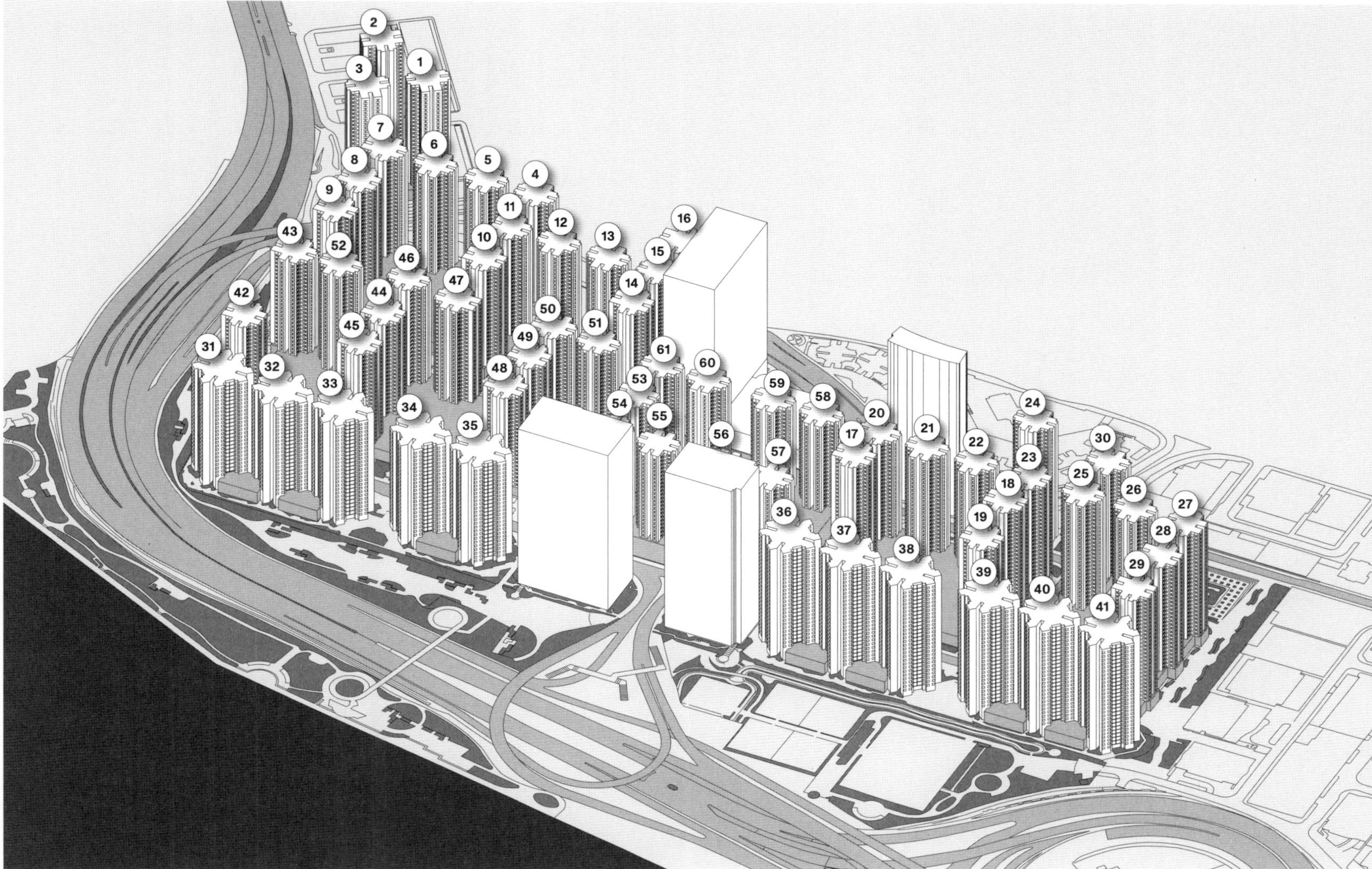

DEVELOPER:	Swire Properties		
ARCHITECT:	Wong Tung & Partners Limited		
LOCATION:	Quarry Bay		
POPULATION:	36,796		
TOWERS:	61	UNIQUE:	29
APARTMENTS:	12,696	UNIT TYPES:	28
CORES:	2	PHASES:	8

Taikoo Shing was built by Swire Properties on the former site of the Taikoo Dockyard after the dockyard was relocated to Tsing Yi Island. The blocks of the estate are well integrated into the surrounding street grid of the larger neighborhood. The estate is located directly adjacent to the MTR subway station and is connected to the city with many bus lines. In the center of the estate are two large office towers and a shopping mall. With the level of connectivity and amenities that Taikoo Shing provides to its residents and the larger community, it is one of the most thoughtfully integrated into its urban context of all the Hong Kong's larger housing estates.

The podium space of Taikoo Shing is quite sophisticated in that it provides mixed-use programs on different levels. A lower layer of commercial space faces outward to integrate the base of the estate into the urban context, and an upper level of commercial space is turned inward toward the courtyards to service the residents of the estate. The ground level of the podium extends to the periphery of the block and forms a consistent street edge. Retail shops and restaurants fill the perimeter of the block on the main roads to provide the neighborhood with an active street life. A second level of publically accessible commercial space on the upper level of the podium provides space for businesses and amenities more closely related to the residential towers. The area includes real estate offices, child-care facilities, as well as dry cleaning and laundry services.

Towers are organized in a gridded formation around large open spaces. These large-scale courtyards provide leisure and recreational spaces for the residential community. On the north edge of the estate is a row of towers rotated forty-five degrees in relation to the grid of the overall estate. These towers, angled to provide sweeping views of Victoria Harbor, house larger, more luxurious units designed to bring in higher rents. This row of towers also has its own exclusive clubhouse and recreational facilities.

Type A

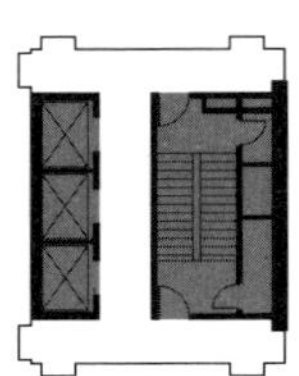

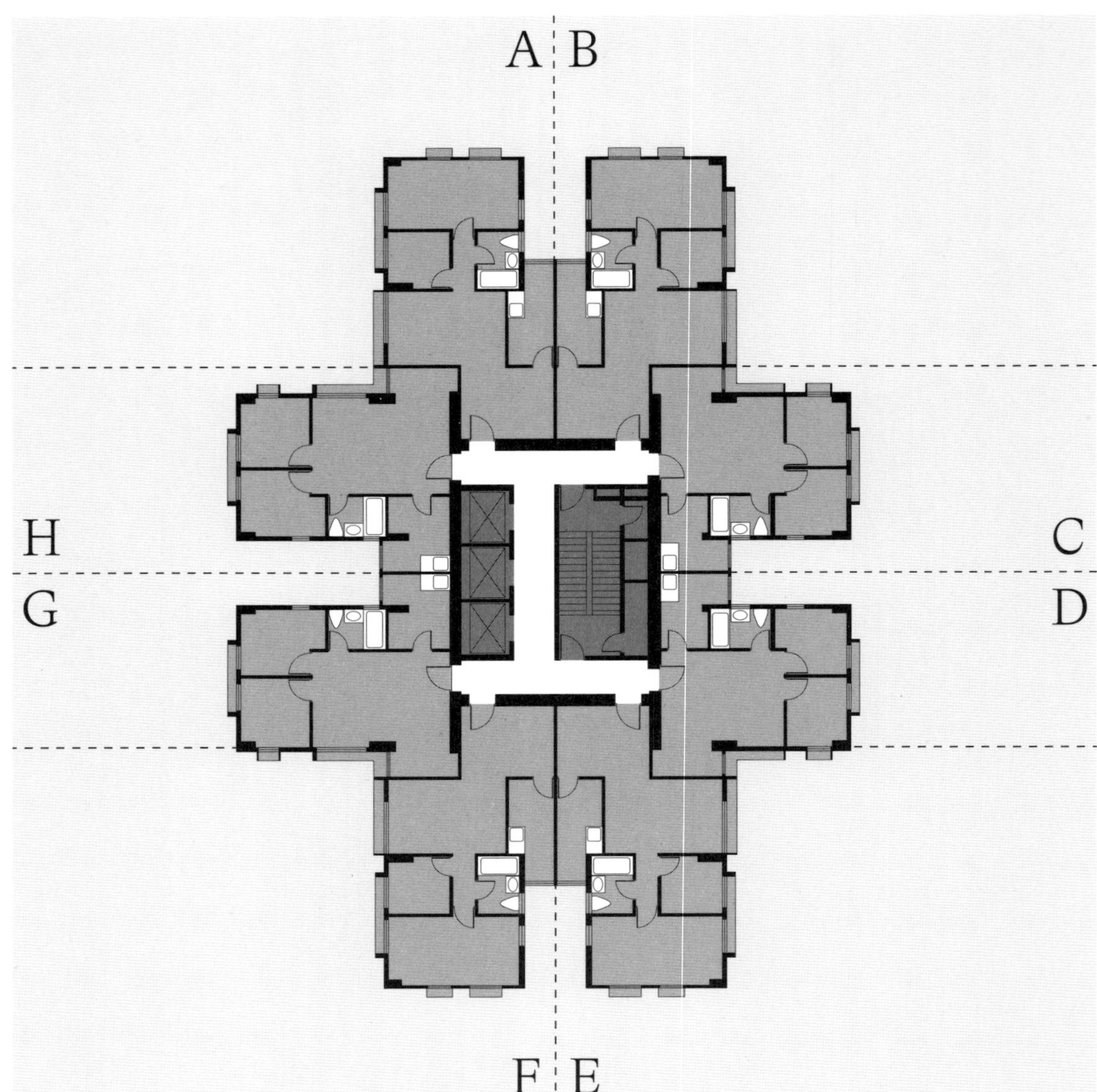

Type B

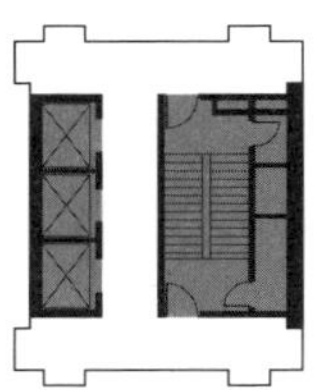

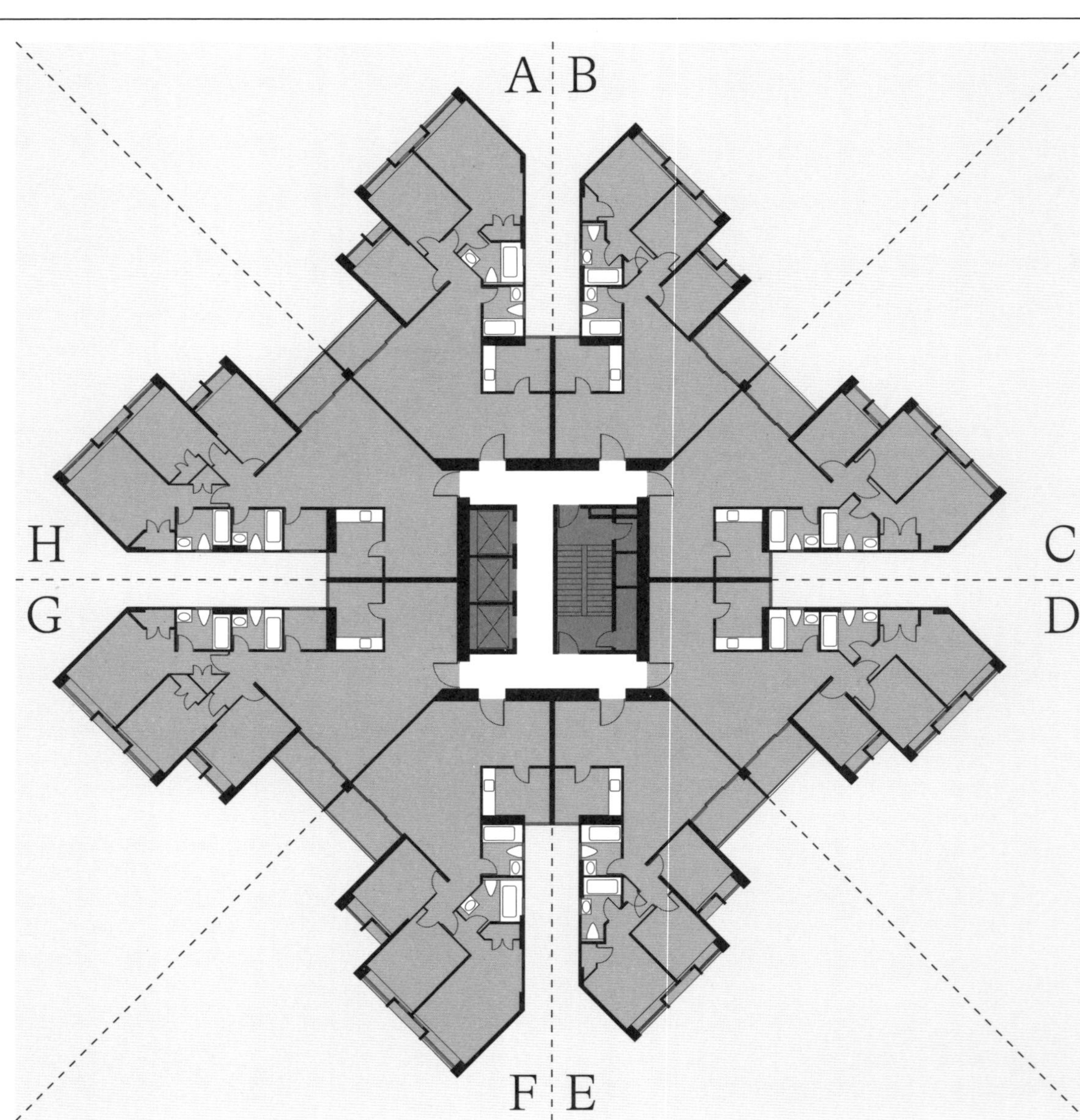

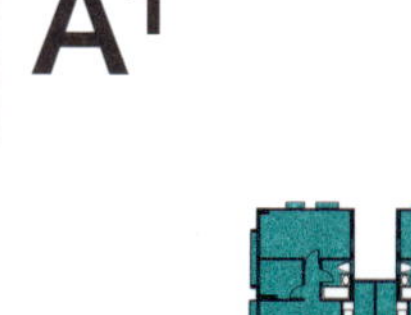

A1

4 Copies

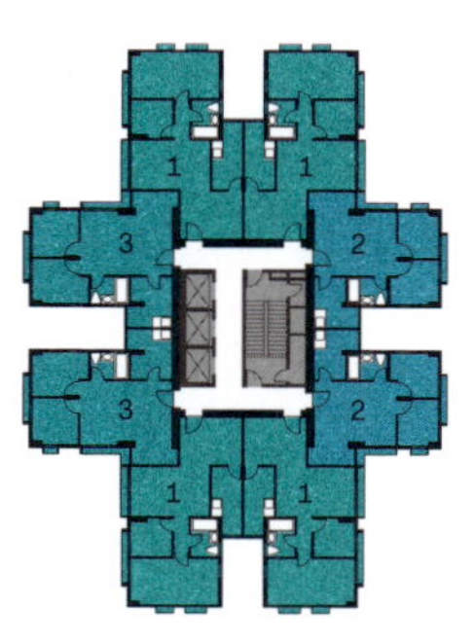

Phase 1 →/ **Block 1–3**
Phase 3 →/ **Block 17**
Phase 6 →/ **Block 52**

A2

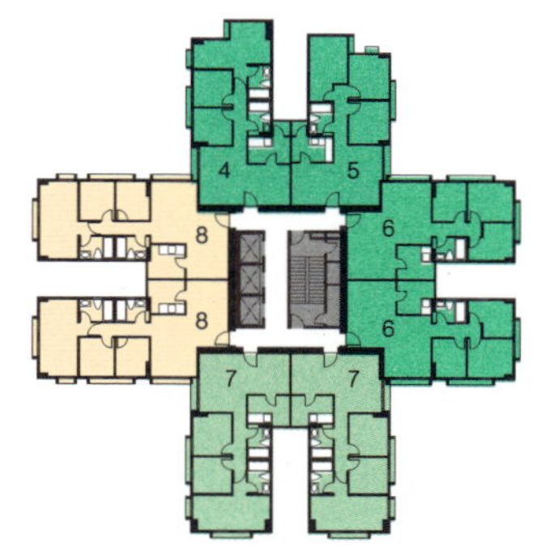

Phase 2 →/ **Block 4**

A3

4 Copies

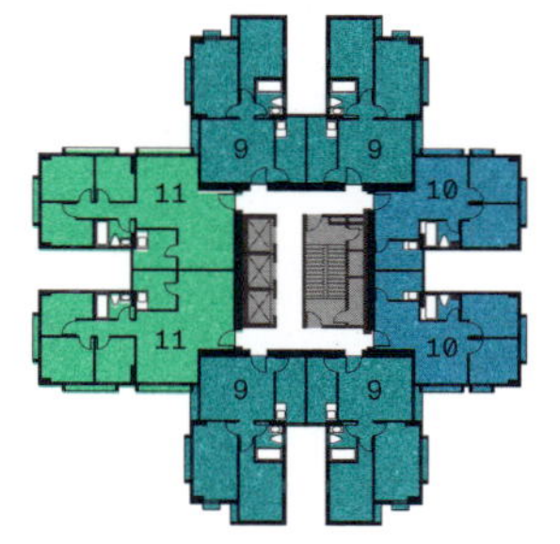

Phase 2 →/ **Block 5, 7, 8, 12, 13**

A4

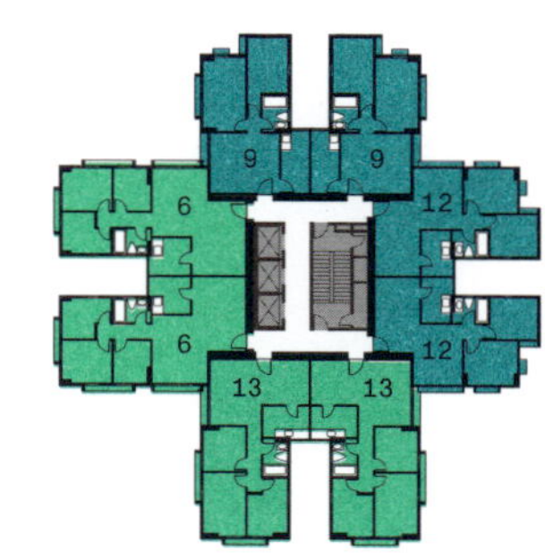

Phase 2 →/ **Block 6**

A5

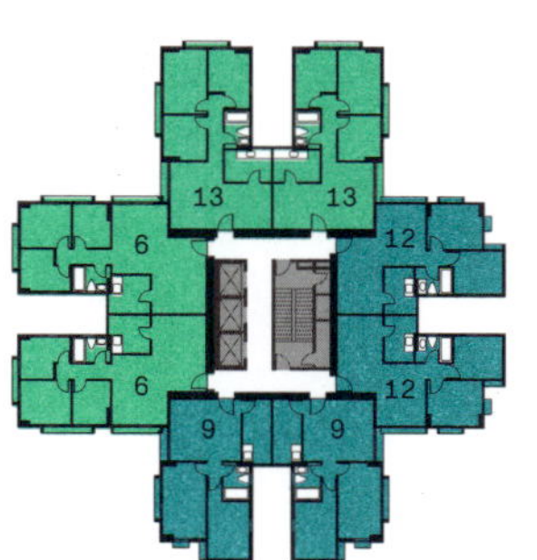

Phase 2 →/ **Block 9**

A6

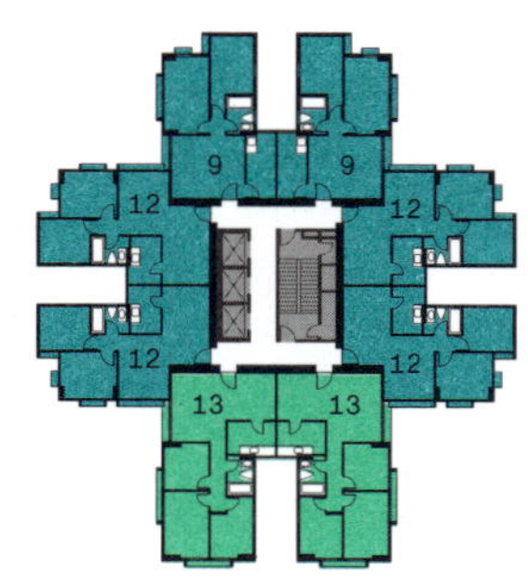

Phase 2 →/ **Block 10**

A7

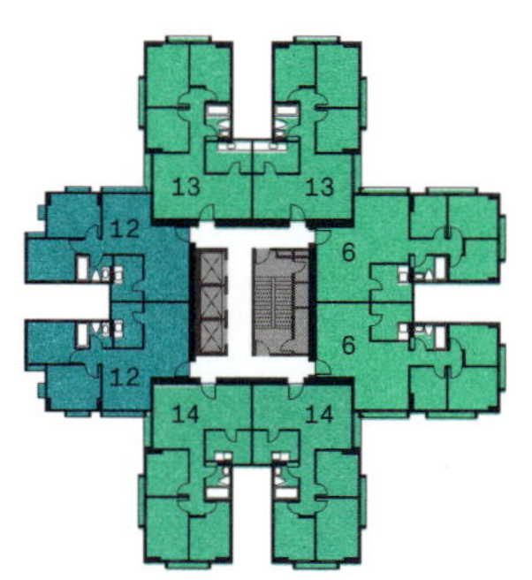

Phase 2 →/ **Block 11**

A8

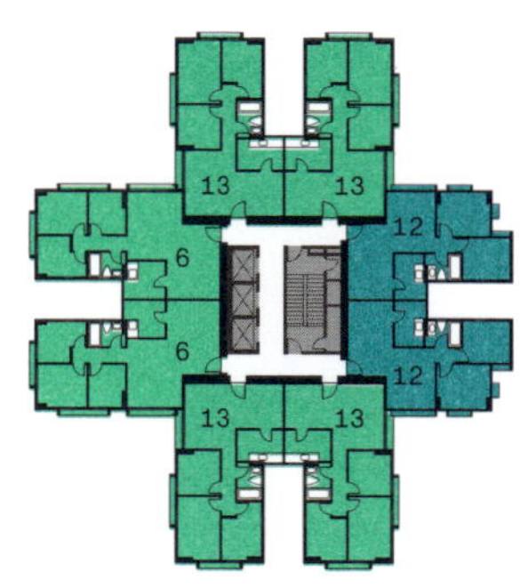

Phase 2 →/ **Block 14**

A9

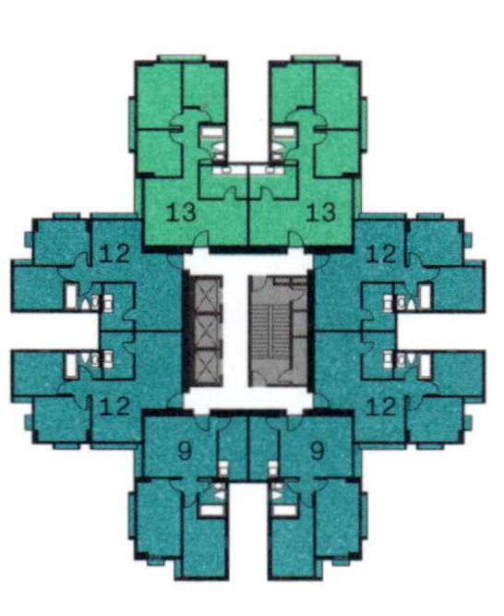

Phase 2 →/ **Block 15**

A10

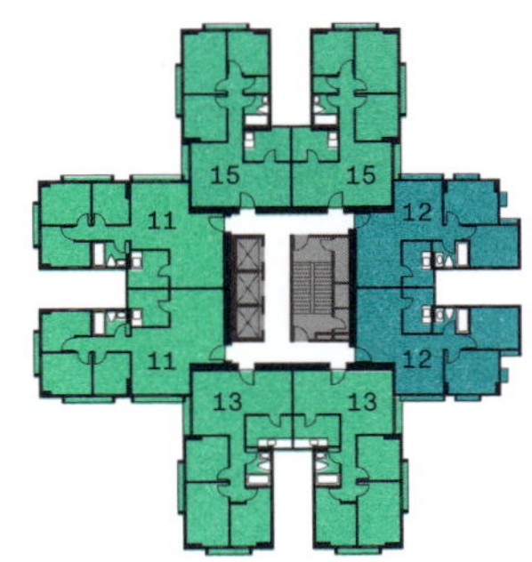

Phase 2 →/ **Block 16**

A11

2 Copies

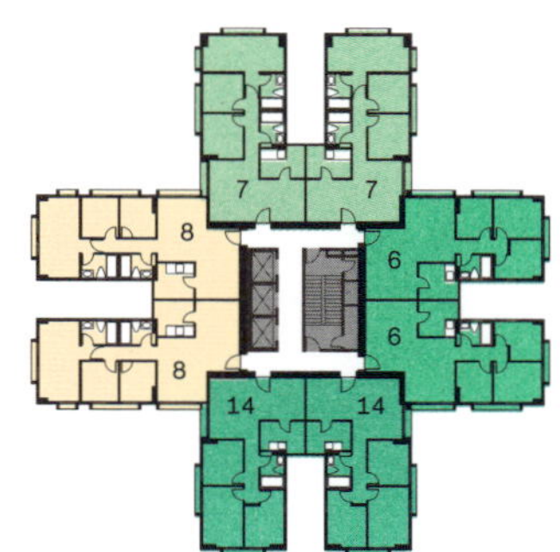

Phase 3 →/ **Block 18, 24**
Phase 7 →/ **Block 44**

A12

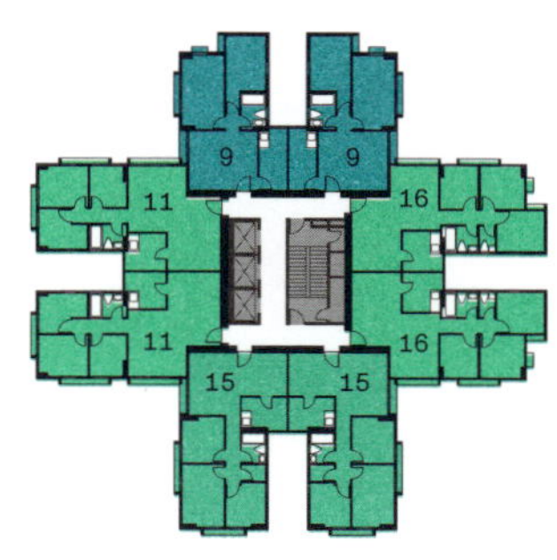

Phase 3 →/ **Block 19**

A13

2 Copies

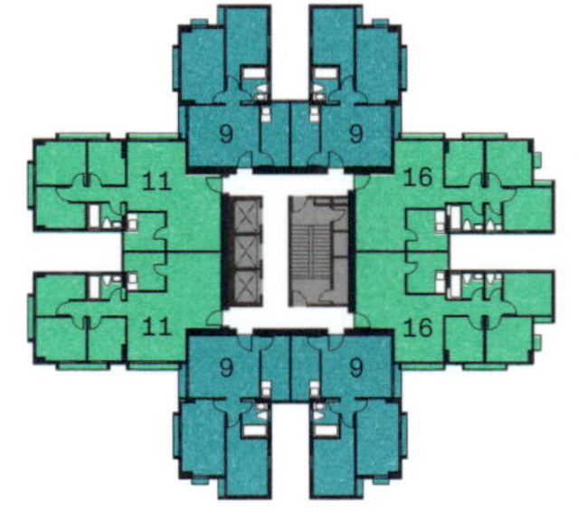

Phase 3 →/ **Block 20, 21, 23**

A14

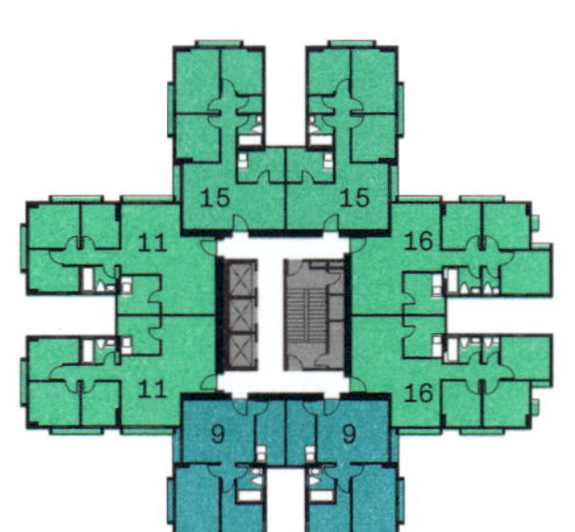

Phase 3 →/ **Block 22**

A15

3 Copies

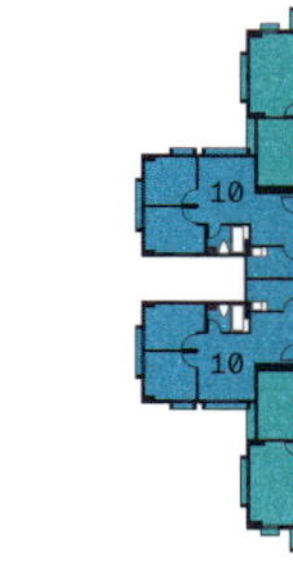
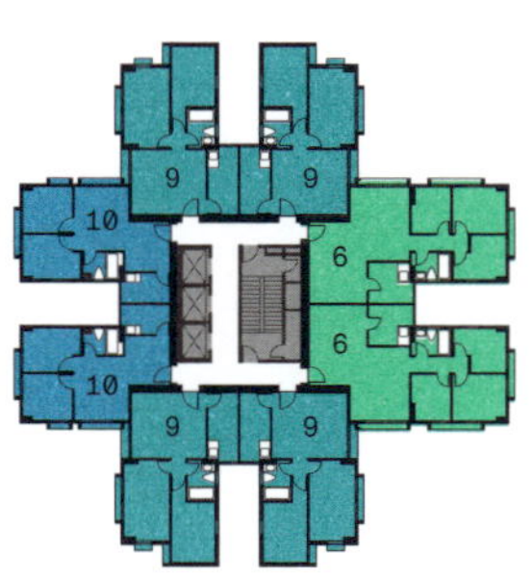

Phase 4 →/ **Block 25, 27, 29, 30**

A16

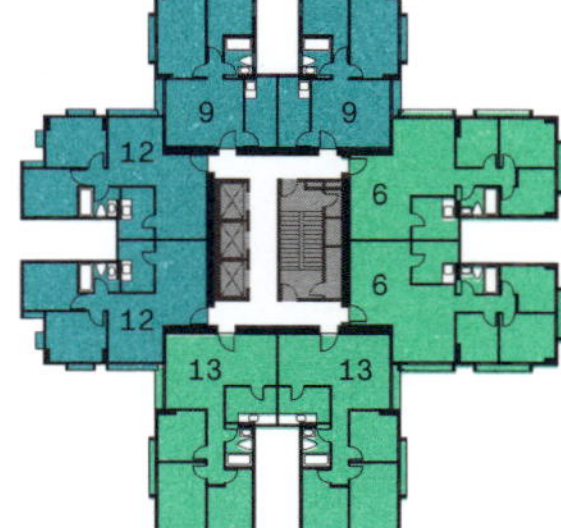

Phase 4 →/ **Block 26**

A^{17}

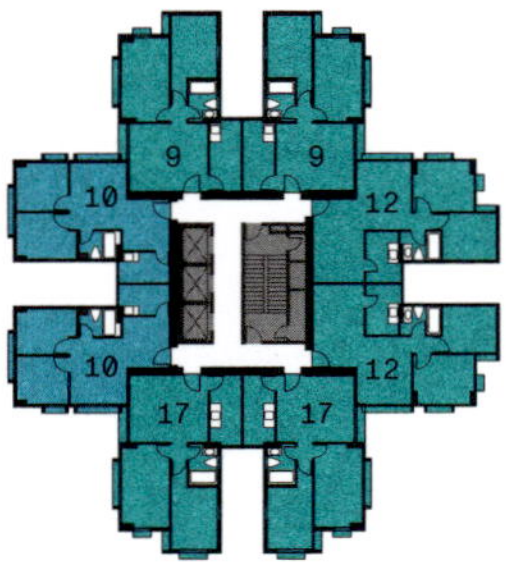

Phase 4 →/ **Block 28**

A^{18}

1 Copy

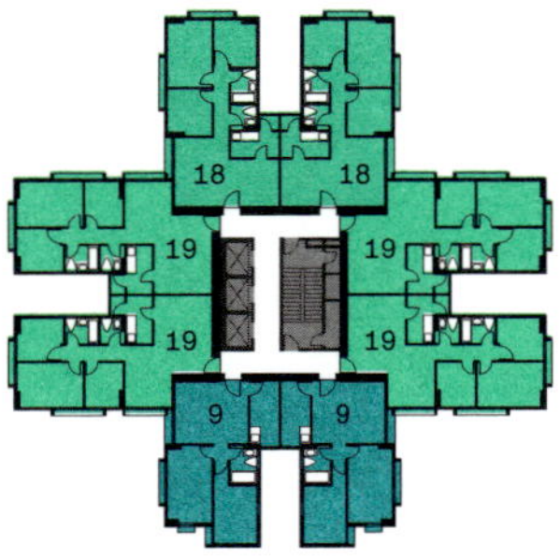

Phase 6 →/ **Block 42, 43**

A^{19}

1 Copy

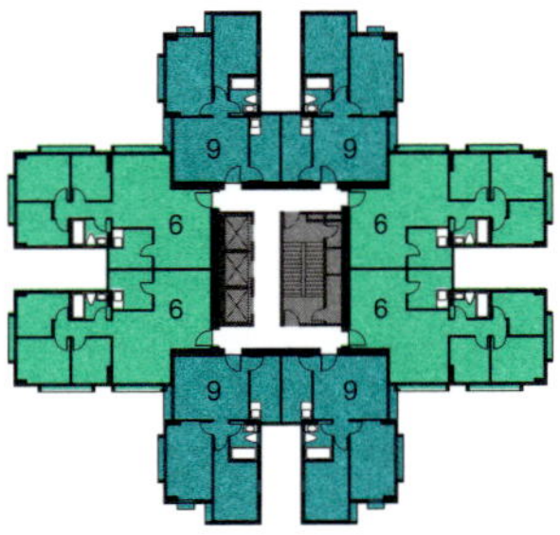

Phase 7 →/ **Block 45, 50**

A^{20}

1 Copy

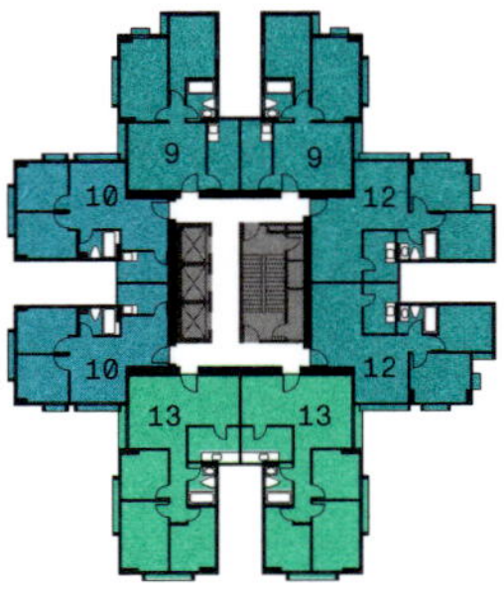

Phase 7 →/ **Block 46, 48**

A^{21}

1 Copy

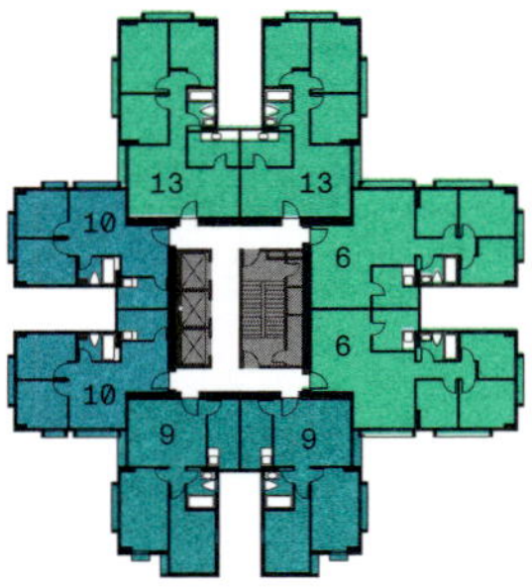

Phase 7 →/ **Block 47, 49**

A^{22}

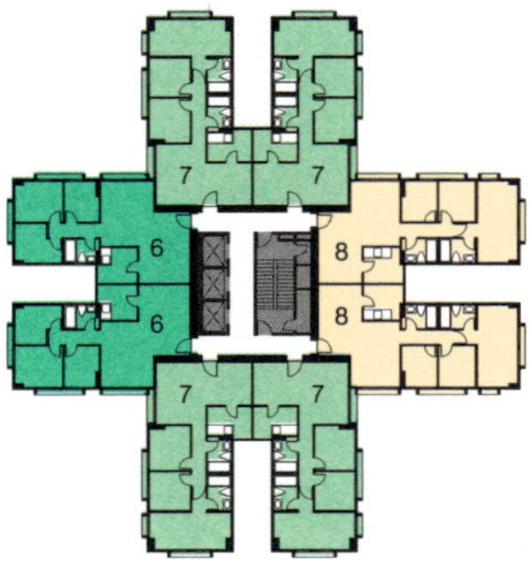

Phase 7 →/ **Block 51**

A^{23}

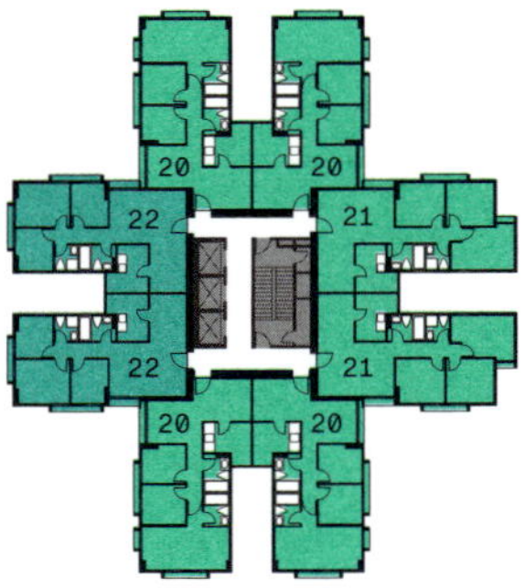

Phase 8 →/ **Block 53**

A^{24}

2 Copies

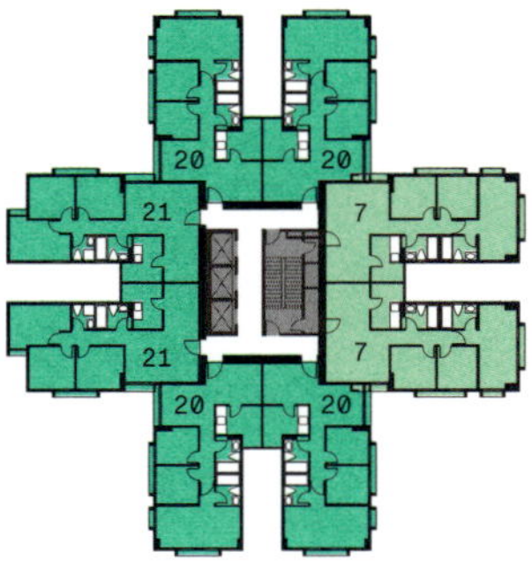

Phase 8 →/ **Block 54–56**

A^{25}

1 Copy

A^{26}

A^{27}

A^{28}

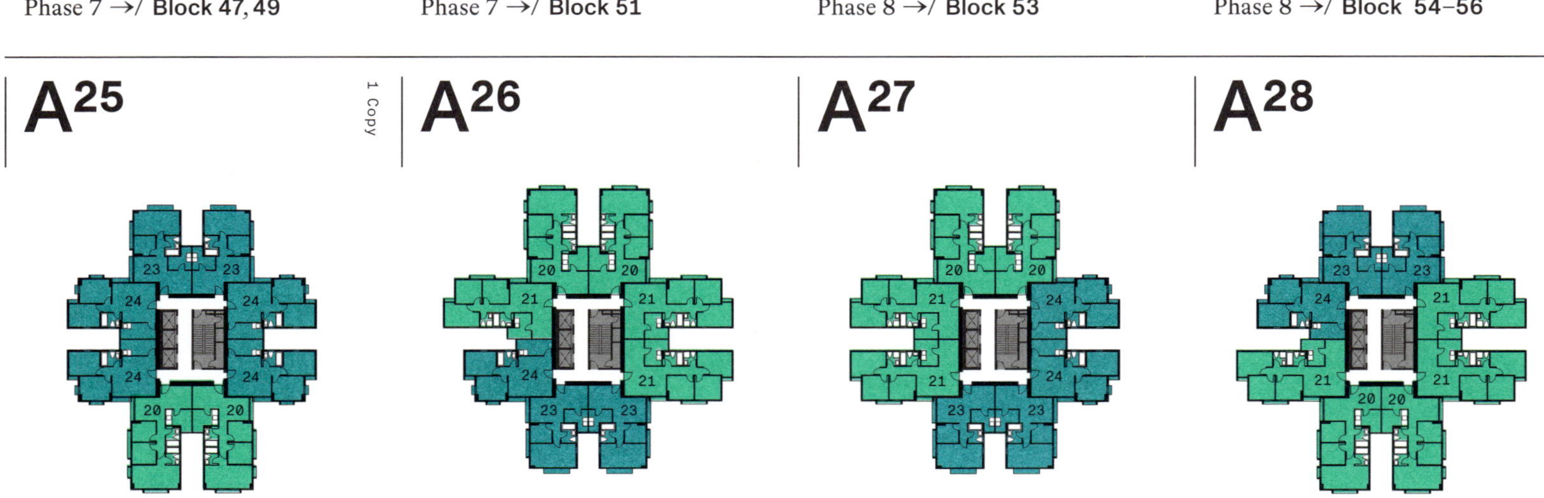

Phase 8 →/ **Block 57, 58**

Phase 8 →/ **Block 59**

Phase 8 →/ **Block 60**

Phase 8 →/ **Block 61**

B^{1}

10 Copies

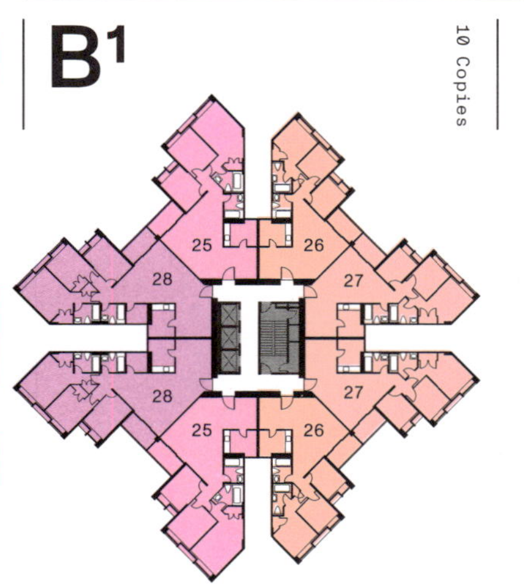

Phase 5 →/ **Block 31–41**

Block	Core	A	B	C	D	E	F	G	H
		Unit							
1	A^1	1 50.1	1 ▶◁	2 42.0	2 ⧗	1 ⧗	1 ↻ 90°	3 46.1	3 ⧗
2		1	1 ▶◁	2	2 ⧗	1 ⧗	1 ↻ 90°	3	3 ⧗
3		1	1 ▶◁	2	2 ⧗	1 ⧗	1 ↻ 90°	3	3 ⧗
17		1	1 ▶◁	2	2 ⧗	1 ⧗	1 ↻ 90°	3	3 ⧗
52		1	1 ▶◁	2	2 ⧗	1 ⧗	1 ↻ 90°	3	3 ⧗
4	A^2	4 58.4	5 58.4	6 59.0	6 ⧗	7 63.3	7 ▶◁	8 66.2	8 ⧗
5	A^3	9 48.7	9 ▶◁	10 42.0	10 ⧗	9 ⧗	9 ↻ 90°	11 59.5	11 ⧗
7		9	9 ▶◁	10	10 ⧗	9 ⧗	9 ↻ 90°	11	11 ⧗
8		9	9 ▶◁	10	10 ⧗	9 ⧗	9 ↻ 90°	11	11 ⧗
12		9	9 ▶◁	10	10 ⧗	9 ⧗	9 ↻ 90°	11	11 ⧗
13		9	9 ▶◁	10	10 ⧗	9 ⧗	9 ↻ 90°	11	11 ⧗
6	A^4	9	9 ▶◁	12 49.1	12 ⧗	13 58.1	13 ▶◁	6 ↻ 90°	6 ▶◁
9	A^5	13 ↻ 90°	13 ⧗	12	12 ⧗	9 ↻ 90°	9 ⧗	6 ↻ 90°	6 ▶◁
10	A^6	9	9 ▶◁	12	12 ⧗	13	13 ▶◁	12 ↻ 90°	12 ▶◁
11	A^7	13 ↻ 90°	13 ⧗	6	6 ⧗	14 57.9	14 ▶◁	12 ↻ 90°	12 ▶◁

Block Core	A	B	C	D	E	F	G	H
	Unit							
14 A[8]	13 ↻ 90°	13 ⧗	12	12 ⧗	13	13 ▶◁	6 ↻ 90°	6 ▶◁
15 A[9]	13 ↻ 90°	13 ⧗	12	12 ⧗	9 ↻ 90°	9 ⧗	12 ↻ 90°	12 ▶◁
16 A[10]	15 57.9	15 ▶◁	12	12 ⧗	13	13 ▶◁	11	11 ⧗
18 A[11]	7 ↻ 90°	7 ↻ 90° ▶◁	6	6 ⧗	14	14 ▶◁	8	8 ⧗
24	7 ↻ 90°	7 ↻ 90° ▶◁	6	6 ⧗	14	14 ▶◁	8	8 ⧗
44	7 ↻ 90°	7 ↻ 90° ▶◁	6	6 ⧗	14	14 ▶◁	8	8 ⧗
19 A[12]	9	9 ▶◁	16 59.1	16 ⧗	15 ↻ 90°	15 ⧗	11	11 ⧗
20 A[13]	9	9 ▶◁	16	16 ⧗	9 ↻ 90°	9 ⧗	11	11 ⧗
21	9	9 ▶◁	16	16 ⧗	9 ↻ 90°	9 ⧗	11	11 ⧗
23	9	9 ▶◁	16	16 ⧗	9 ↻ 90°	9 ⧗	11	11 ⧗
22 A[14]	15	15 ▶◁	16	16 ⧗	9 ↻ 90°	9 ⧗	11	11 ⧗
25 A[15]	9	9 ▶◁	6	6 ⧗	9 ↻ 90°	9 ⧗	10 ↻ 90°	10 ▶◁
27	9	9 ▶◁	6	6 ⧗	9 ↻ 90°	9 ⧗	10 ↻ 90°	10 ▶◁
29	9	9 ▶◁	6	6 ⧗	9 ↻ 90°	9 ⧗	10 ↻ 90°	10 ▶◁
30	9	9 ▶◁	6	6 ⧗	9 ↻ 90°	9 ⧗	10 ↻ 90°	10 ▶◁

Block Core	Unit A	B	C	D	E	F	G	H
26 A^{16}	9	9 ▶◁	6	6 ⧗	13	13 ▶◁	12 ↻ 90°	12 ▶◁
28 A^{17}	9	9 ▶◁	12	12 ⧗	17 49.2	17 ▶◁	10 ↻ 90°	10 ▶◁
42 A^{18}	18 58.7	18 ▶◁	19 60.3	19 ⧗	9 ↻ 90°	9 ⧗	19 ↻ 180°	19 ▶◁
43	18	18 ▶◁	19	19 ⧗	9 ↻ 90°	9 ⧗	19 ↻ 180°	19 ▶◁
45 A^{19}	9	9 ▶◁	6	6 ⧗	9 ↻ 90°	9 ⧗	6 ↻ 90°	6 ▶◁
50	9	9 ▶◁	6	6 ⧗	9 ↻ 90°	9 ⧗	6 ↻ 90°	6 ▶◁
46 A^{20}	9	9 ▶◁	12	12 ⧗	13	13 ▶◁	10 ↻ 90°	10 ▶◁
48	9	9 ▶◁	12	12 ⧗	13	13 ▶◁	10 ↻ 90°	10 ▶◁
47 A^{21}	13 ↻ 90°	13 ⧗	6	6 ⧗	9 ↻ 90°	9 ⧗	10 ↻ 90°	10 ▶◁
49	13 ↻ 90°	13 ⧗	6	6 ⧗	9 ↻ 90°	9 ⧗	10 ↻ 90°	10 ▶◁
51 A^{22}	7 ↻ 90°	7 ↻ 90° ▶◁	8 ↻ 90°	8 ▶◁	7	7 ▶◁	6 ↻ 90°	6 ▶◁
53 A^{23}	20 58.3	20 ▶◁	21 58.8	21 ⧗	20 ↻ 90°	20 ⧗	22 54.9	22 ⧗
54 A^{24}	20	20 ▶◁	7 ↻ 270°	7 ↻ 90° ▶◁	20 ↻ 90°	20 ⧗	21 ↻ 90°	21 ▶◁
55	20	20 ▶◁	7 ↻ 270°	7 ↻ 90° ▶◁	20 ↻ 90°	20 ⧗	21 ↻ 90°	21 ▶◁
56	20	20 ▶◁	7 ↻ 270°	7 ↻ 90° ▶◁	20 ↻ 90°	20 ⧗	21 ↻ 90°	21 ▶◁

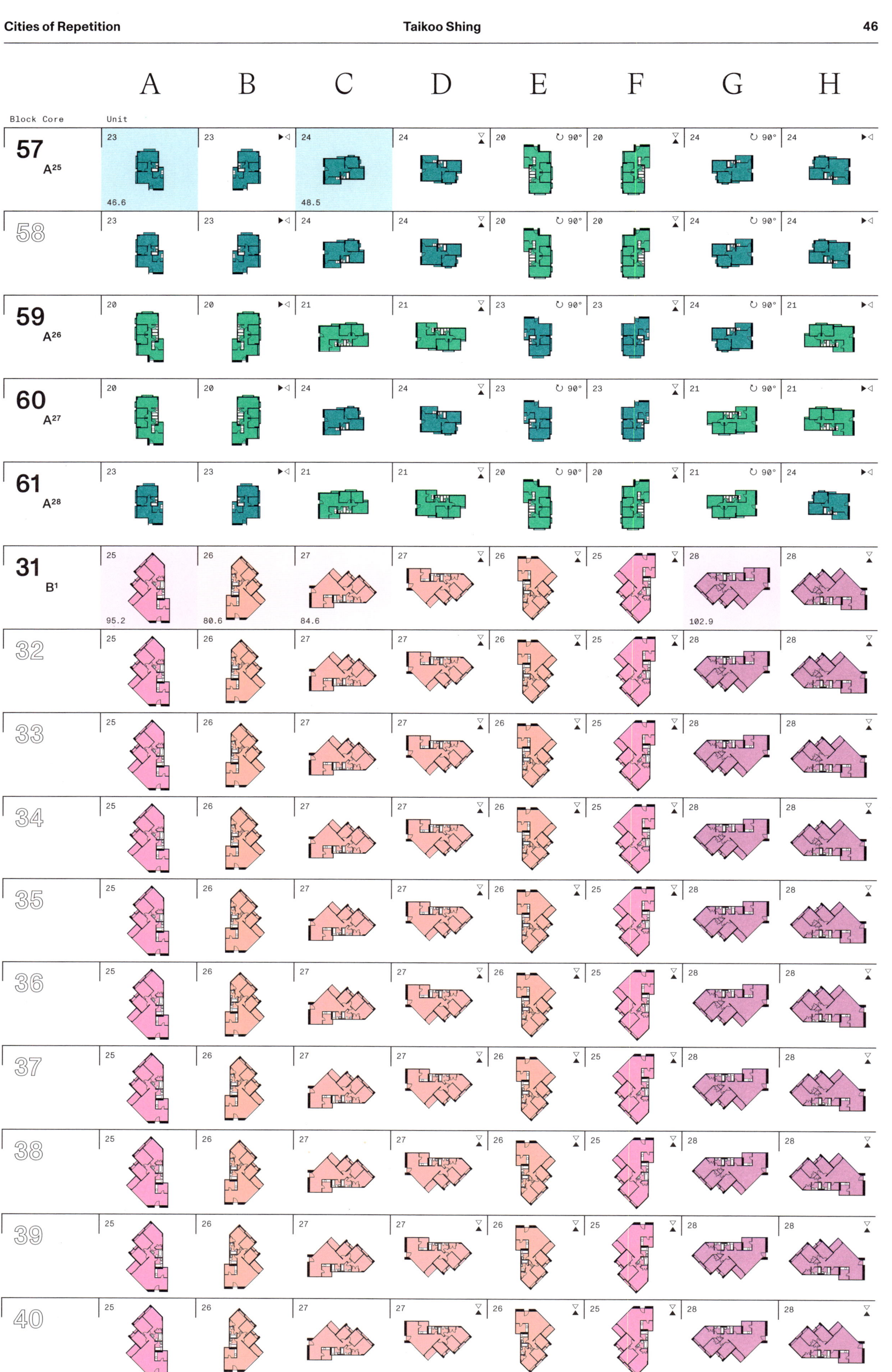
A B C D E F G H
Block Core
Unit
57 A25 23 23 24 24 20 ↻ 90° 20 24 ↻ 90° 24
46.6 48.5
58 23 23 24 24 20 ↻ 90° 20 24 ↻ 90° 24
59 A26 20 20 21 21 23 ↻ 90° 23 24 ↻ 90° 21
60 A27 20 20 24 24 23 ↻ 90° 23 21 ↻ 90° 21
61 A28 23 23 21 21 20 ↻ 90° 20 21 ↻ 90° 24
31 B1 25 26 27 27 26 25 28 28
95.2 80.6 84.6 102.9
32 25 26 27 27 26 25 28 28
33 25 26 27 27 26 25 28 28
34 25 26 27 27 26 25 28 28
35 25 26 27 27 26 25 28 28
36 25 26 27 27 26 25 28 28
37 25 26 27 27 26 25 28 28
38 25 26 27 27 26 25 28 28
39 25 26 27 27 26 25 28 28
40 25 26 27 27 26 25 28 28

Heng Fa Chuen 1978–89

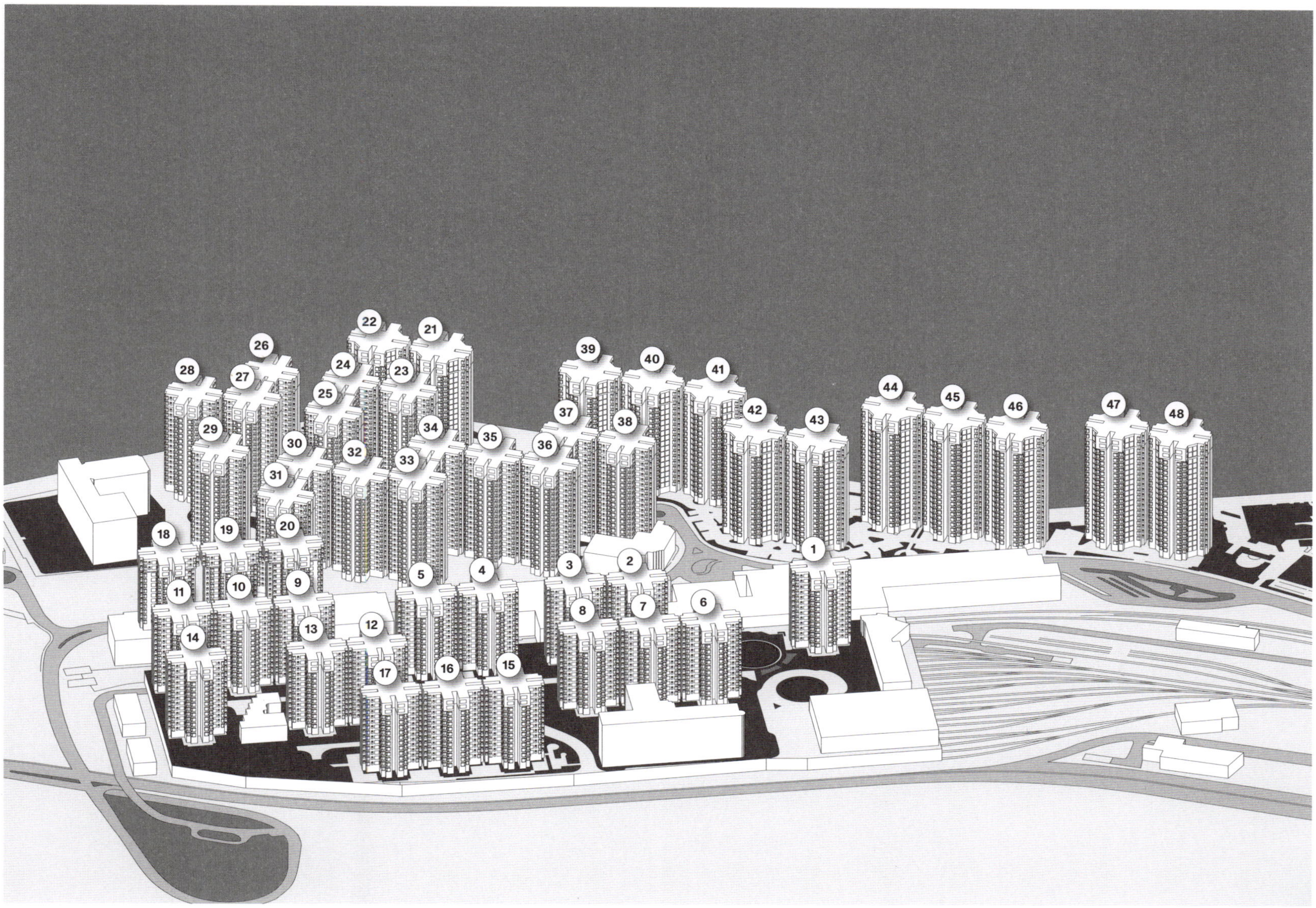

DEVELOPER:	MTR Corporation, Kerry Properties		
ARCHITECT:	Simon Kwan and Associates		
LOCATION:	Chai Wan		
POPULATION:	18,921		
TOWERS:	48	UNIQUE:	22
APARTMENTS:	6,505	UNIT TYPES:	15
CORES:	3	PHASES:	1

Heng Fa Chuen is built on an area of land reclaimed from Victoria Harbor in the Eastern District of Hong Kong Island. The waterfront estate is surrounded by hillsides and is relatively isolated from the dense urban context of central Hong Kong Island. Like many housing estates in Hong Kong, the development of the Heng Fa Chuen is closely linked to the development of the city's transportation infrastructure. The forty-eight-tower estate was jointly developed by Kerry Properties together with the MTR (Mass Transit Railway) Corporation. The project was built together with the MTR Island Line and accommodates a covered rail depot, which is used for storage and maintenance of MTR trains.

Clusters of tower blocks within the estate are organized into two main areas, upper and lower. The two sections are laid out on two grids set at an angle to each other. The tower blocks on the upper level are built on top of the MTR train depot and a parking garage. The tower blocks on the lower section are set directly on the ground level. A long linear shopping mall and MTR station divides the upper and lower sections and provides commercial, retail, and communal space and amenities for the estate.

Taikoo Shing B1

Block Core Unit

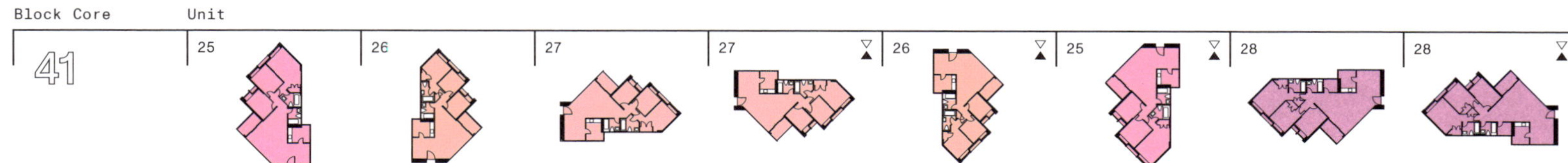

Type A

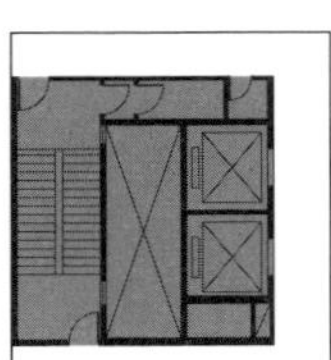

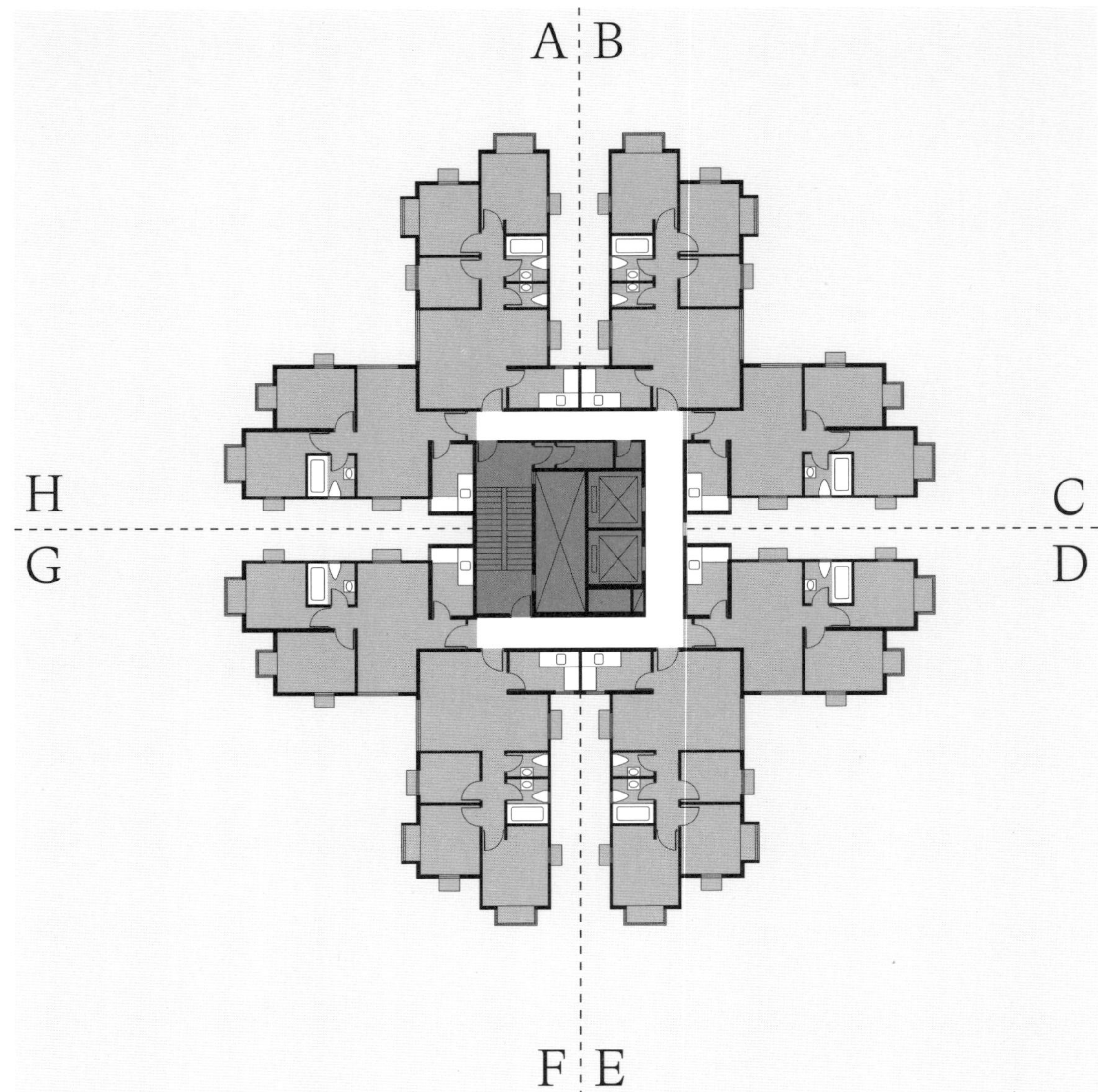

Type B

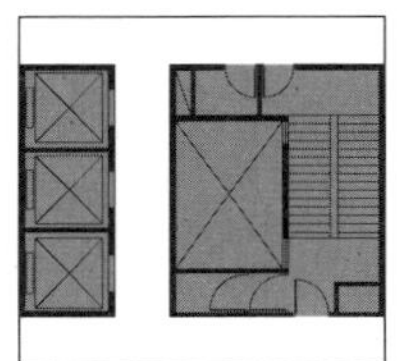

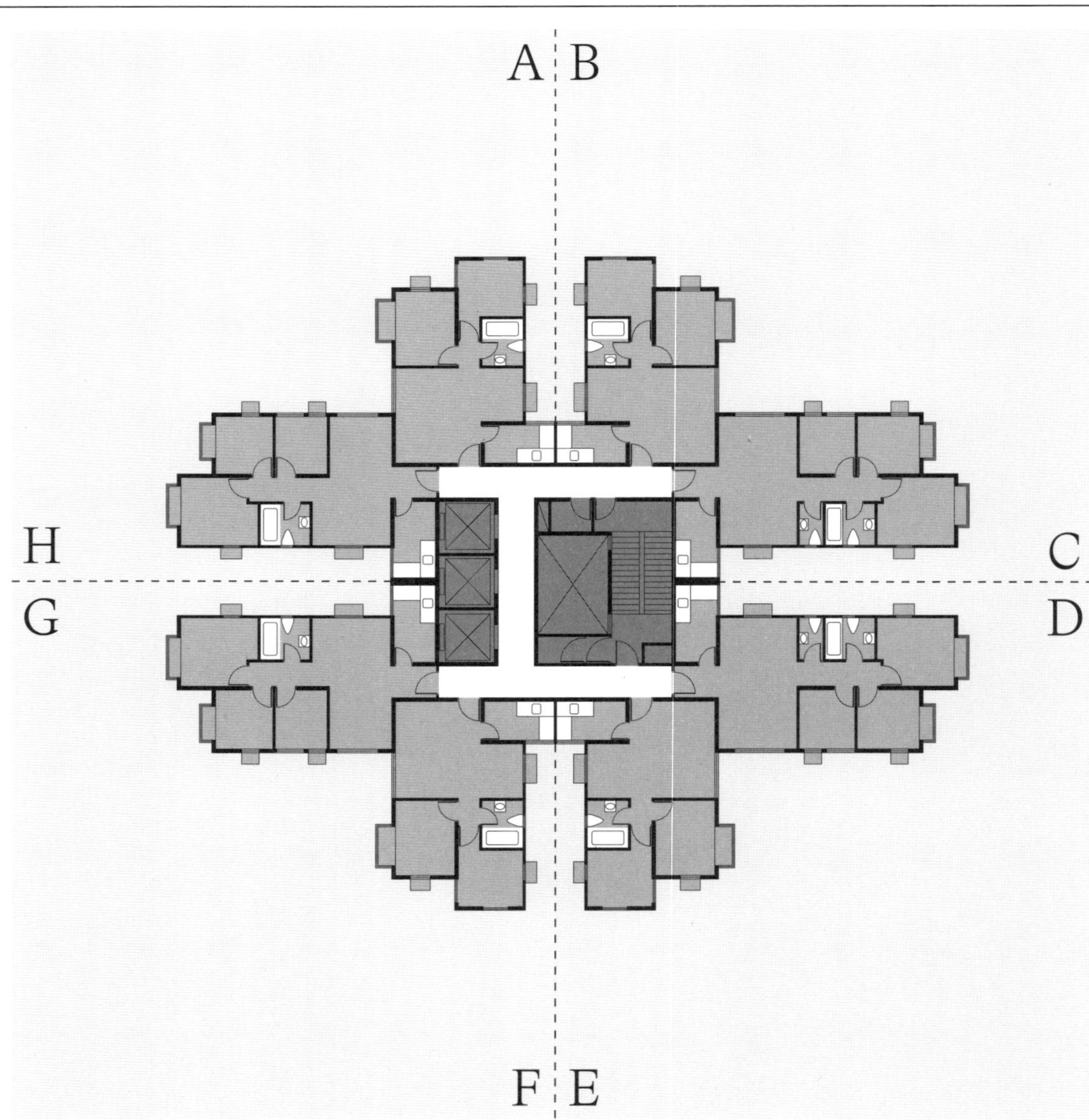

Type C

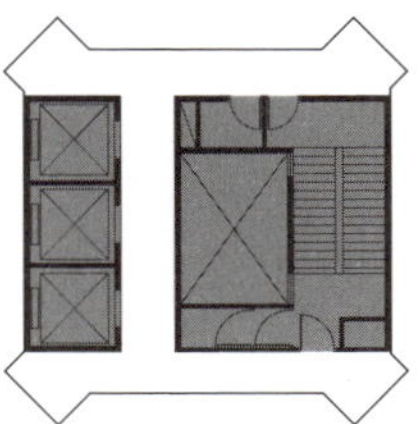

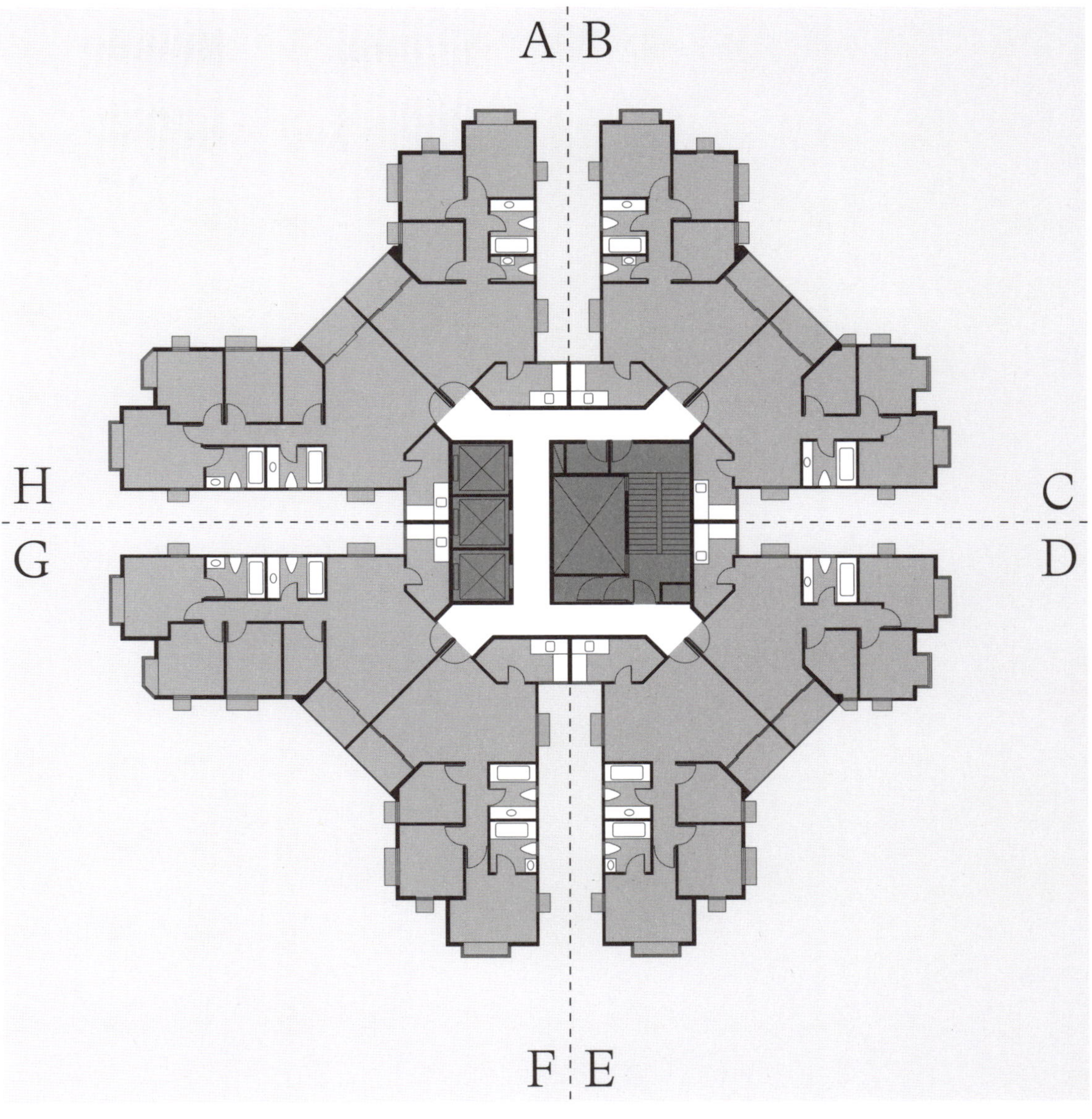

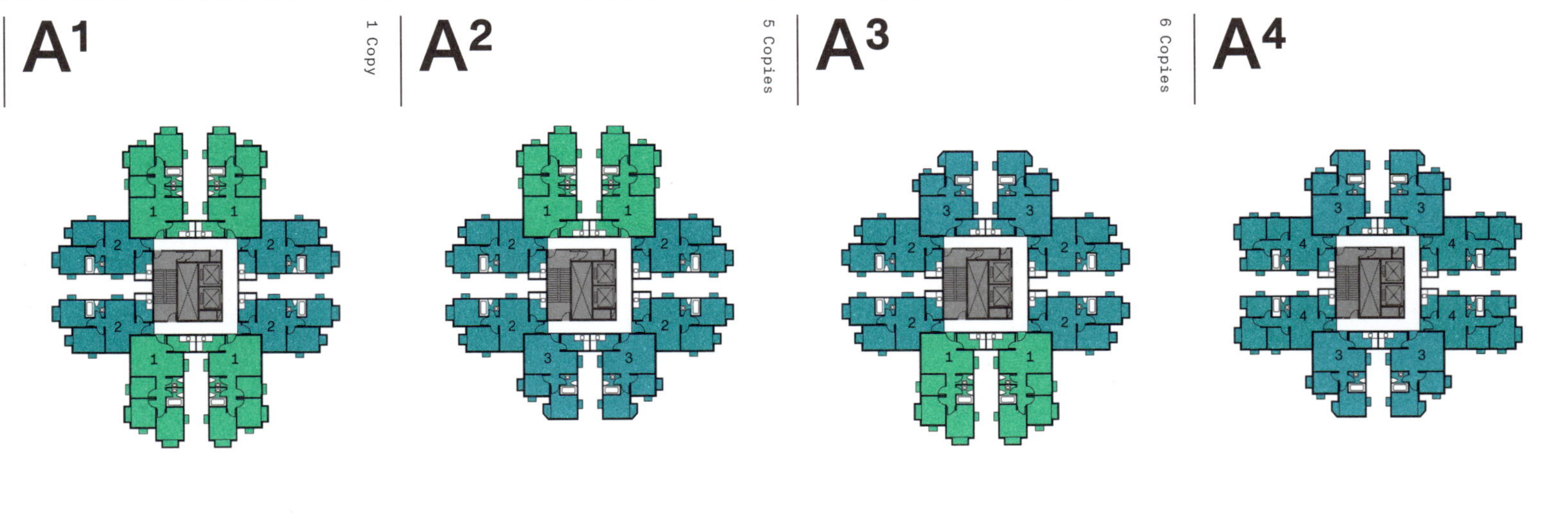

Phase 1 →/ **Block 1, 14**

Phase 1 →/ **Block 2, 4, 8, 9, 13, 17**

Phase 1 →/ **Block 3, 5, 6, 11, 12, 15, 18**

Phase 1 →/ **Block 7, 10, 16**

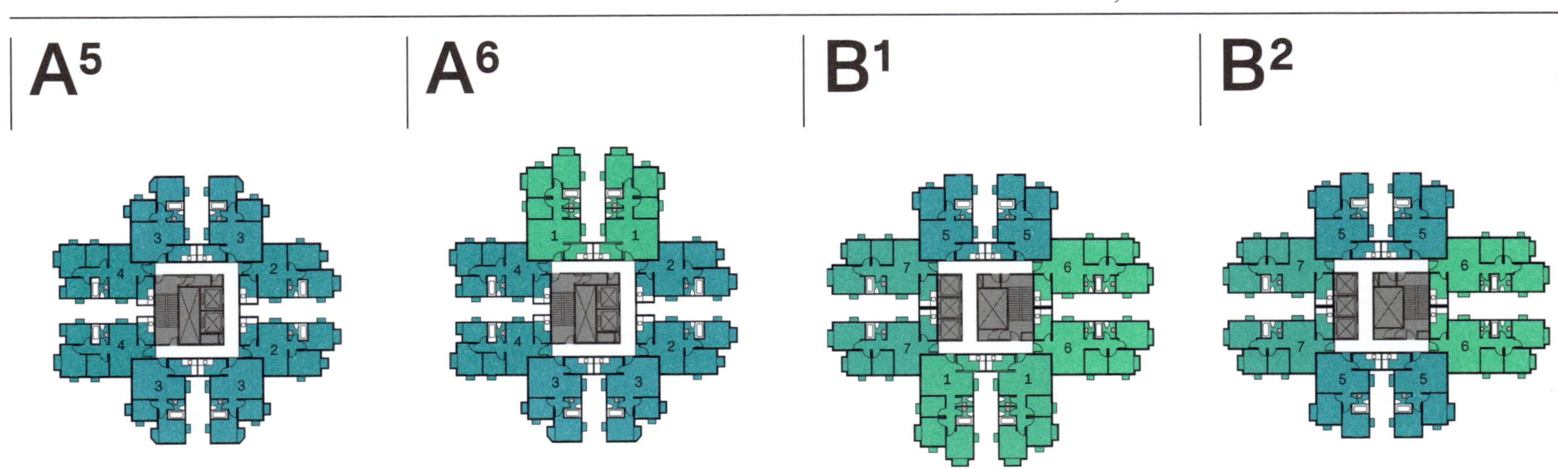

Phase 1 →/ **Block 19**

Phase 1 →/ **Block 20**

Phase 1 →/ **Block 23**

Phase 1 →/ **Block 24, 30**

B^3

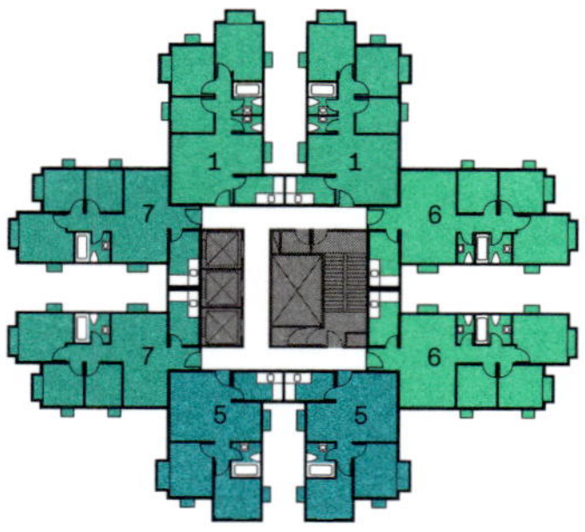

Phase 1 →/ **Block 25**

B^4

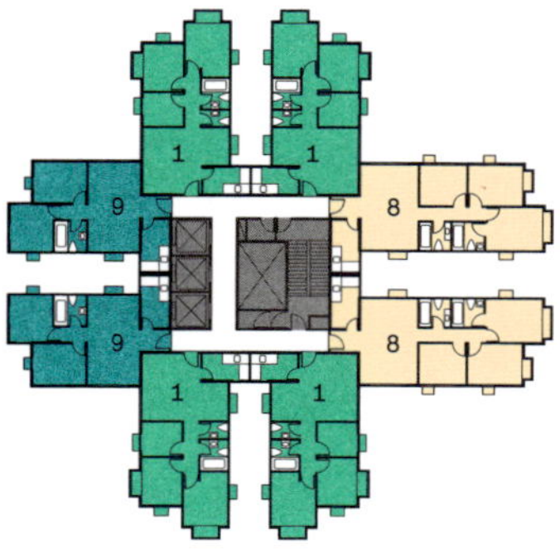

Phase 1 →/ **Block 26**

B^5

1 Copy

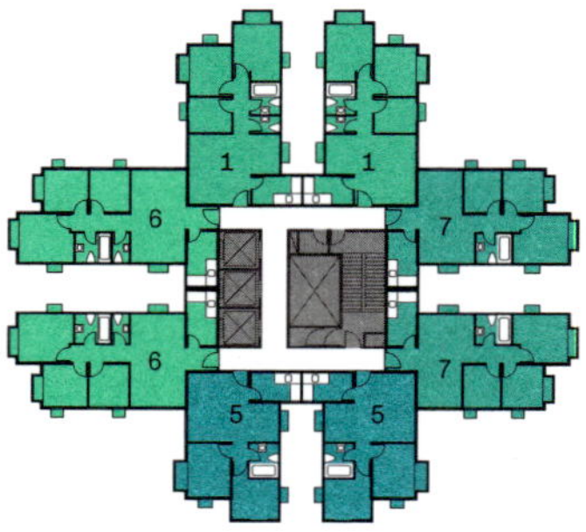

Phase 1 →/ **Block 27, 33**

B^6

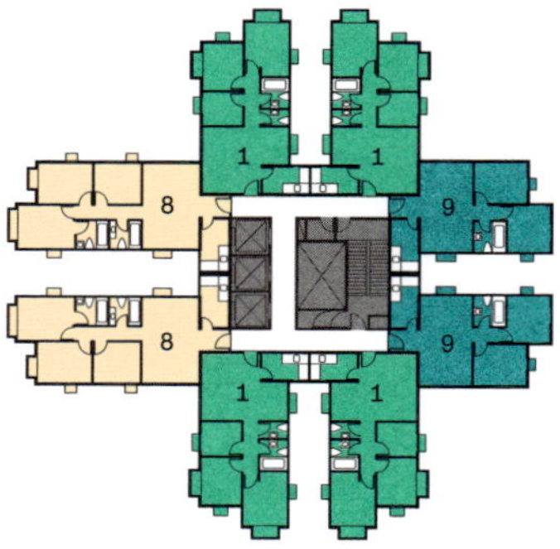

Phase 1 →/ **Block 28**

B^7

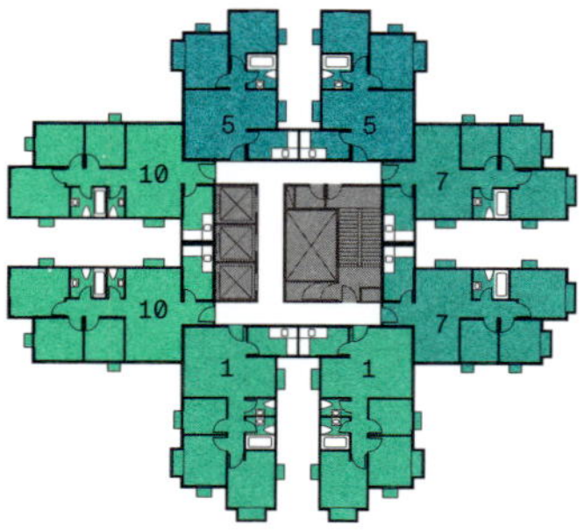

Phase 1 →/ **Block 29**

B^8

5 Copies

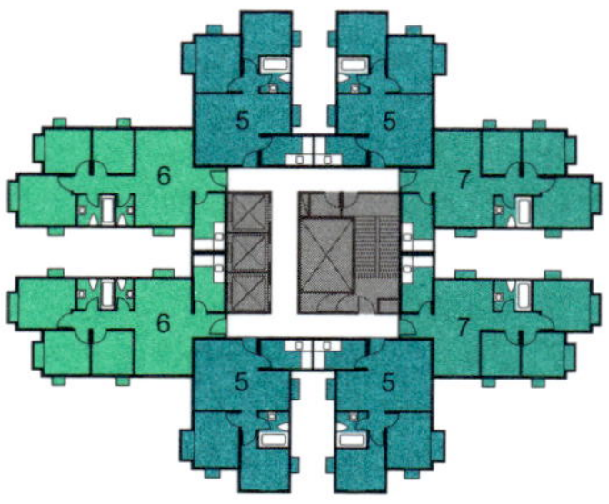

Phase 1 →/ **Block 31,32,35–38**

B^9

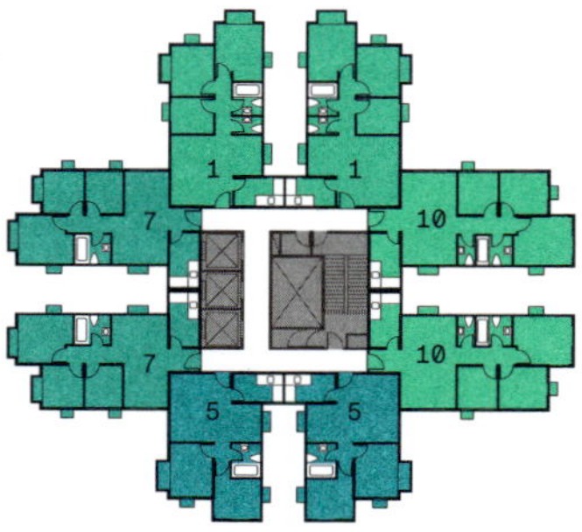

Phase 1 →/ **Block 34**

C^1

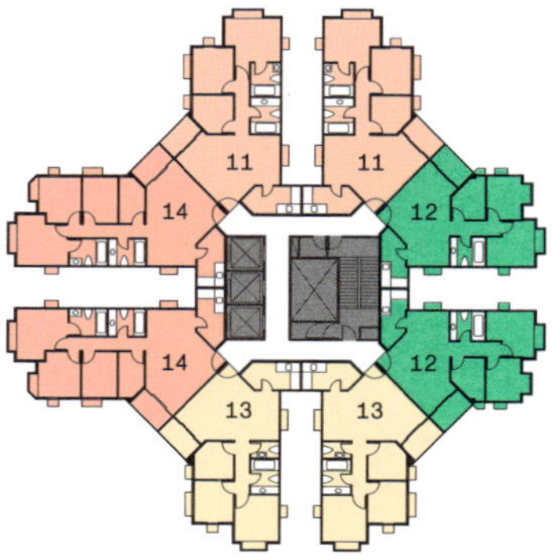

Phase 1 →/ **Block 21**

C^2

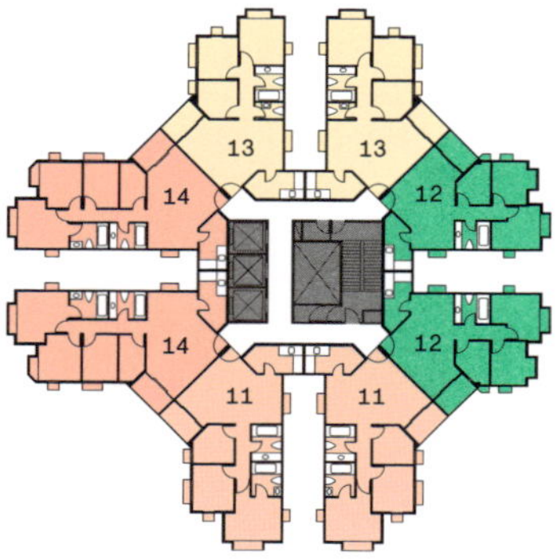

Phase 1 →/ **Block 22**

C^3

1 Copy

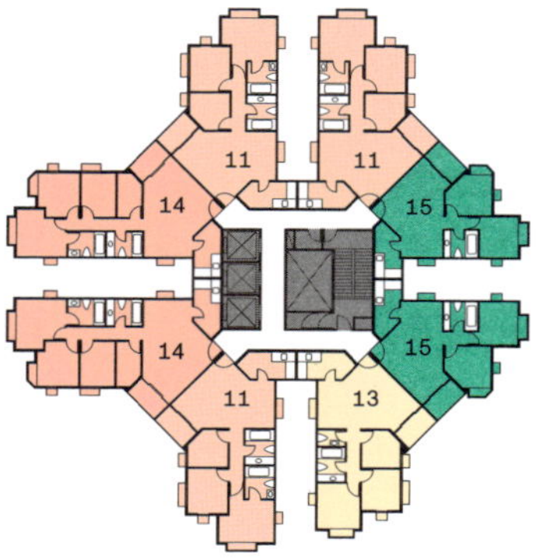

Phase 1 →/ **Block 39, 42**

C^4

1 Copy

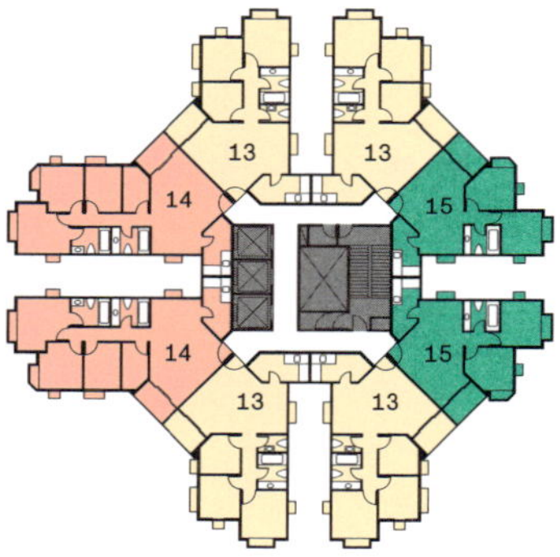

Phase 1 →/ **Block 40, 45**

C^5

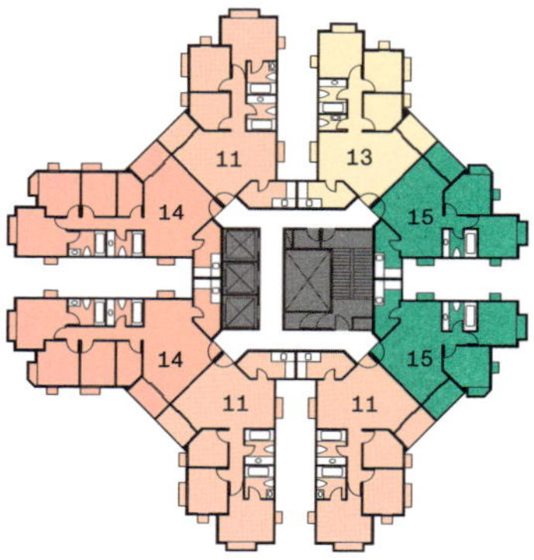

Phase 1 →/ **Block 41**

C^6

2 Copies

C^7

1 Copy

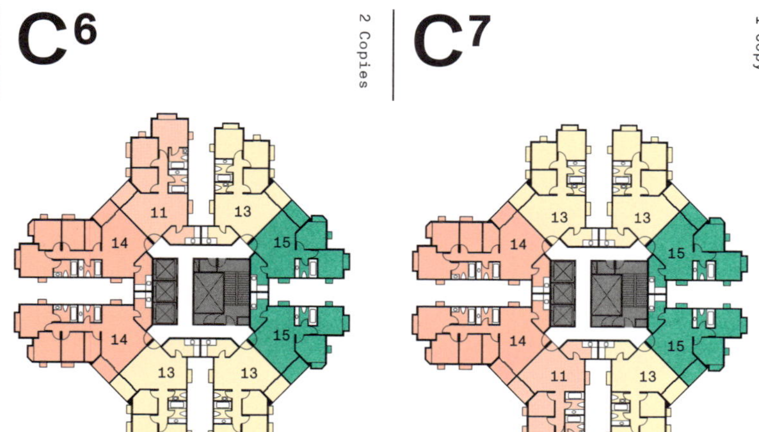

Phase 1 →/ **Block 43,46,48**

Phase 1 →/ **Block 44,47**

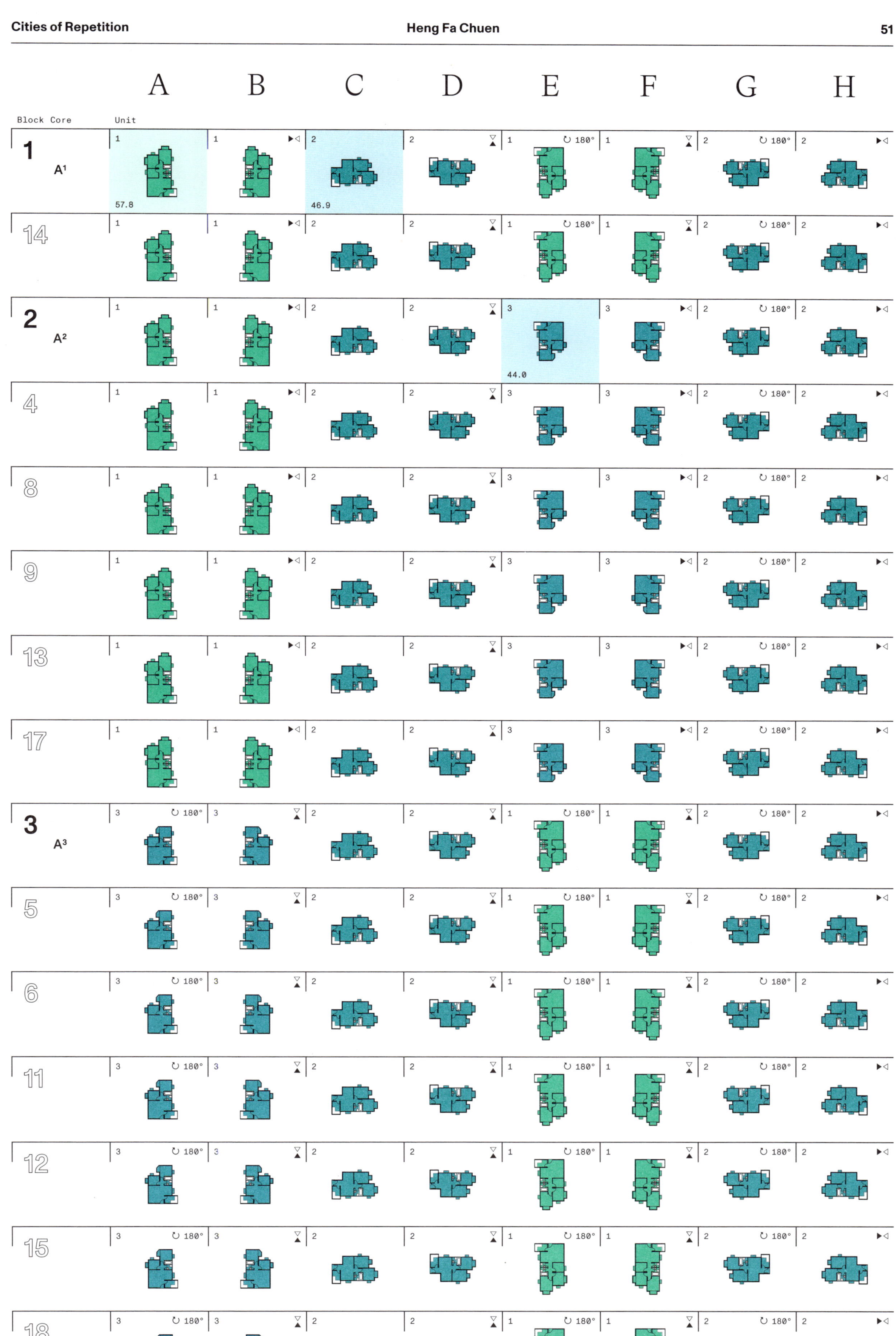
A
B
C
D
E
F
G
H
Block Core
Unit
1
A1
14
2
A2
4
8
9
13
17
3
A3
5
6
11
12
15
18
57.8
46.9
44.0
180°

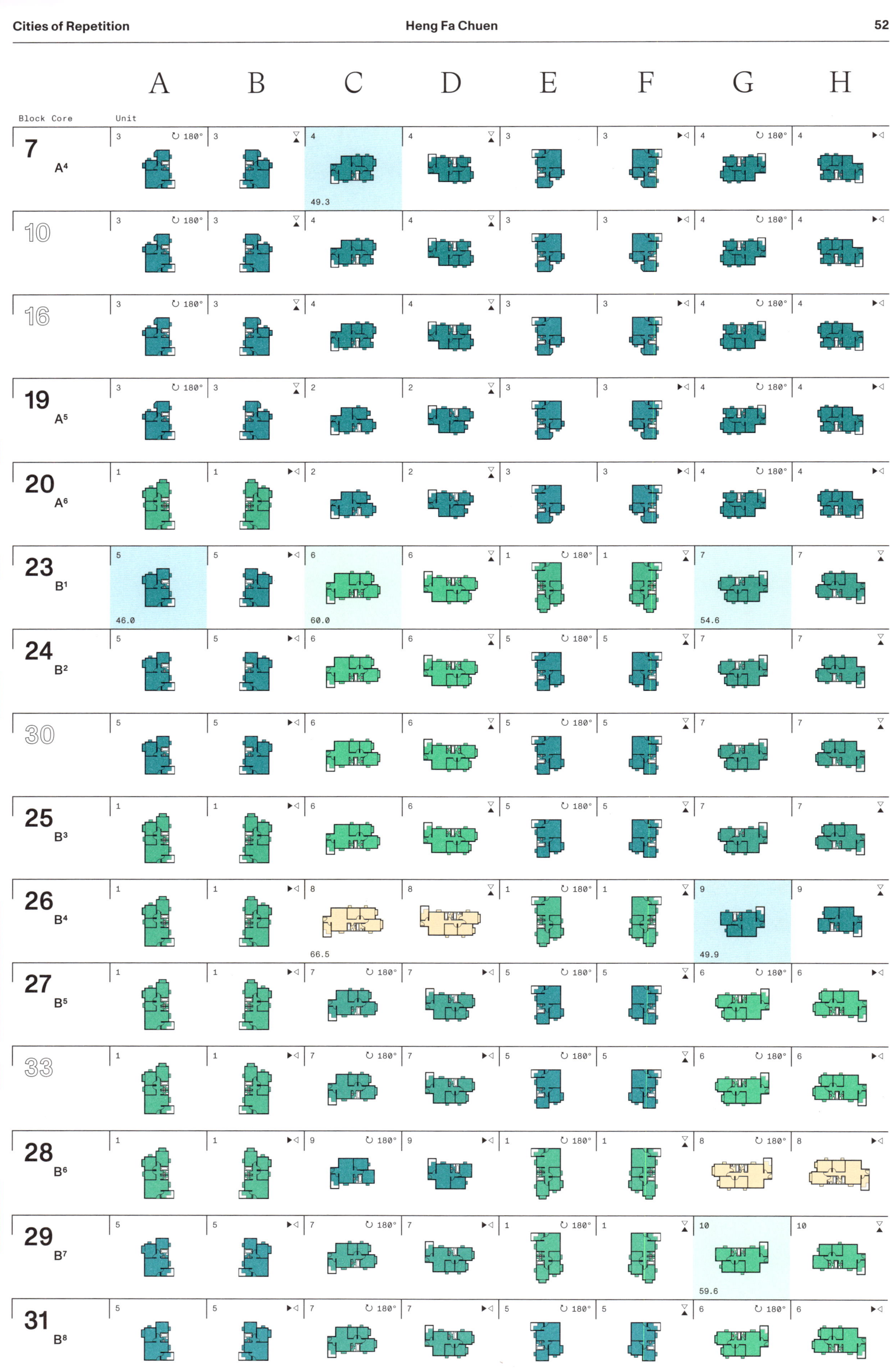
A B C D E F G H
Block Core
Unit
7 A4
3 180° 3 4 4 3 3 4 180° 4
49.3
10
3 180° 3 4 4 3 3 4 180° 4
16
3 180° 3 4 4 3 3 4 180° 4
19 A5
3 180° 3 2 2 3 3 4 180° 4
20 A6
1 1 2 2 3 3 4 180° 4
23 B1
5 5 6 6 1 180° 1 7 7
46.0
60.0
54.6
24 B2
5 5 6 6 5 180° 5 7 7
30
5 5 6 6 5 180° 5 7 7
25 B3
1 1 6 6 5 180° 5 7 7
26 B4
1 1 8 8 1 180° 1 9 9
66.5
49.9
27 B5
1 1 7 180° 7 5 180° 5 6 180° 6
33
1 1 7 180° 7 5 180° 5 6 180° 6
28 B6
1 1 9 180° 9 1 180° 1 8 180° 8
29 B7
5 5 7 180° 7 1 180° 1 10 10
59.6
31 B8
5 5 7 180° 7 5 180° 5 6 180° 6

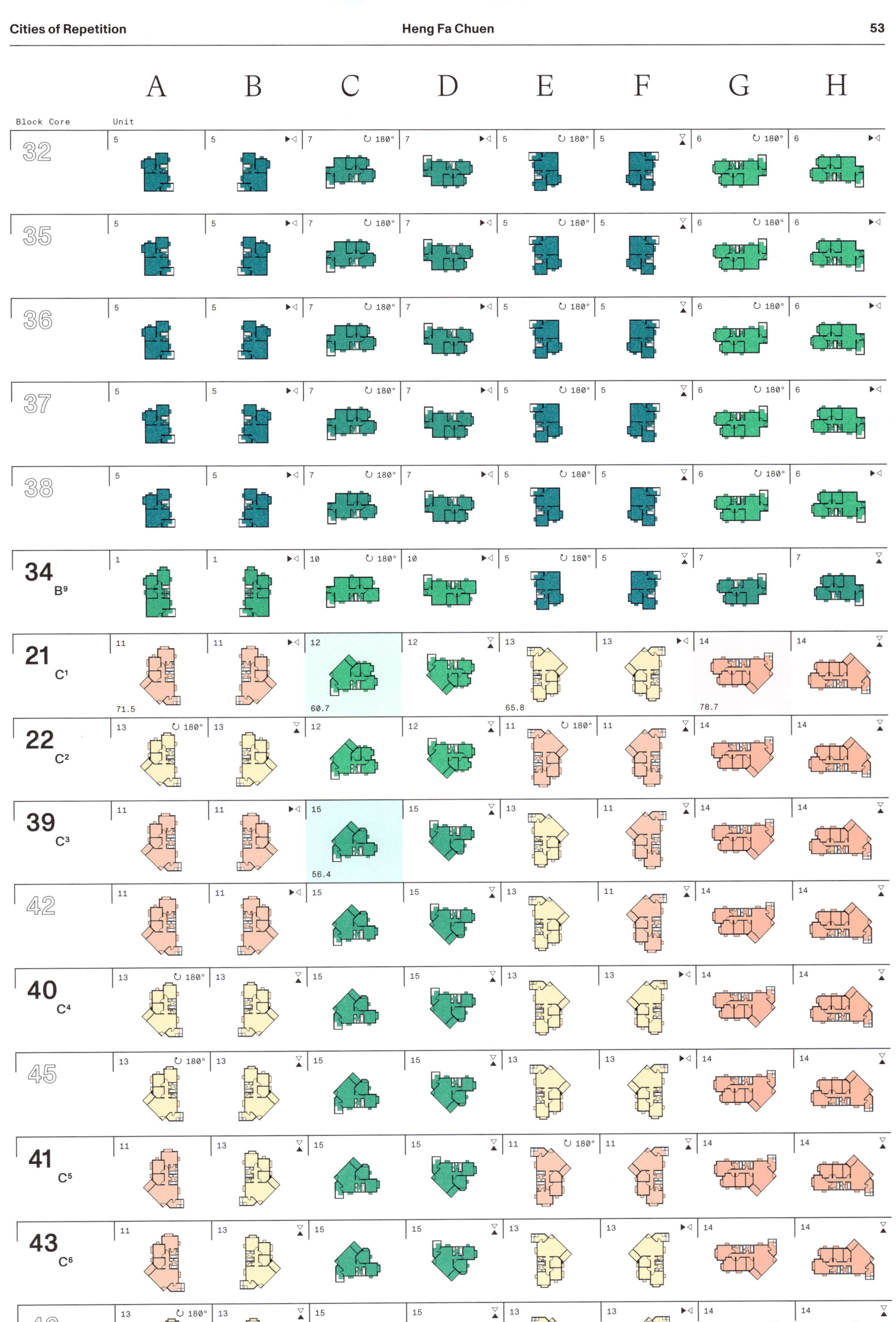
A B C D E F G H
Block Core
Unit
32
5 5 7 180° 7 5 180° 5 6 180° 6
35
5 5 7 180° 7 5 180° 5 6 180° 6
36
5 5 7 180° 7 5 180° 5 6 180° 6
37
5 5 7 180° 7 5 180° 5 6 180° 6
38
5 5 7 180° 7 5 180° 5 6 180° 6
34 B9
1 1 10 180° 10 5 180° 5 7 7
21 C1
11 11 12 12 13 13 14 14
71.5 60.7 65.8 78.7
22 C2
13 180° 13 12 12 11 180° 11 14 14
39 C3
11 11 15 15 13 11 14 14
56.4
42
11 11 15 15 13 11 14 14
40 C4
13 180° 13 15 15 13 13 14 14
45
13 180° 13 15 15 13 13 14 14
41 C5
11 13 15 15 11 180° 11 14 14
43 C6
11 13 15 15 13 13 14 14
46
13 180° 13 15 15 13 13 14 14

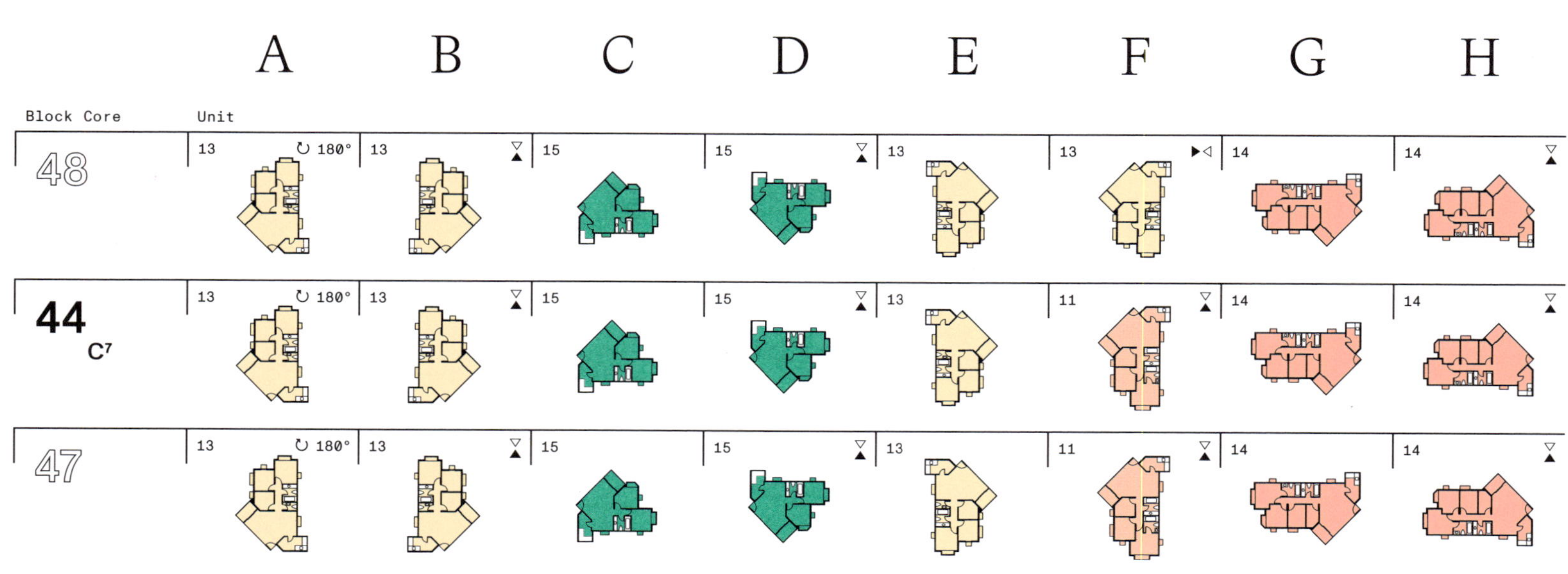
A
B
C
D
E
F
G
H
Block Core
Unit
48
13 180°
13
15
15
13
13
14
14
44
C7
13 180°
13
15
15
13
11
14
14
47
13 180°
13
15
15
13
11
14
14

Kornhill 1980–87

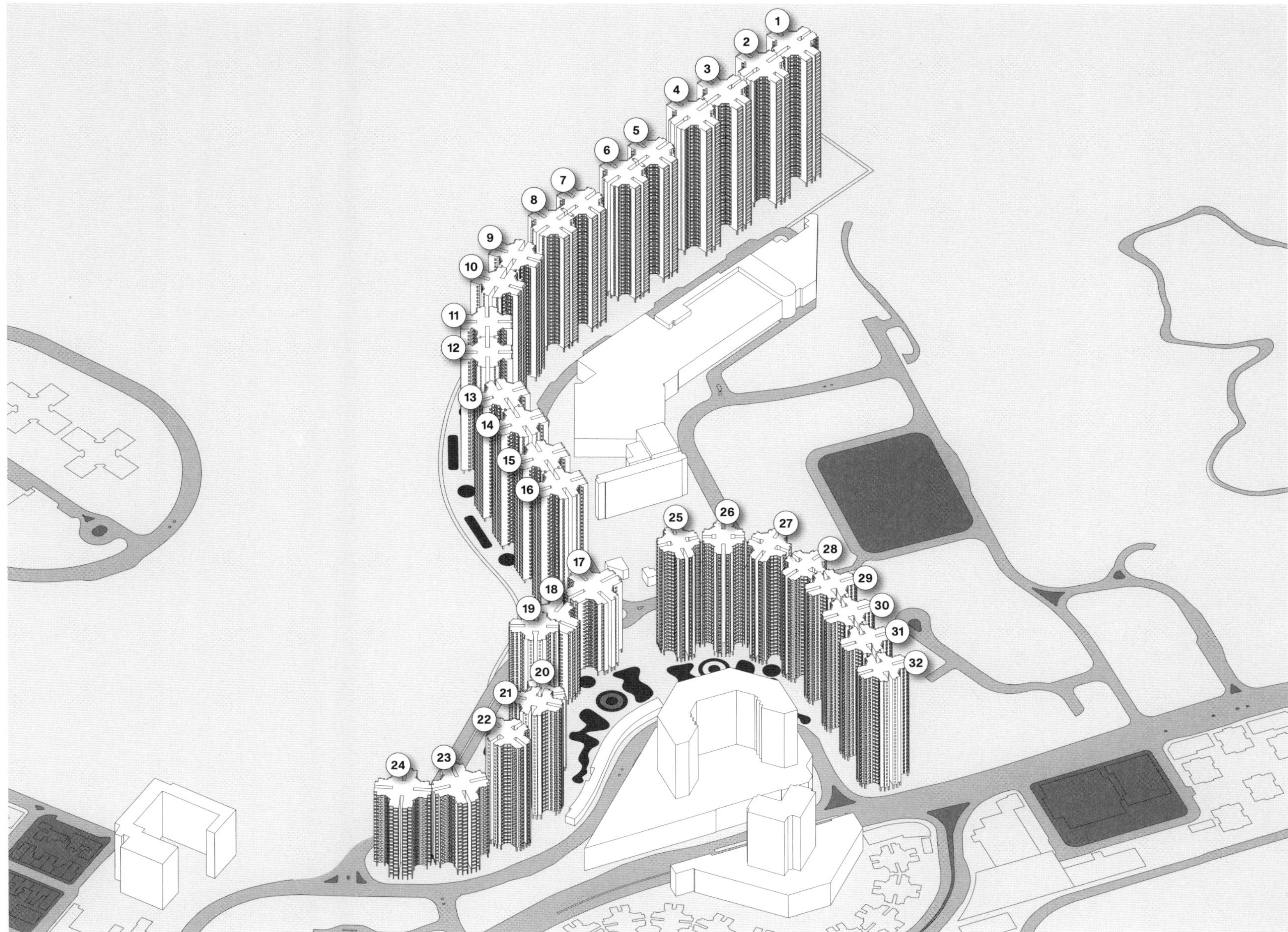

DEVELOPER:	Hang Lung, New World Development, MTR Property		
ARCHITECT:	DLN Architects & Engineers		
LOCATION:	Quarry Bay		
POPULATION:	18,450		
TOWERS:	32	UNIQUE:	9
APARTMENTS:	6,651	UNIT TYPES:	29
CORES:	3	PHASES:	1

The **Kornhill** Estate planning resembles a Y-shaped series of linear strings of tower blocks that follow a gentle slope of the site's topography. The linear organization of the blocks creates open views toward the mountains or the city for many of the units. Located near the Tai Koo Shing estate and Tai Koo MTR transit station, Kornhill was developed and is maintained by the MTR Property, the real estate arm of the company that provides rapid rail transit to Hong Kong. The housing at Kornhill was built in parallel with the development of the MTR's Island Line in the 1980s.

Type A

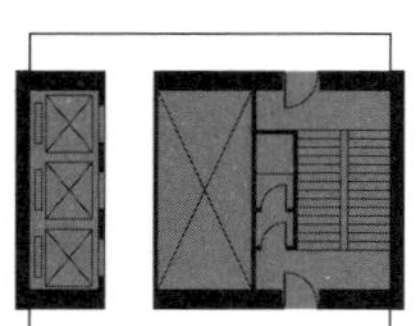

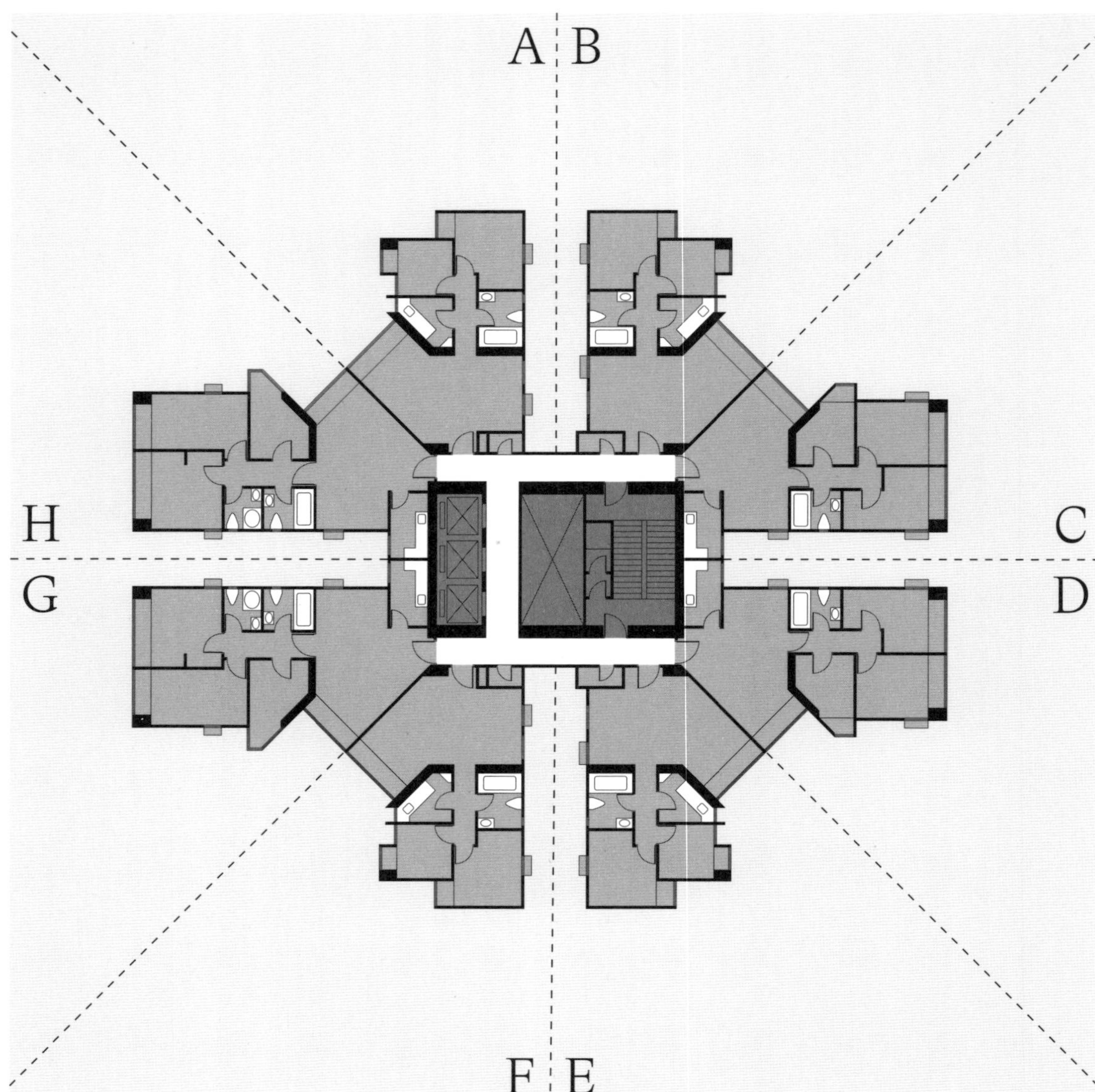

Type B

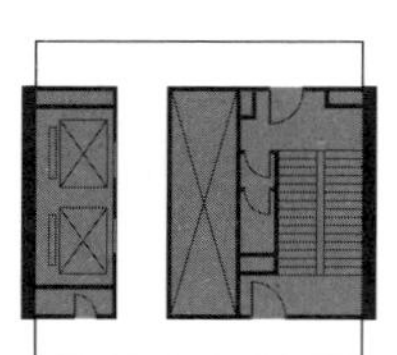

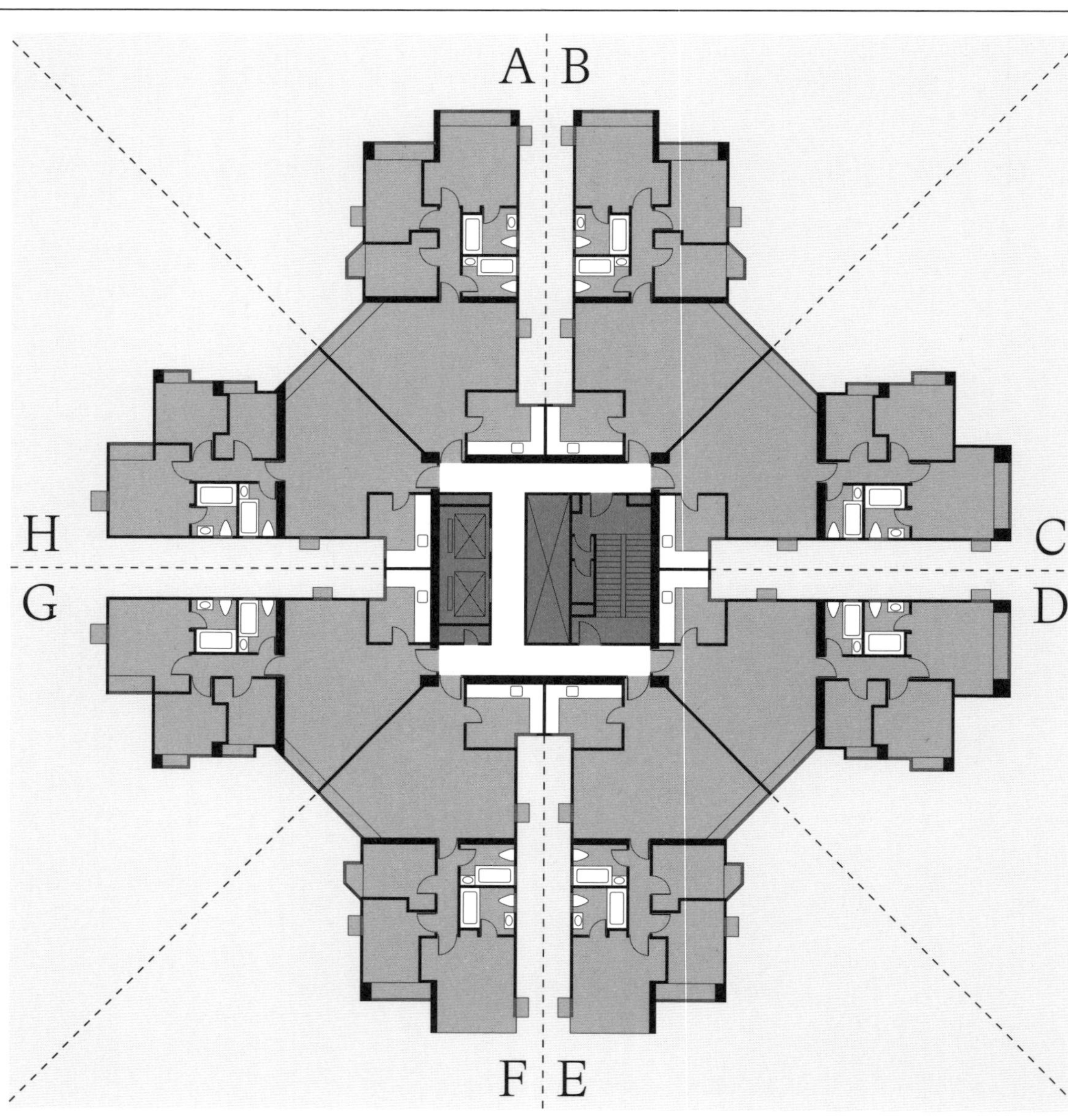

Type C

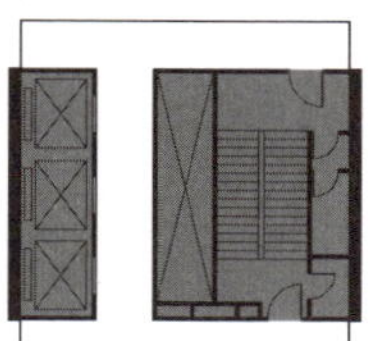

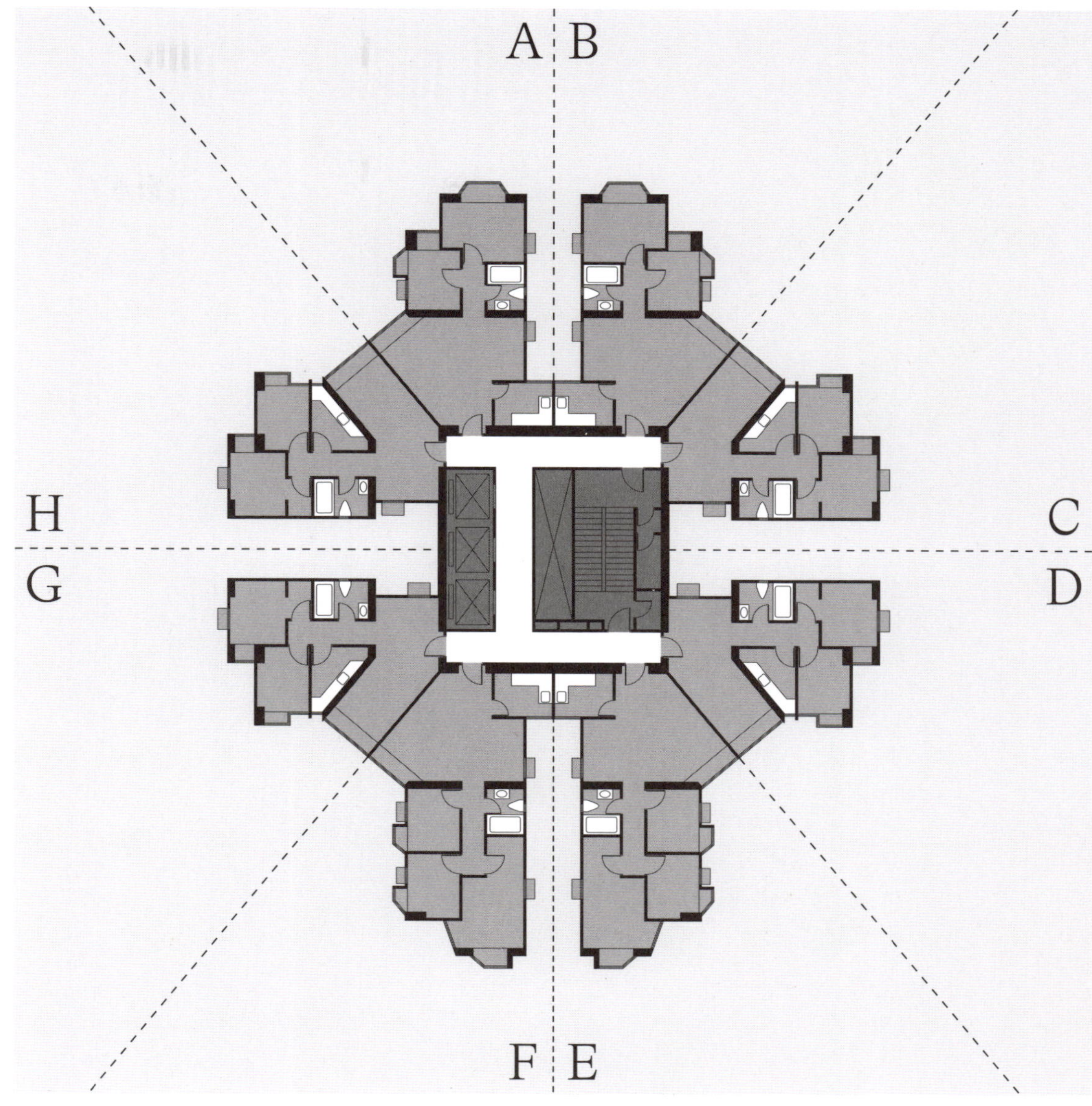

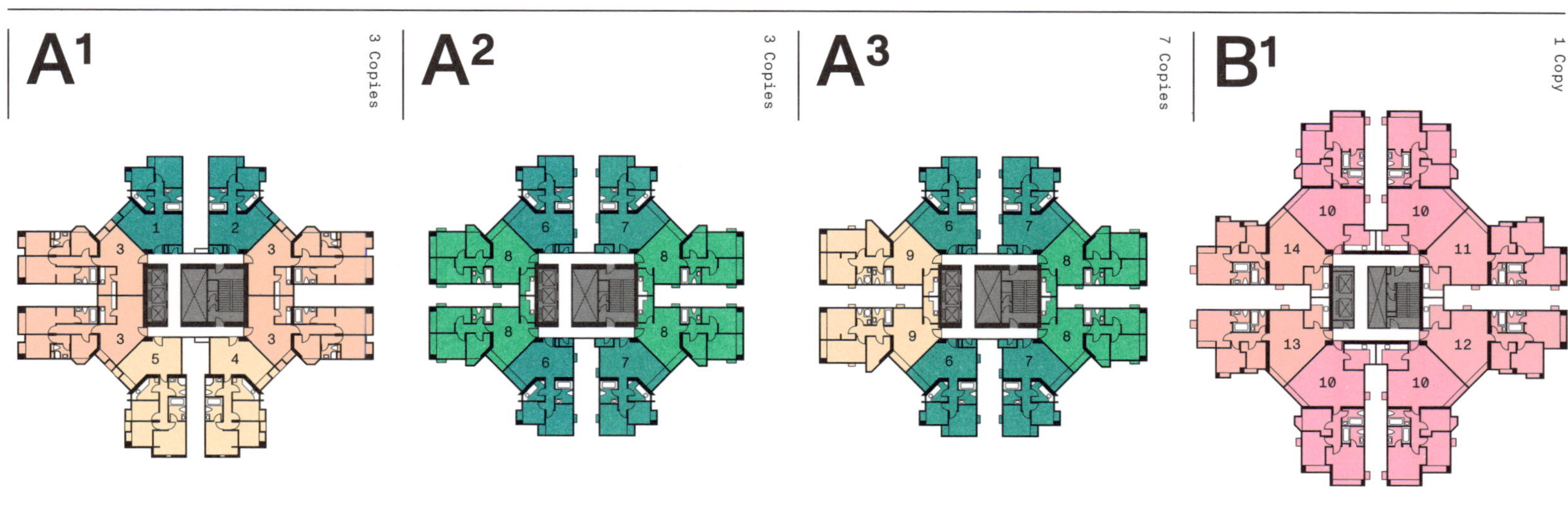

Phase 1 →/ **Block 1–4**

Phase 1 →/ **Block 5–8**

Phase 1 →/ **Block 9–16**

Phase 1 →/ **Block 17, 24**

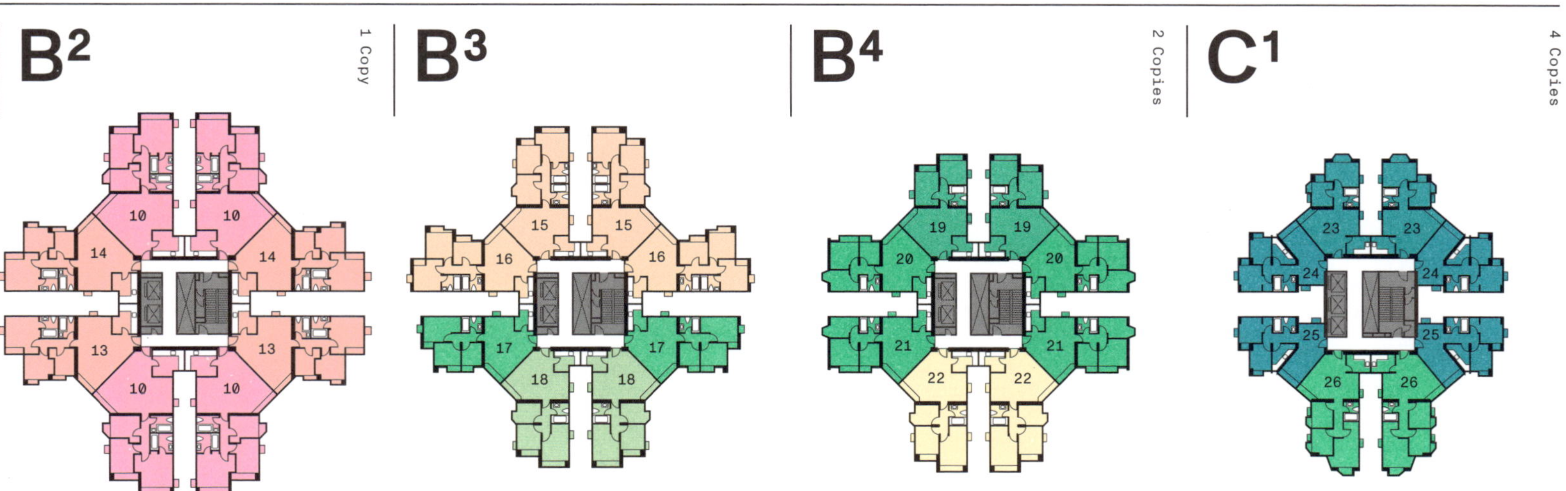

Phase 1 →/ **Block 18, 23**

Phase 1 →/ **Block 19**

Phase 1 →/ **Block 20–22**

Phase 1 →/ **Block 25–28, 31**

C²

2 Copies

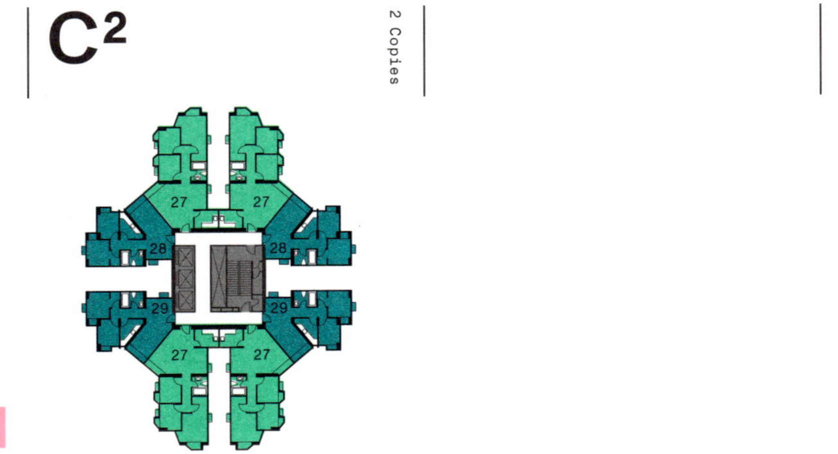

Phase 1 →/ Block 29, 30, 32

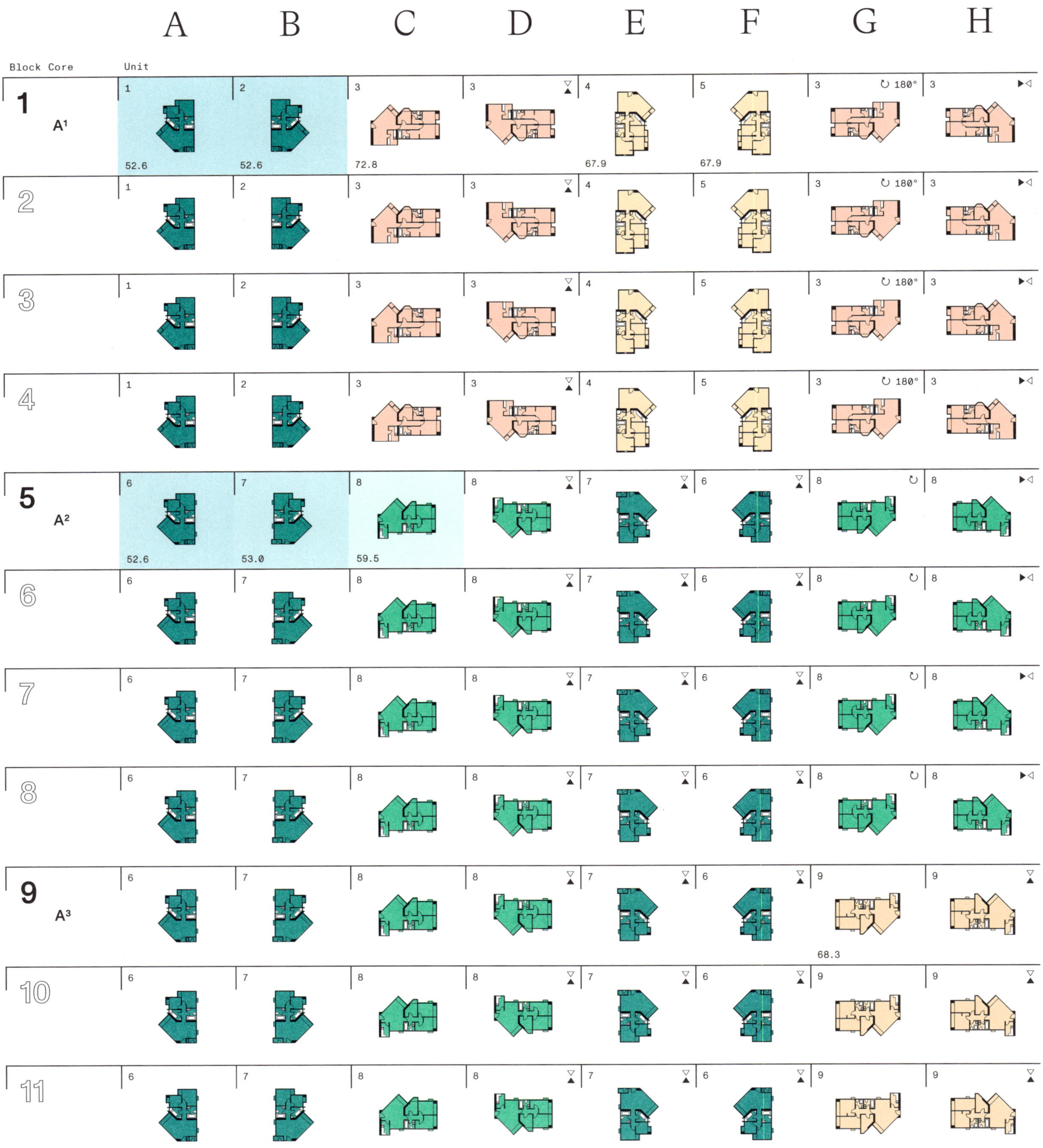

Block Core	A (Unit)	B	C	D	E	F	G	H
1 A¹	1 52.6	2 52.6	3 72.8	3 ⧖	4 67.9	5 67.9	3 ↻ 180°	3 ▶◁
2	1	2	3	3 ⧖	4	5	3 ↻ 180°	3 ▶◁
3	1	2	3	3 ⧖	4	5	3 ↻ 180°	3 ▶◁
4	1	2	3	3 ⧖	4	5	3 ↻ 180°	3 ▶◁
5 A²	6 52.6	7 53.0	8 59.5	8 ⧖	7 ⧖	6 ⧖	8 ↻	8 ▶◁
6	6	7	8	8 ⧖	7 ⧖	6 ⧖	8 ↻	8 ▶◁
7	6	7	8	8 ⧖	7 ⧖	6 ⧖	8 ↻	8 ▶◁
8	6	7	8	8 ⧖	7 ⧖	6 ⧖	8 ↻	8 ▶◁
9 A³	6	7	8	8 ⧖	7 ⧖	6 ⧖	9 68.3	9 ⧖
10	6	7	8	8 ⧖	7 ⧖	6 ⧖	9	9 ⧖
11	6	7	8	8 ⧖	7 ⧖	6 ⧖	9	9 ⧖

Block Core	A	B	C	D	E	F	G	H
	Unit							
12	6	7	8	8	7	6	9	9
13	6	7	8	8	7	6	9	9
14	6	7	8	8	7	6	9	9
15	6	7	8	8	7	6	9	9
16	6	7	8	8	7	6	9	9
17 B^1	10 91.0	10	11 87.9	12 87.9	10 ↻ 180°	10	13 81.5	14 81.5
24	10	10	11	12	10 ↻ 180°	10	13	14
18 B^2	10	10	14	13	10 ↻ 180°	10	13	14
23	10	10	14	13	10 ↻ 180°	10	13	14
19 B^3	15 70.9	15	16 69.1	17 61.3	18 63.8	18	17	16
20 B^4	19 56.0	19	20 59.7	21 59.7	22 65.9	22	21	20
21	19	19	20	21	22	22	21	20
22	19	19	20	21	22	22	21	20
25 C^1	23 50.4	23	24 46.4	25 46.4	26 60.3	26	25	24
26	23	23	24	25	26	26	25	24

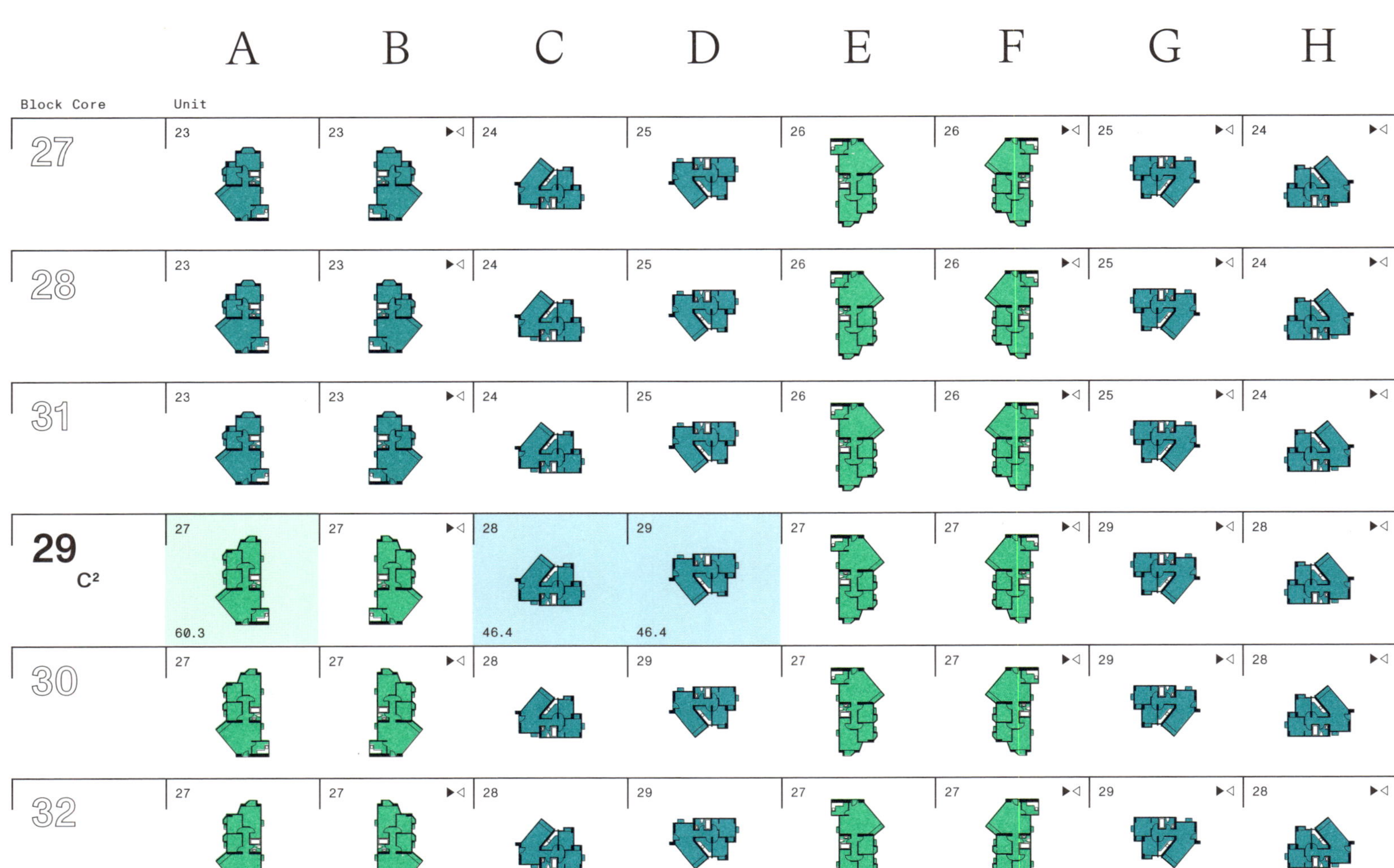
A B C D E F G H
Block Core
Unit
27
23 23 ▶◁ 24 25 26 26 ▶◁ 25 ▶◁ 24 ▶◁
28
23 23 ▶◁ 24 25 26 26 ▶◁ 25 ▶◁ 24 ▶◁
31
23 23 ▶◁ 24 25 26 26 ▶◁ 25 ▶◁ 24 ▶◁
29 C^2
27 27 ▶◁ 28 29 27 27 ▶◁ 29 ▶◁ 28 ▶◁
60.3 46.4 46.4
30
27 27 ▶◁ 28 29 27 27 ▶◁ 29 ▶◁ 28 ▶◁
32
27 27 ▶◁ 28 29 27 27 ▶◁ 29 ▶◁ 28 ▶◁

City One 1980–88

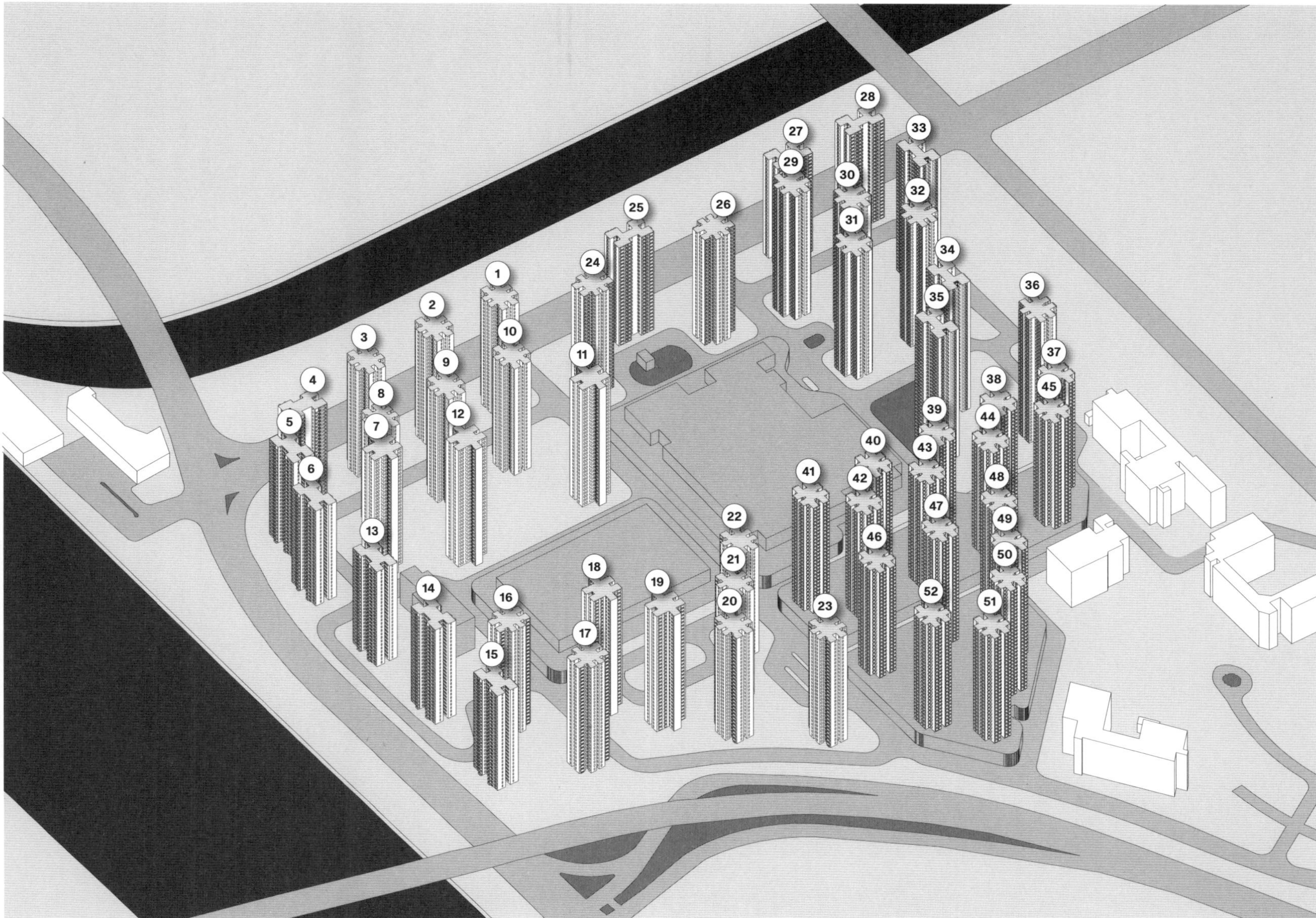

DEVELOPER:	New World Development, Henderson Land Development, Sun Hung Kai Properties, Cheung Kong Holdings		
ARCHITECT:	Wong and Ouyang (HK) Ltd.		
LOCATION:	Sha Tin		
POPULATION:	24,758		
TOWERS:	52	UNIQUE:	10
APARTMENTS:	10,643	UNIT TYPES:	18
CORES:	6	PHASES:	7

City One is located in Sha Tin, a district of Hong Kong's New Territories. A large part of the urban area of Sha Tin is built on land reclaimed from a large, shallow tidal bay during the development of the Sha Tin New Town. Hong Kong's colonial British government built a number of satellite towns, or New Towns, in underdeveloped areas in the outskirts of the colony. These new urban centers were built as part of a planning strategy initiated in the 1950s to decentralize the rapidly growing population of Hong Kong. City One took its name from being the first large site to be sold to the private sector for development.

While some of the towers at the center of the City One estate sit atop a podium with commercial and retail spaces, most of the towers are organized in an offset grid pattern and sit directly on the ground.

Type A

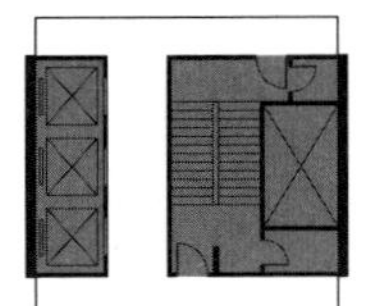

A'

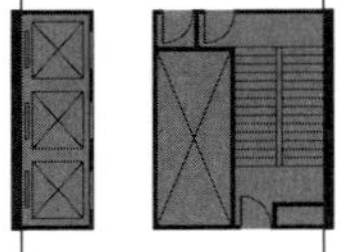

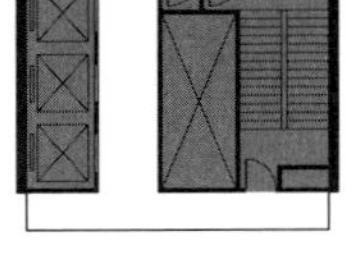

A''

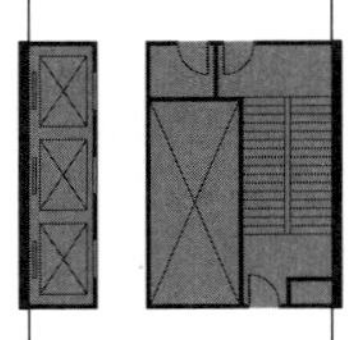

A'''

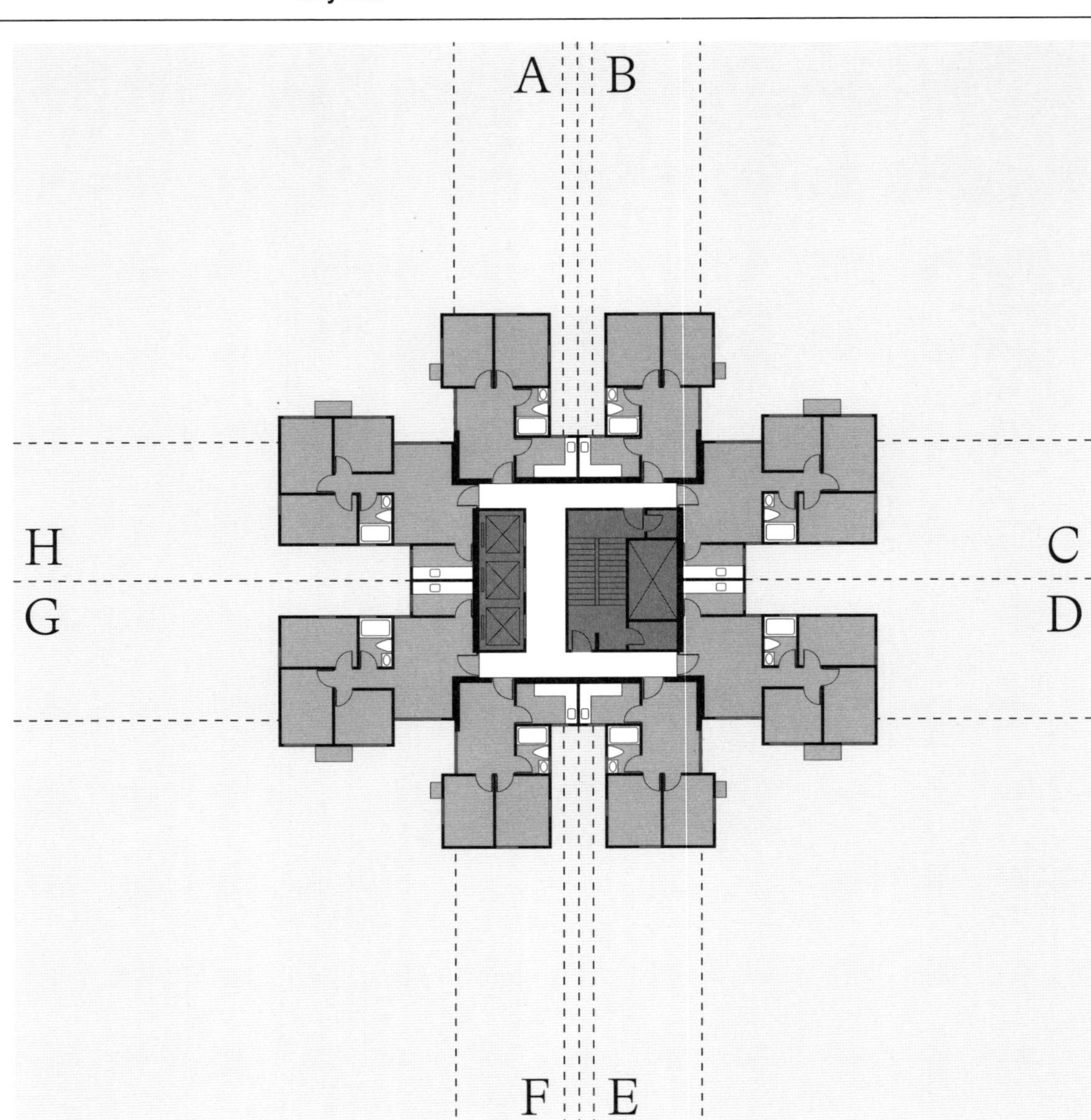

Type B

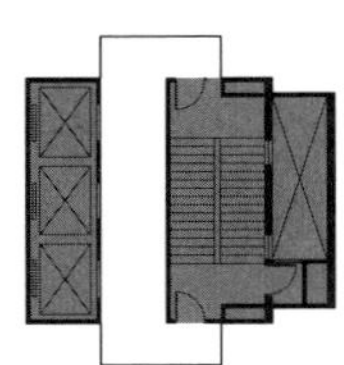

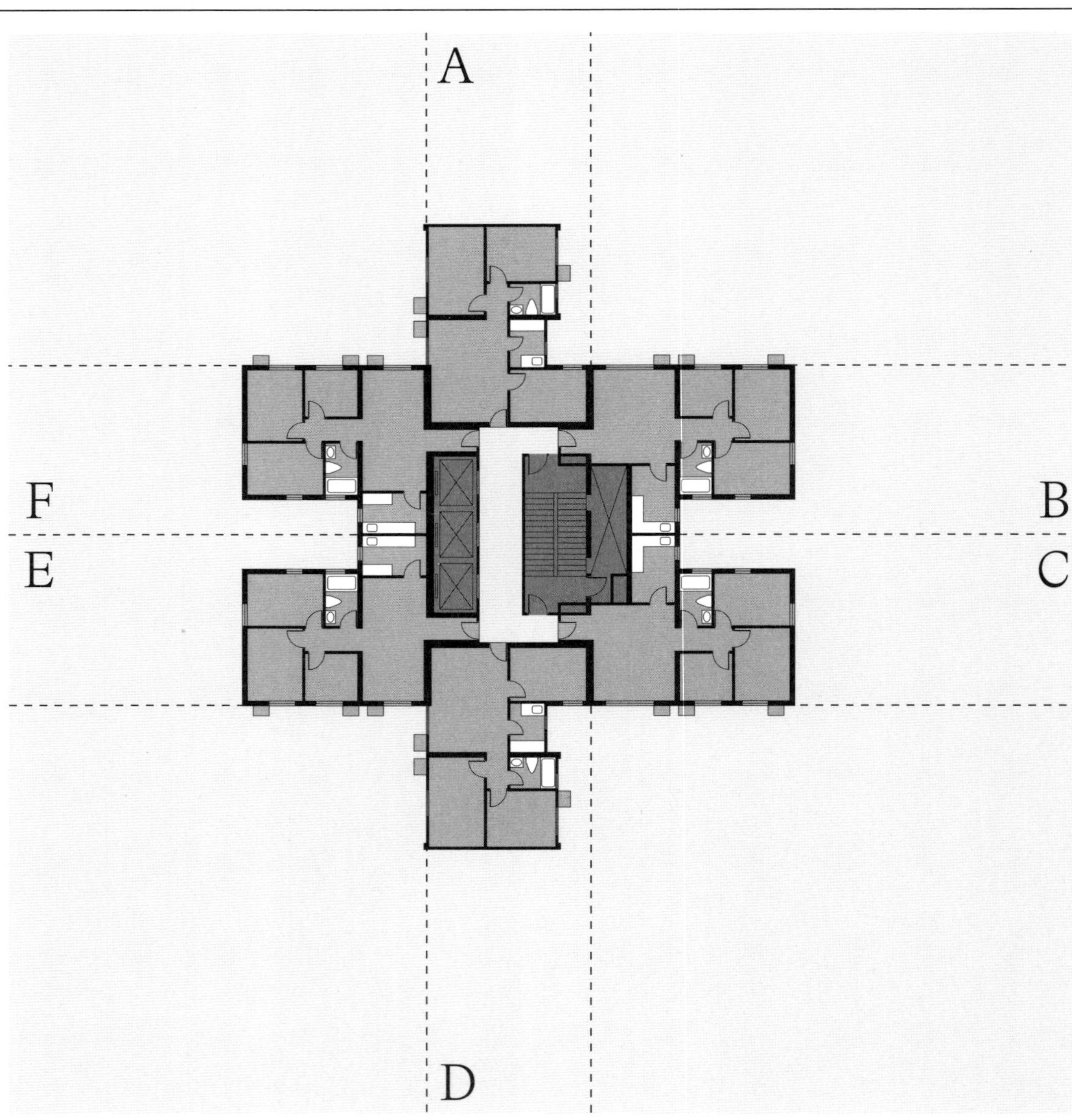

Type C

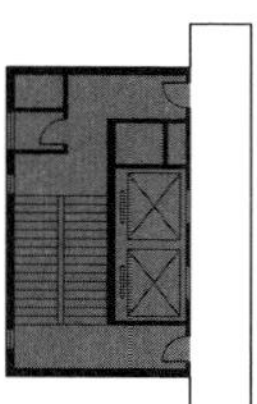

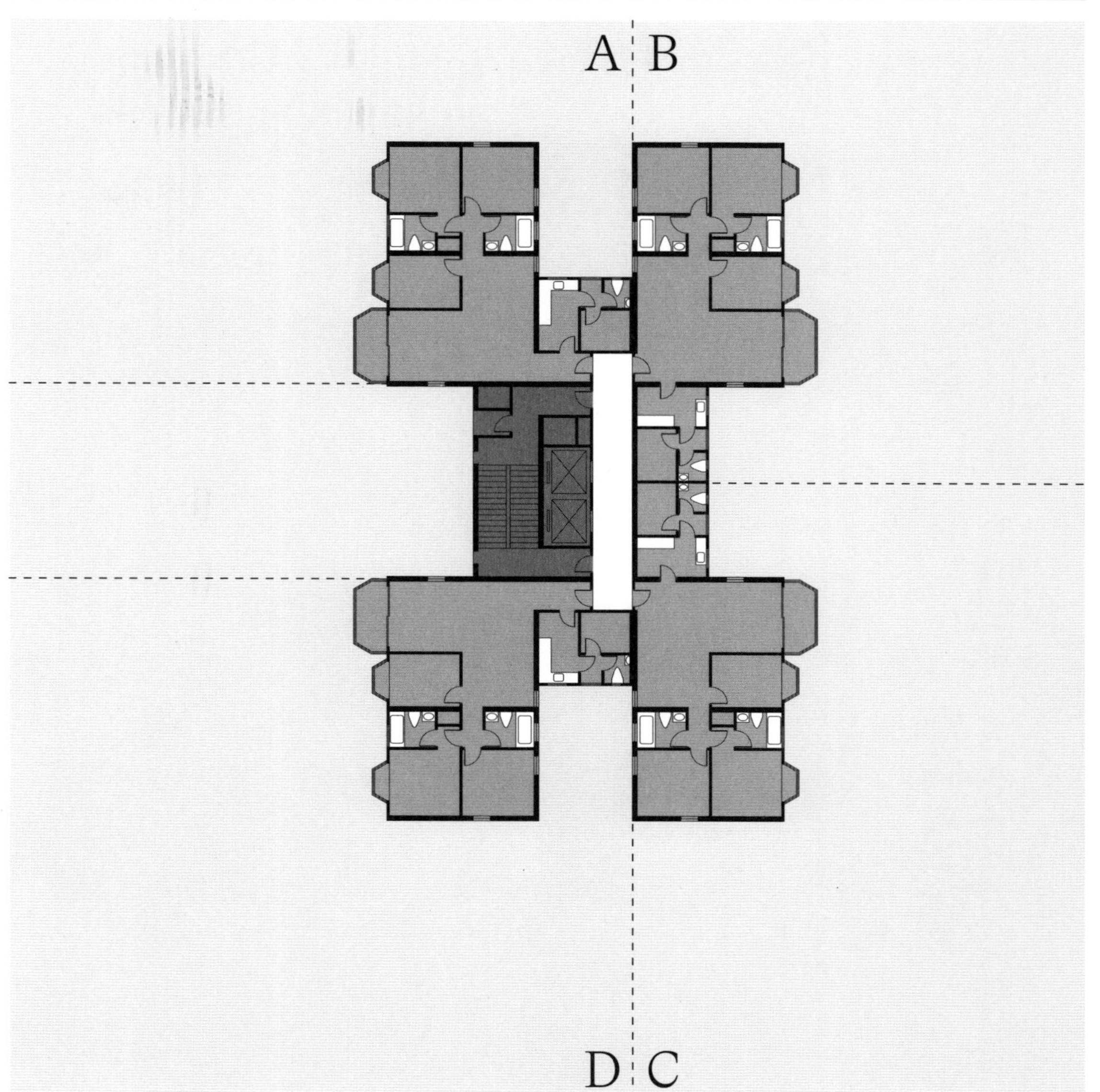

Type D

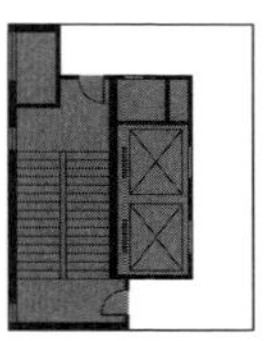

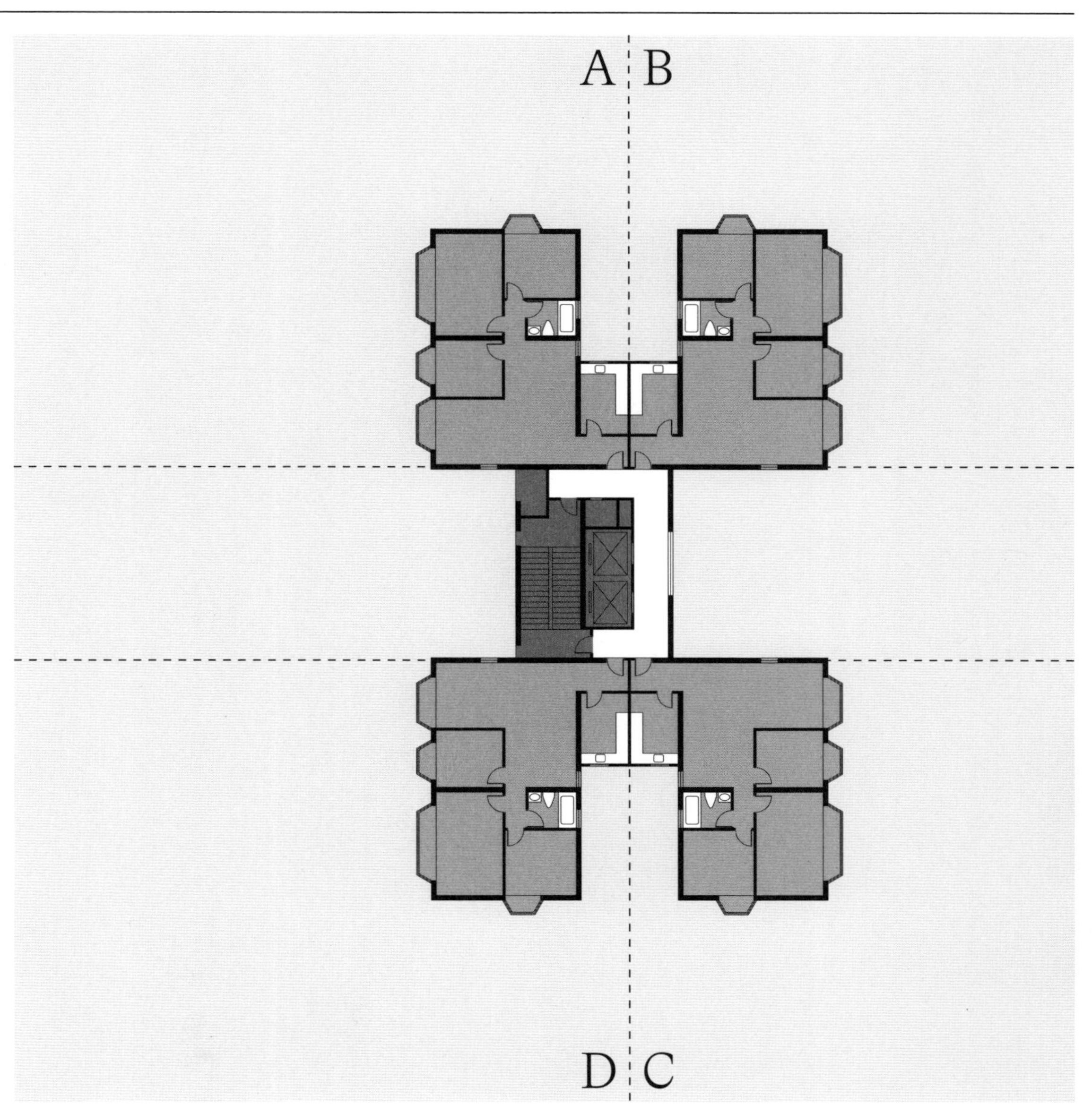

A'1

5 Copies

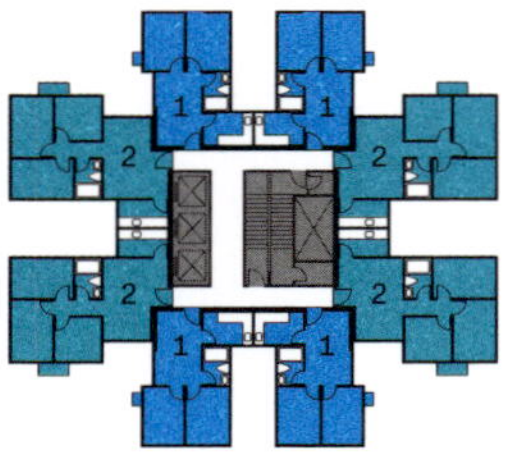

Phase 1 →/ **Block 1, 2, 3, 8, 9, 10**

A'2

9 Copies

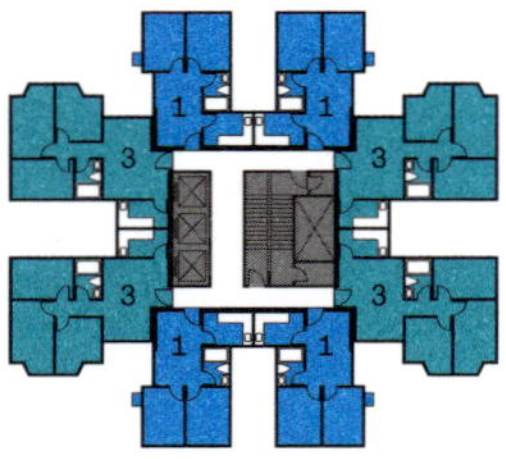

Phase 2 →/ **Block 16, 17, 19–21, 23**
Phase 3 →/ **Block 29–32**

A'3

1 Copy

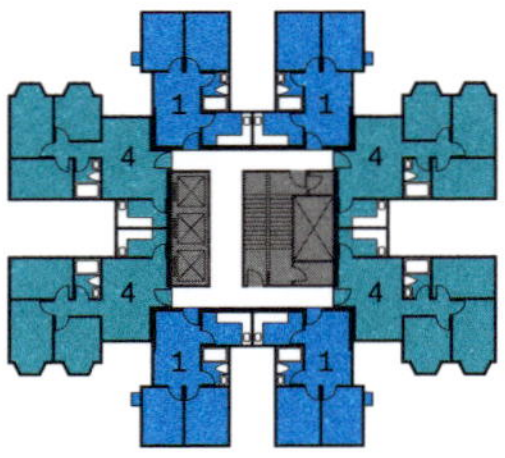

Phase 6 →/ **Block 24, 26**

A''4

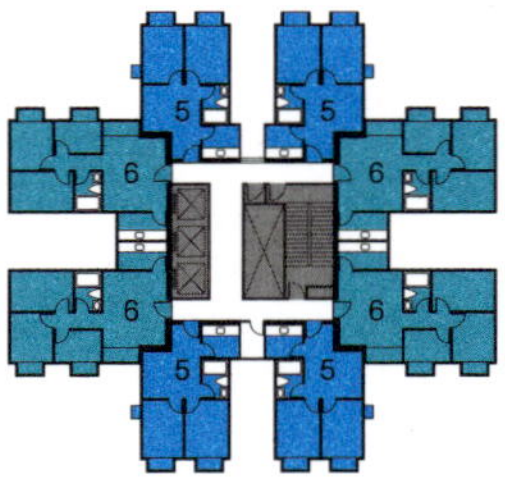

Phase 7 →/ **Block 36**

A'''5

15 Copies

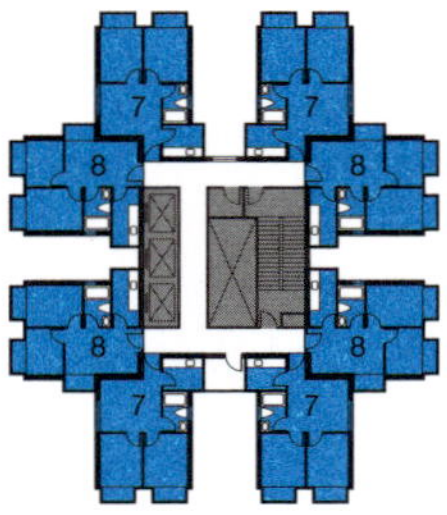

Phase 4 →/ **Block 37–45**
Phase 5→/ **Block 46–52**

B1

2 Copies

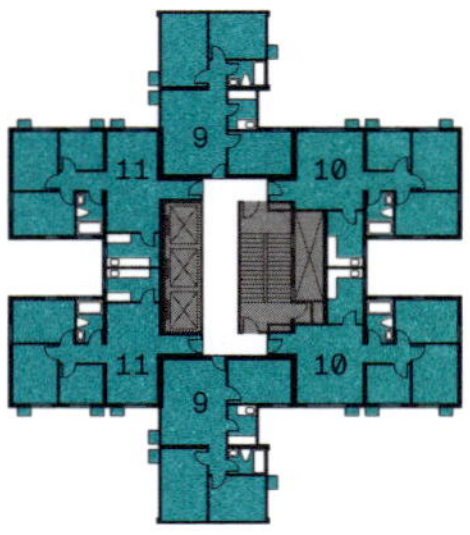

Phase 1 →/ **Block 7, 11, 12**

B2

1 Copy

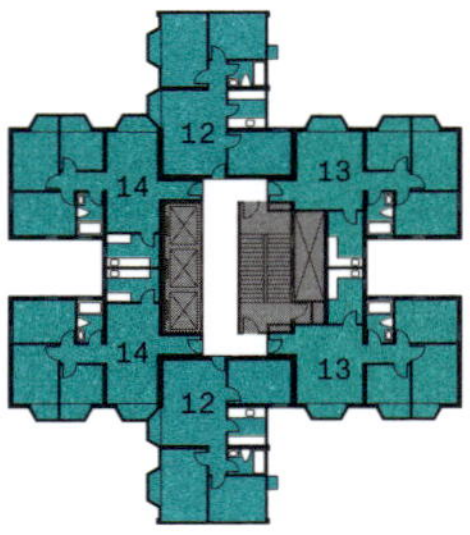

Phase 2 →/ **Block 18, 22**

C1

6 Copies

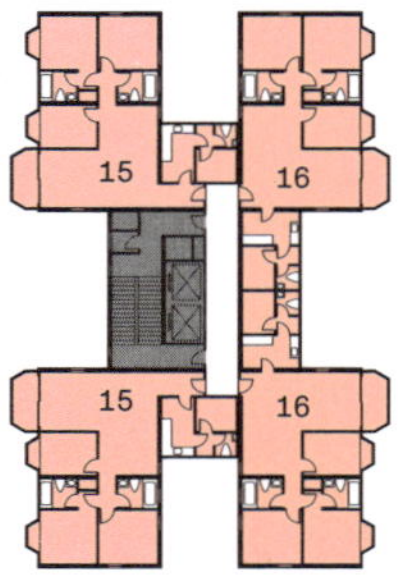

Phase 1 →/ **Block 4, 5, 6, 13, 14**
Phase 2 →/ **Block 15**
Phase 3 →/ **Block 33**

D1

2 Copies

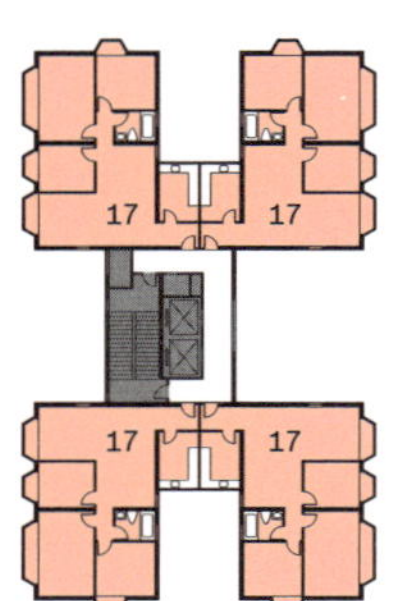

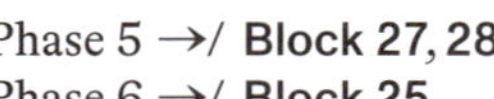

Phase 5 →/ **Block 27, 28**
Phase 6 →/ **Block 25**

D2

1 Copy

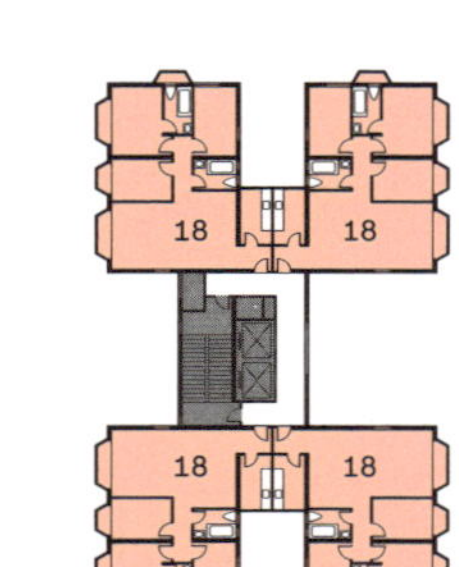

Phase 7 →/ **Block 34, 35**

A B C D E F G H
Block Core
Unit
1
A'1
30.6
41.2
2
3
8
9
10
16
A'2
42.9
17
19
20
21
23
29
30
31
180°

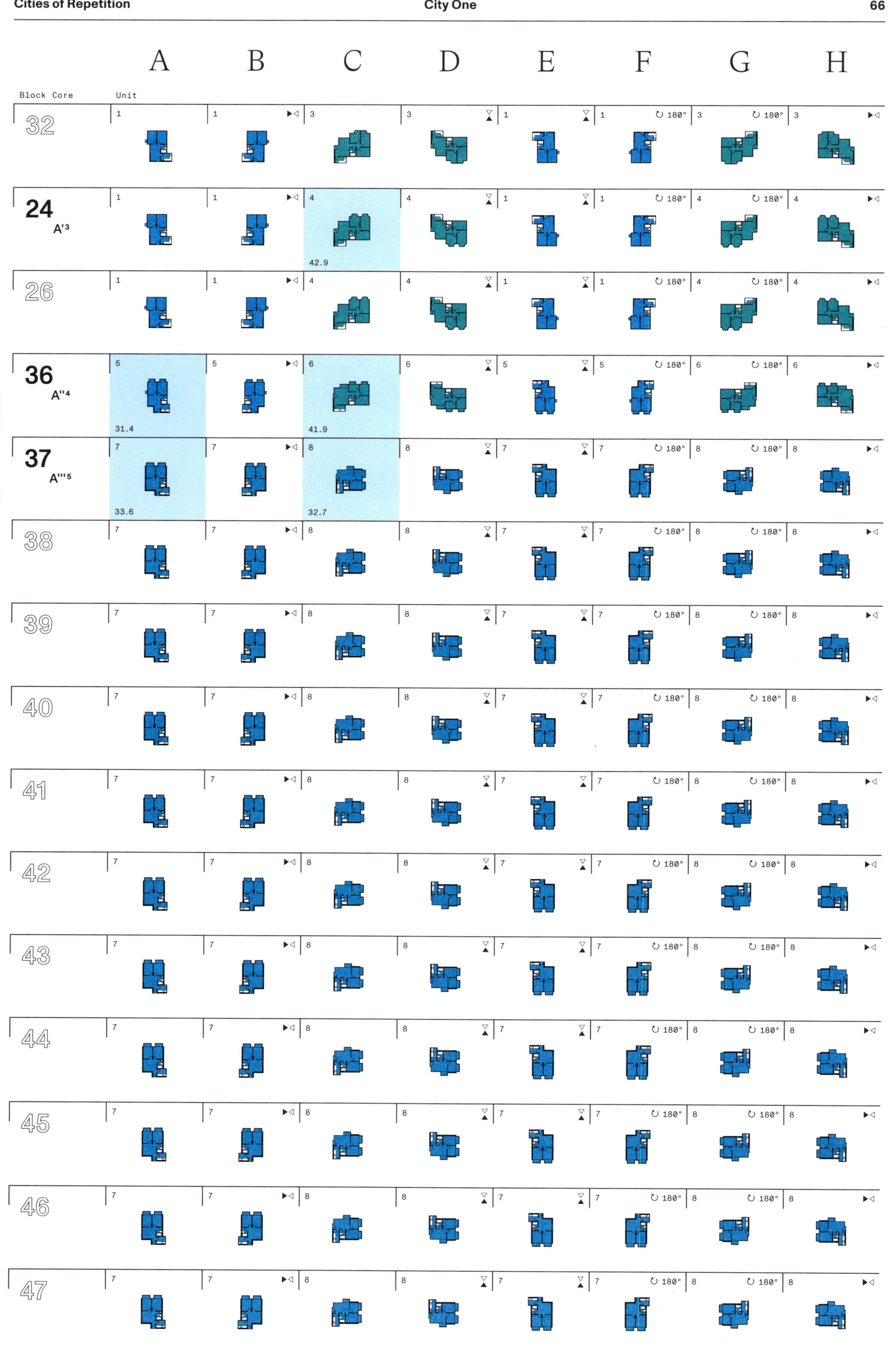
A B C D E F G H
Block Core
Unit
32 1 1 3 3 1 1 180° 3 180° 3
24 A'3 1 1 4 42.9 4 1 1 180° 4 180° 4
26 1 1 4 4 1 1 180° 4 180° 4
36 A''4 5 31.4 5 6 41.9 6 5 5 180° 6 180° 6
37 A'''5 7 33.6 7 8 32.7 8 7 7 180° 8 180° 8
38 7 7 8 8 7 7 180° 8 180° 8
39 7 7 8 8 7 7 180° 8 180° 8
40 7 7 8 8 7 7 180° 8 180° 8
41 7 7 8 8 7 7 180° 8 180° 8
42 7 7 8 8 7 7 180° 8 180° 8
43 7 7 8 8 7 7 180° 8 180° 8
44 7 7 8 8 7 7 180° 8 180° 8
45 7 7 8 8 7 7 180° 8 180° 8
46 7 7 8 8 7 7 180° 8 180° 8
47 7 7 8 8 7 7 180° 8 180° 8

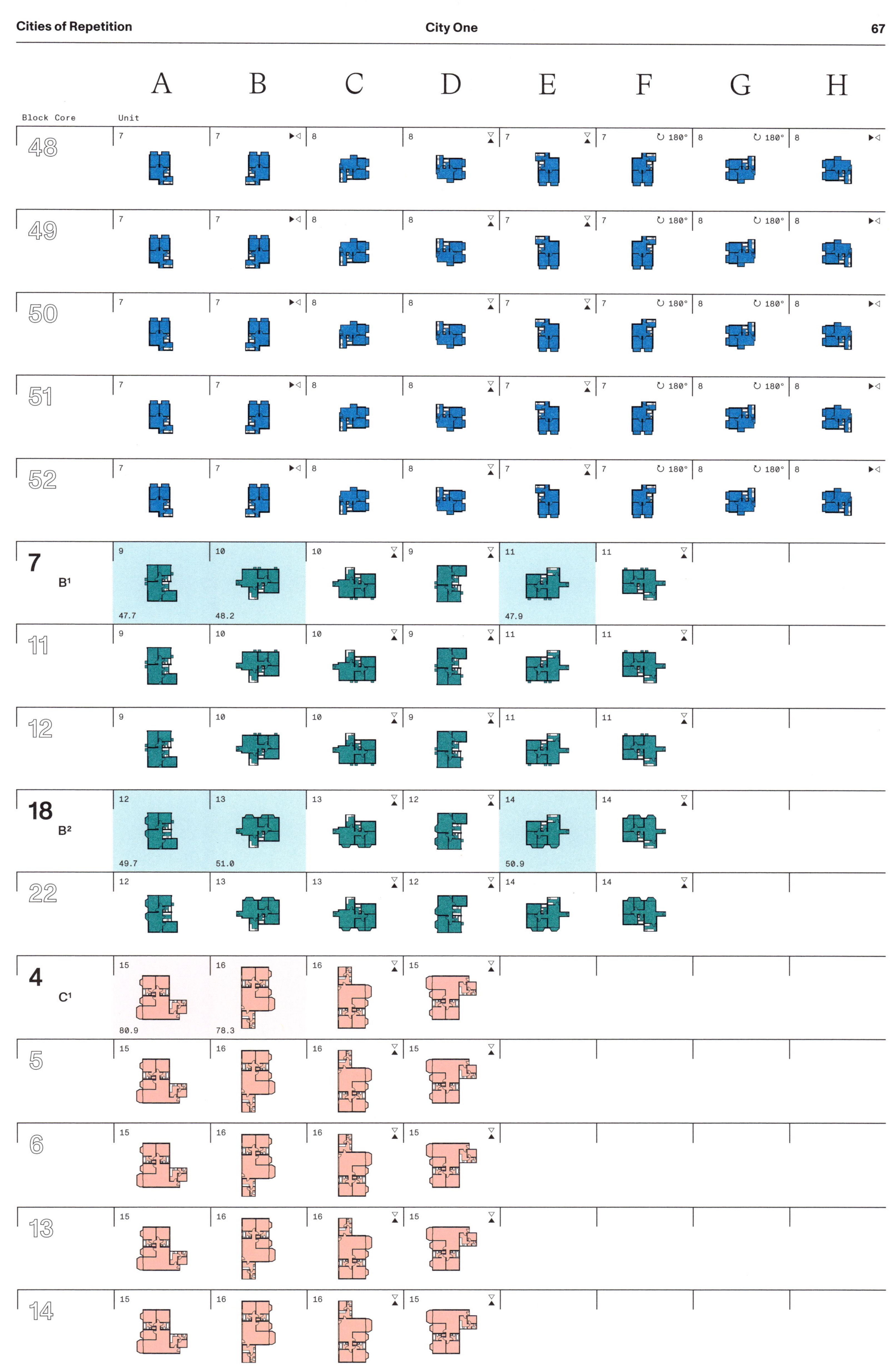
A
B
C
D
E
F
G
H
Block Core
Unit
48
49
50
51
52
7
B¹
47.7
48.2
47.9
11
12
18
B²
49.7
51.0
50.9
22
4
C¹
80.9
78.3
5
6
13
14
180°

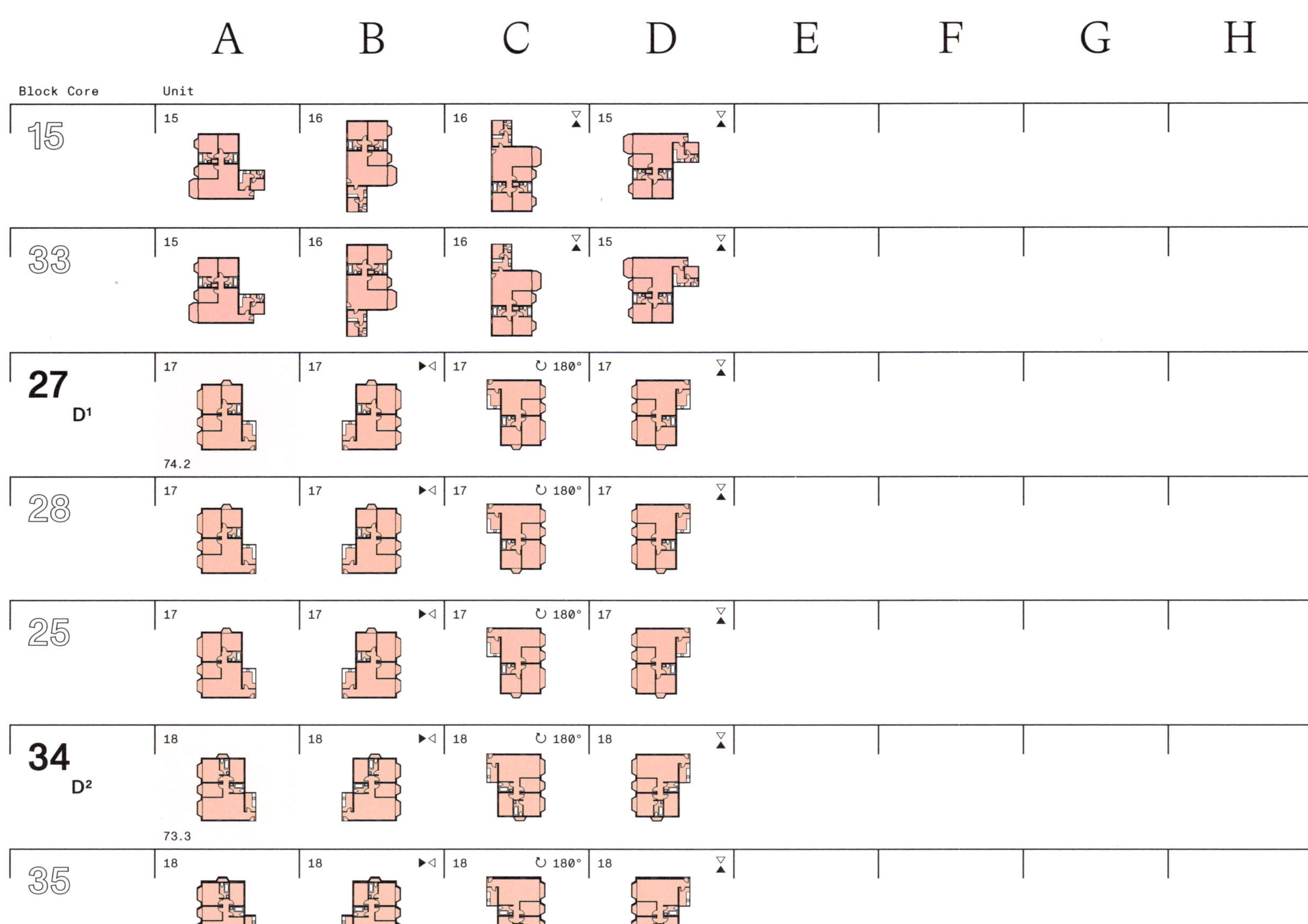
A
B
C
D
E
F
G
H
Block Core
Unit
15
33
27
D1
74.2
28
25
34
D2
73.3
35
180°

Whampoa Garden 1985–91

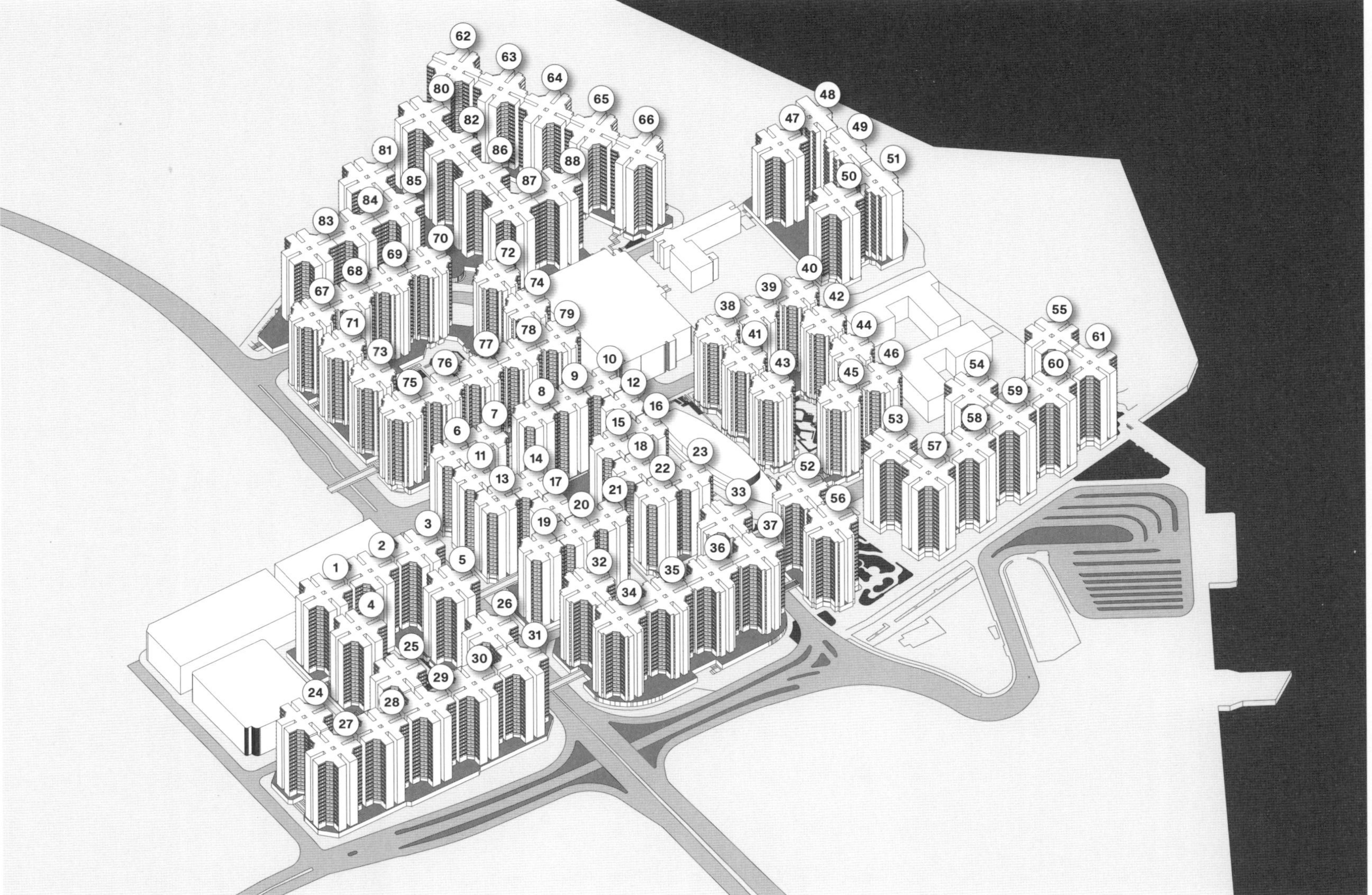

DEVELOPER:	Hutchison Whampoa Limited		
ARCHITECT:	Wong and Ouyang (HK) Ltd., DLN Architects & Engineers.		
LOCATION:	Kowloon City		
POPULATION:	31,613		
TOWERS:	88	UNIQUE:	39
APARTMENTS:	10,441	UNIT TYPES:	49
CORES:	10	PHASES:	12

Like the Taikoo Shing housing estate, **Whampoa Garden** was built on a site formerly occupied by dockyards. The site on the waterfront in Kowloon was comprised of dry docks, boat slips, and a group of industrial buildings and warehouses owned by the Hong Kong Kowloon and Whampoa Dock Company. Whampoa refers to the Chinese name "Huangpu," an area of Guangzhou in mainland China where the company had previously owned another dockyard. In 1973, the Taikoo Dockyard and the Hong Kong Whampoa Dock companies merged and built a new dockyard on Tsing Yi Island, to the northwest of the Kowloon Peninsula. Through this merger, ship builders effectively became real estate developers, and over the following decade, the site was transformed into Hong Kong's second largest private housing estate—after Mei Foo—in Hong Kong.

The estate is well integrated into the surrounding urban context, as the development is built on an extension of the existing street grid. Towers are built on top of podium space and grouped by construction phase. The overall population density of the estate was originally constrained by building height limitations due to the site's proximity to the Kai Tak airport. Large, publically accessible courtyards are formed by the positioning of towers at the periphery of each block. The courtyards provide residents with a variety of landscape and water features as well as recreational facilities such as basketball and badminton courts. Podium spaces and urban courtyards are connected throughout the estate by a network of elevated pedestrian bridges. Podium spaces host retail shops, large shopping centers, and parking below the towers. In the middle of the estate, a shopping center disguised as a ship marks the spot of a former dry dock.

Type A

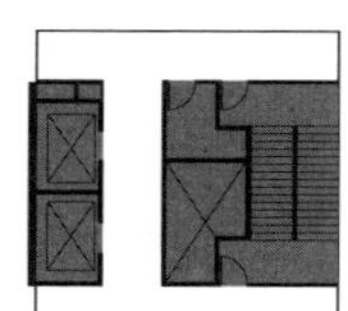

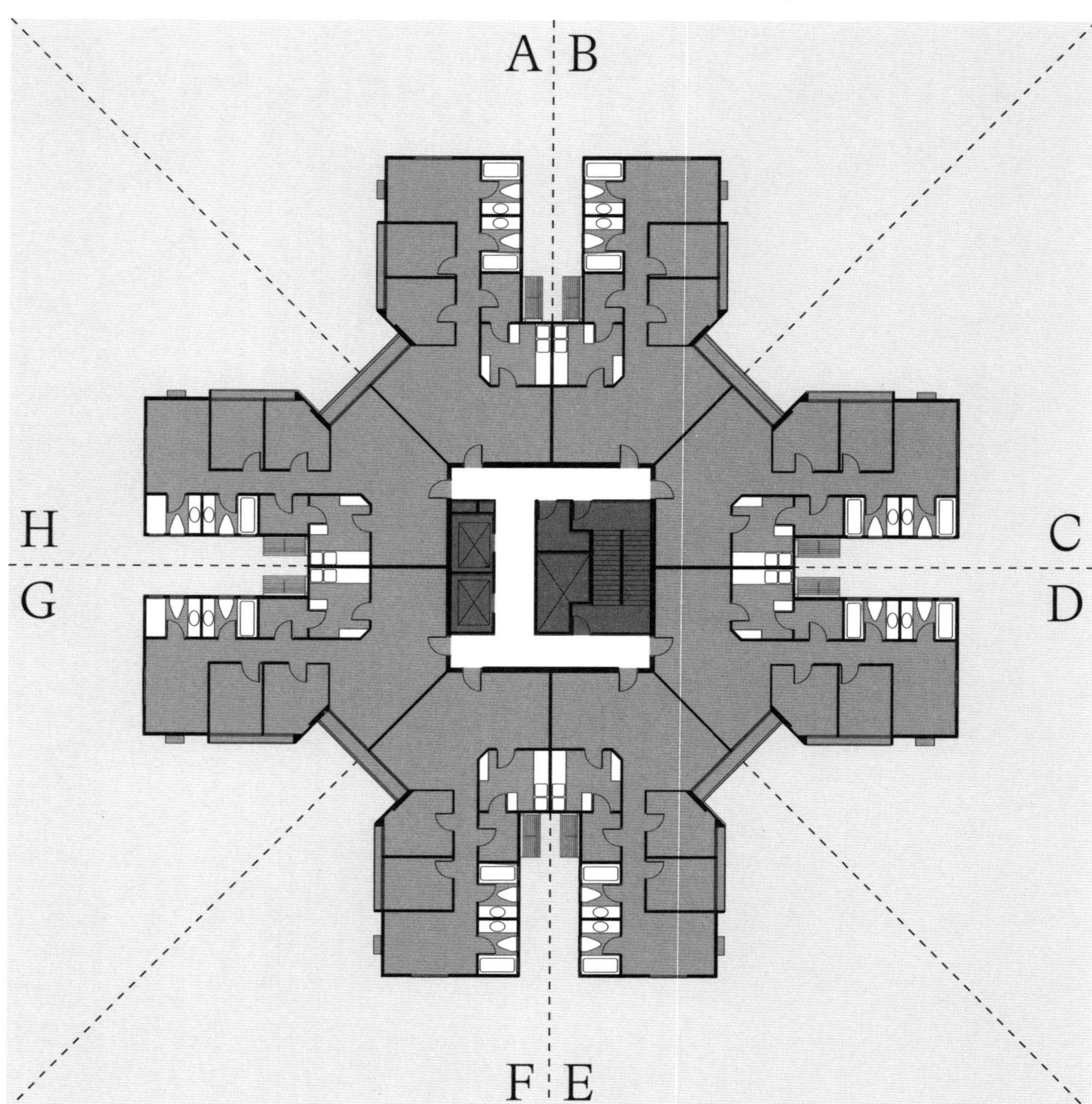

Type B

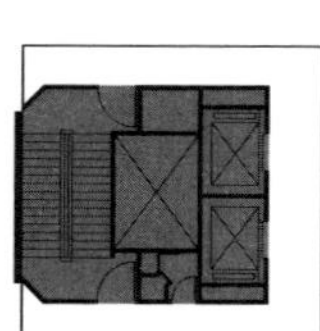

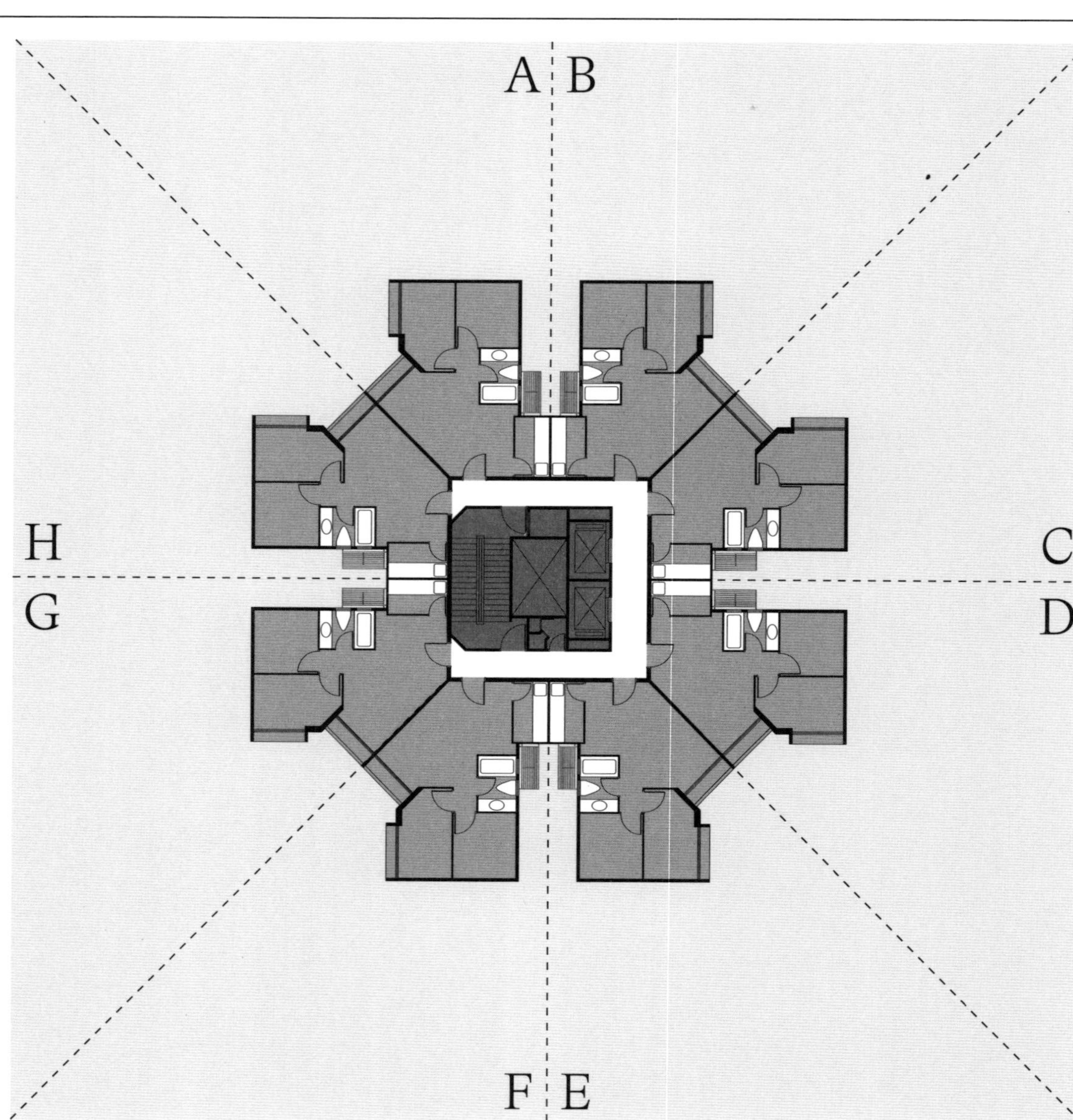

Type C

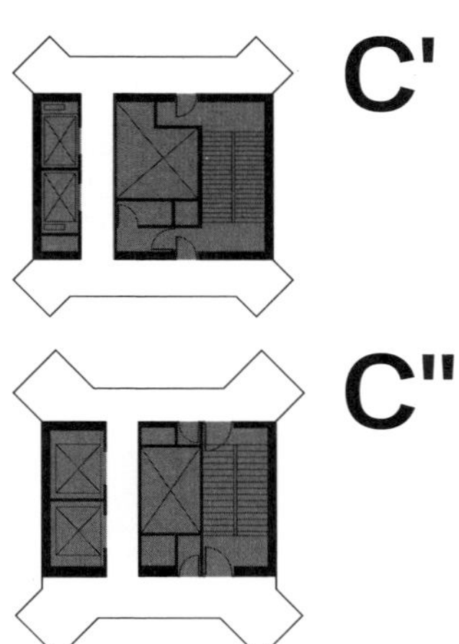

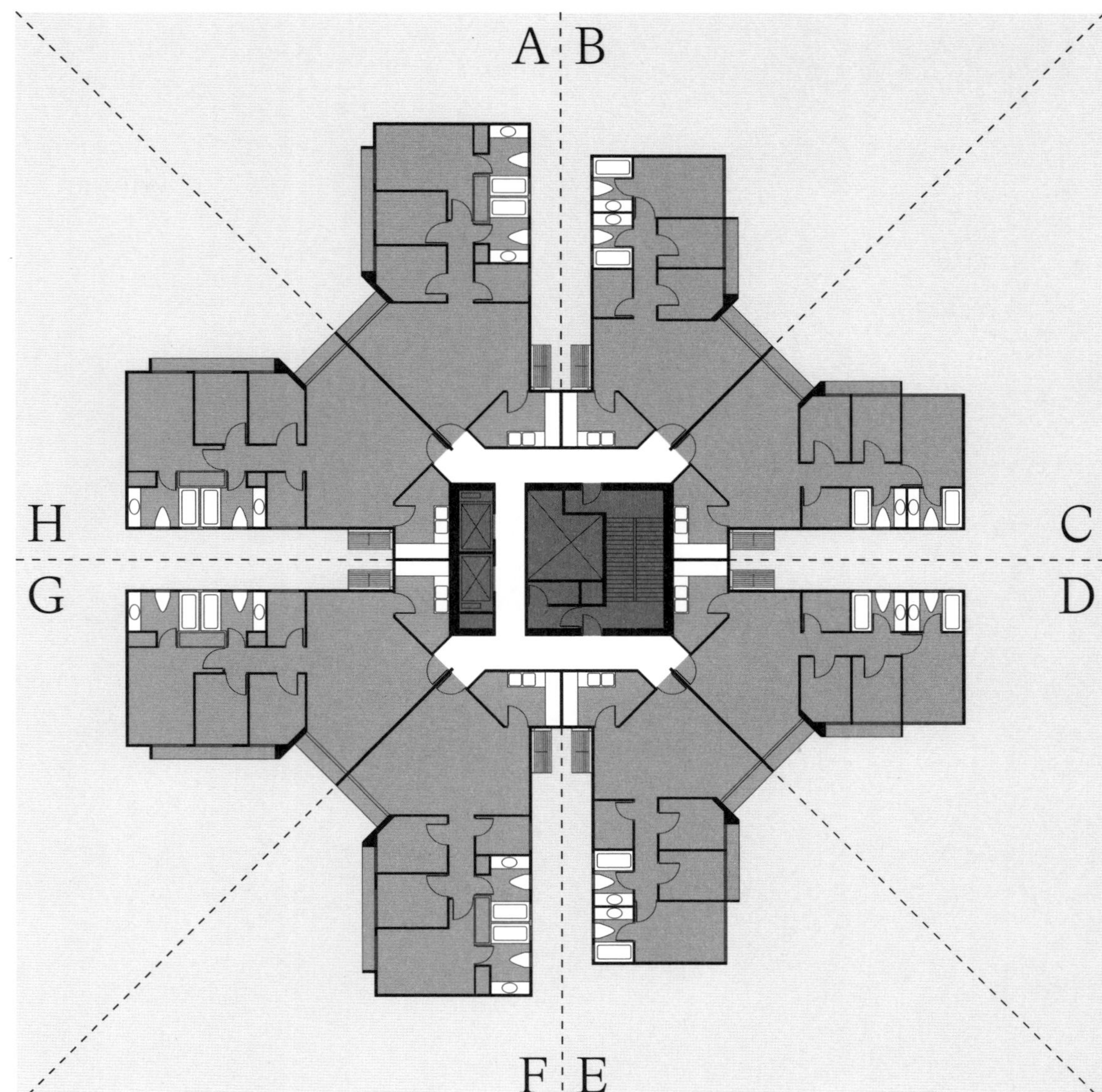

Type D

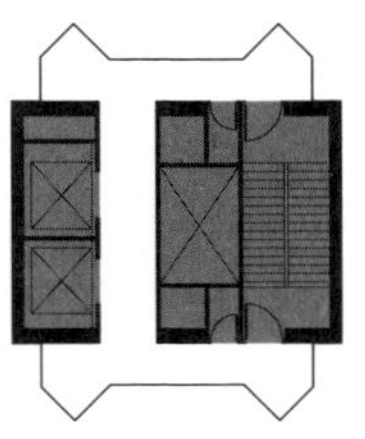

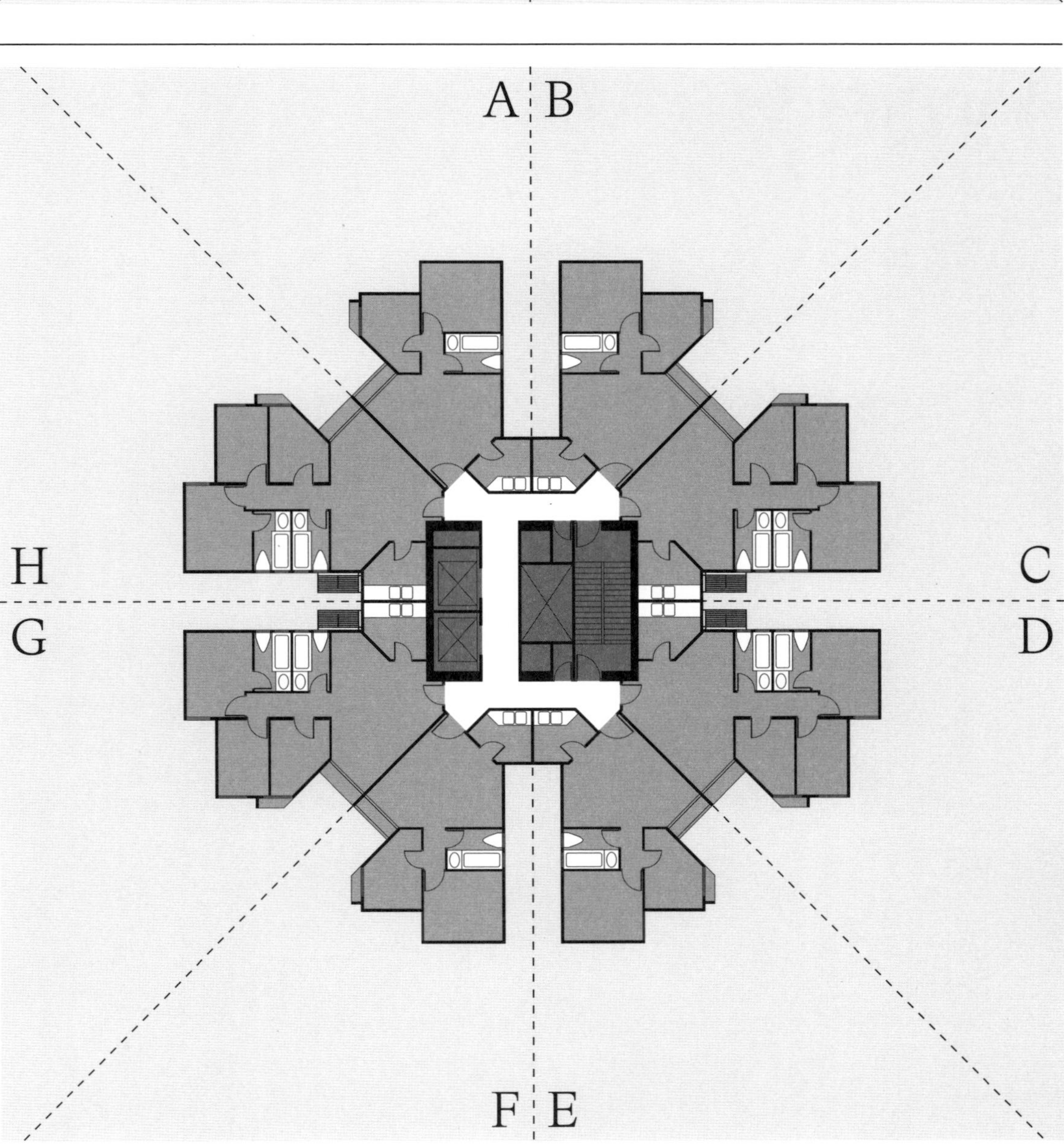

Type E

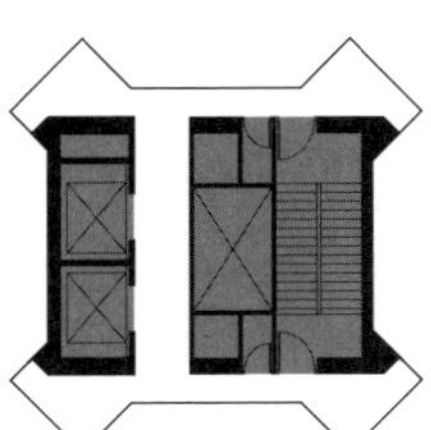

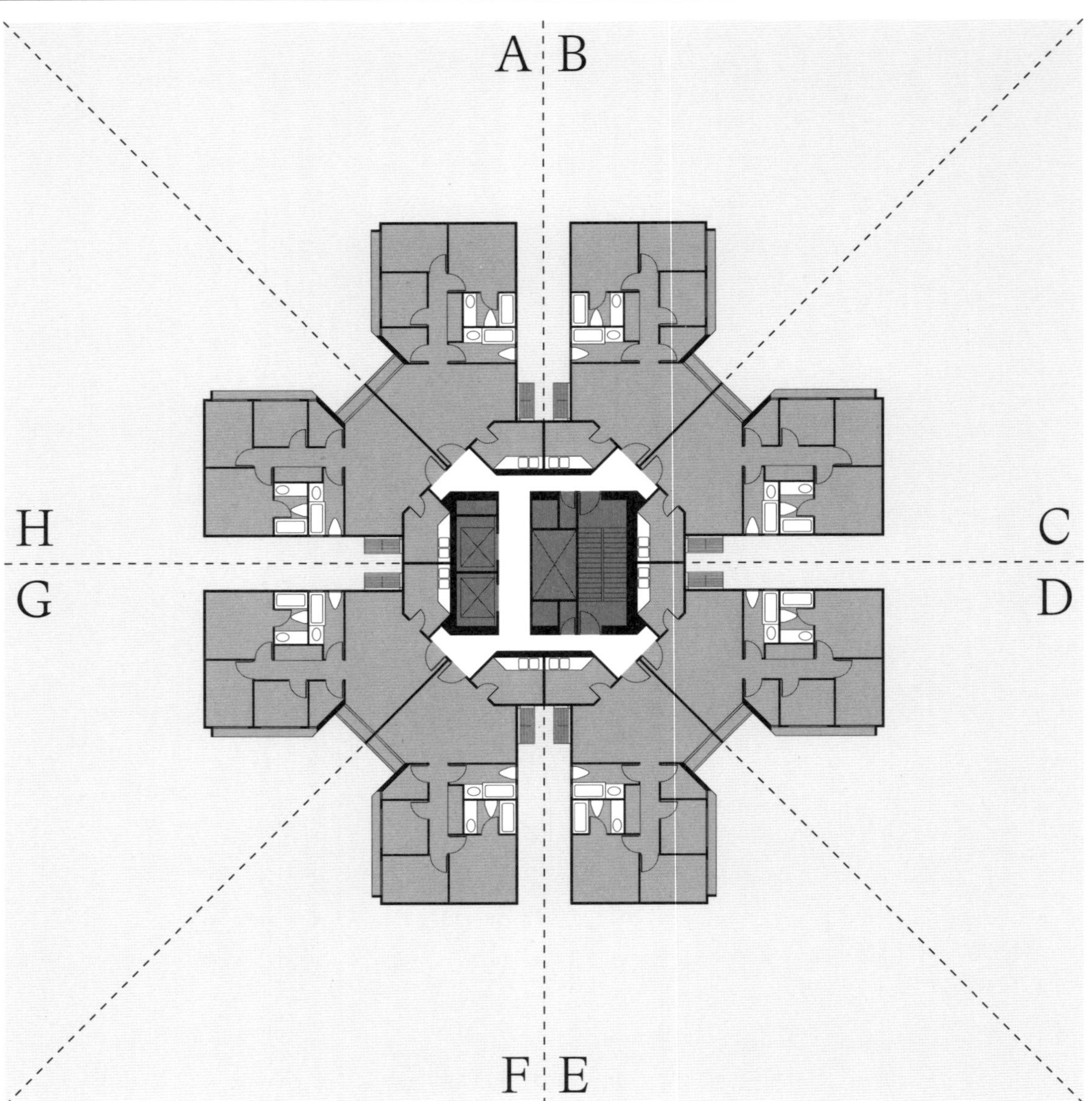

Type F

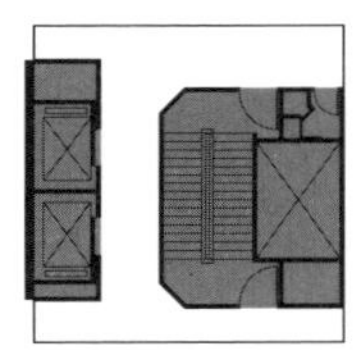

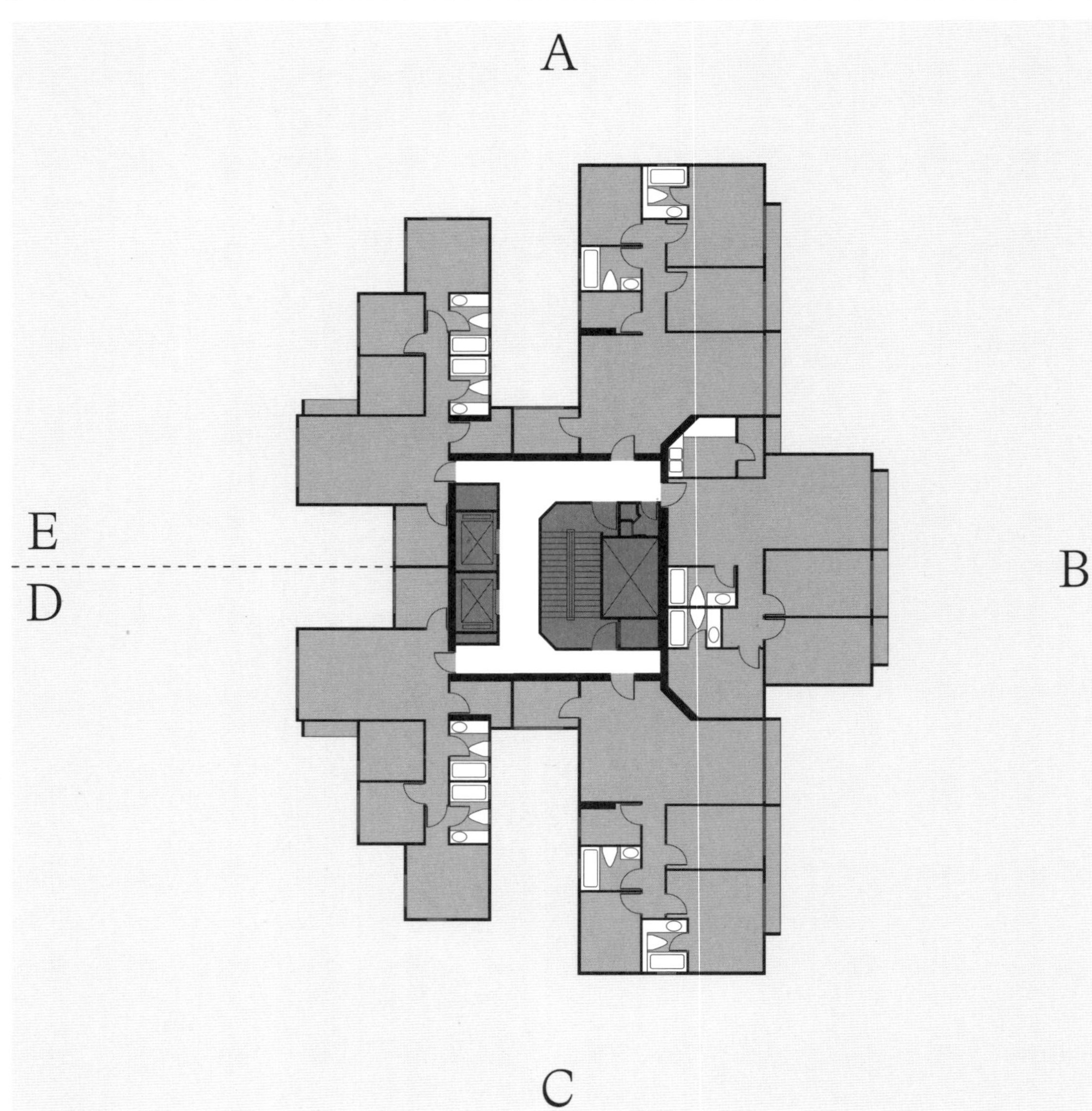

Type G

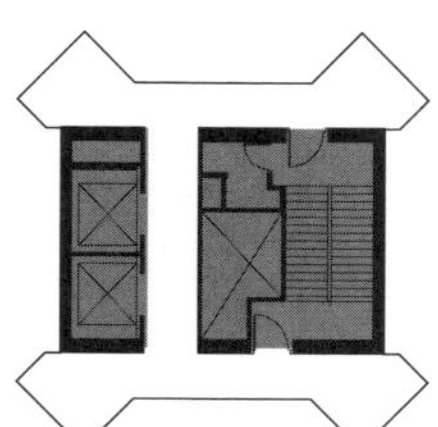

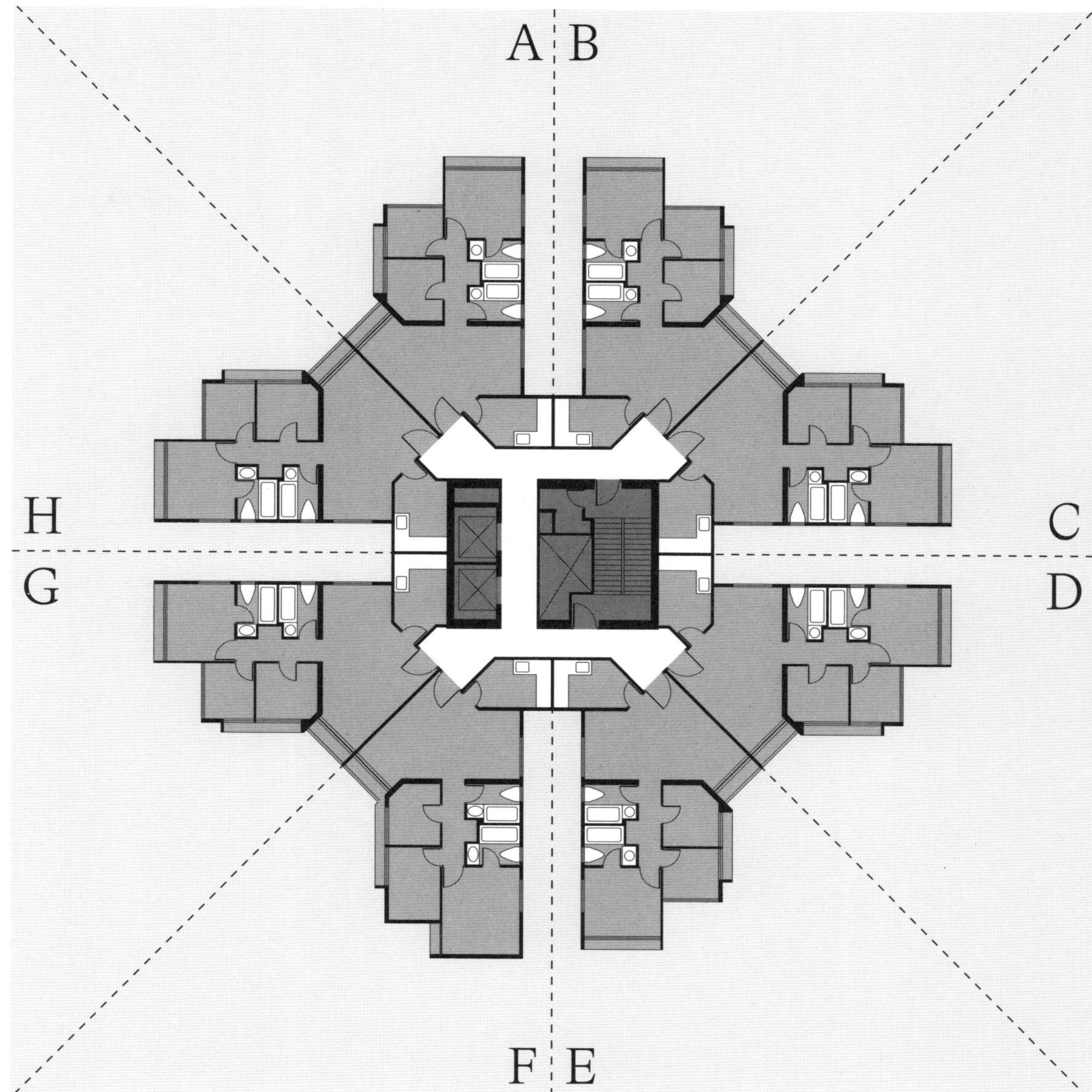

Type H

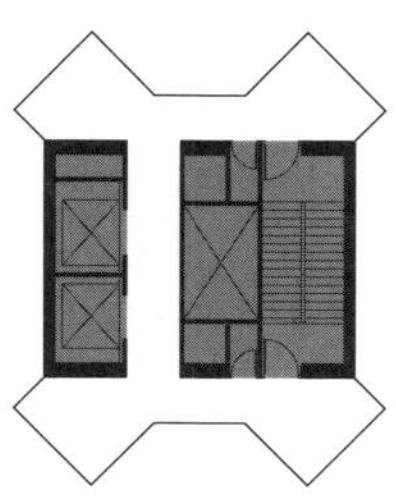

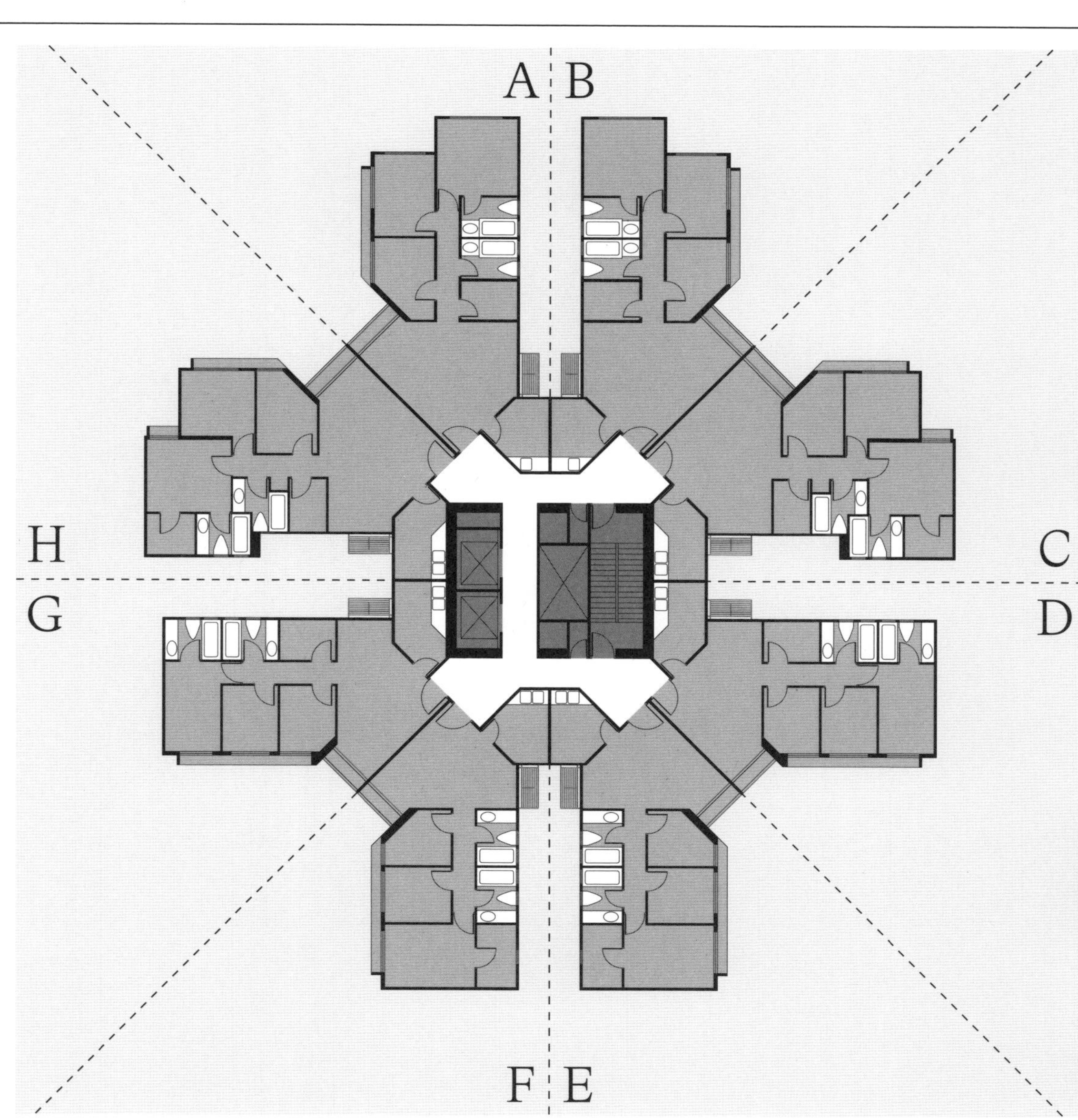

Type I

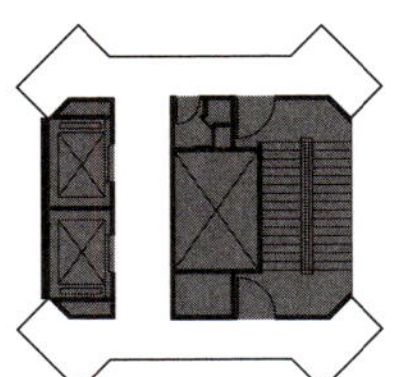

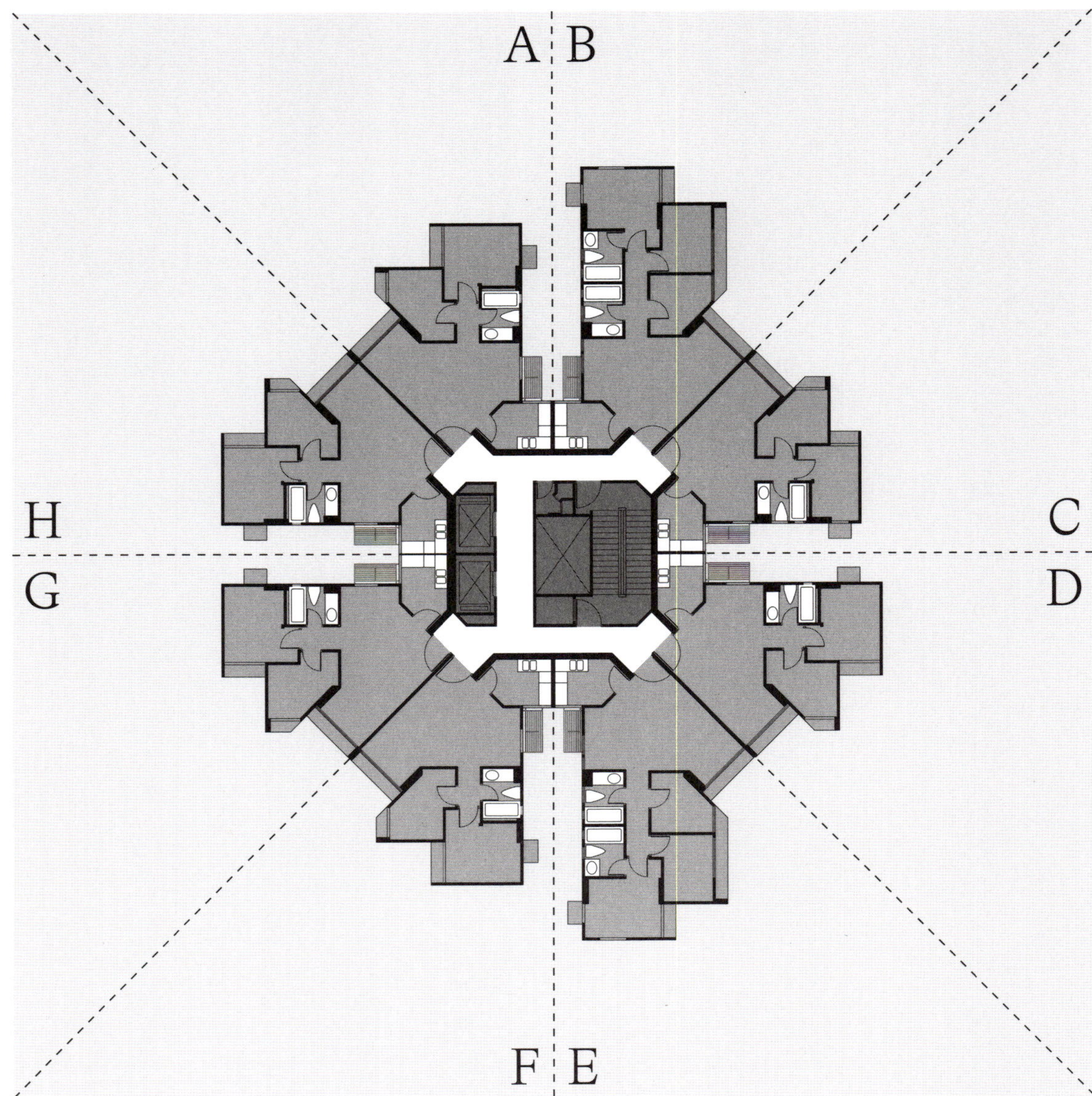

A¹

4 Copies

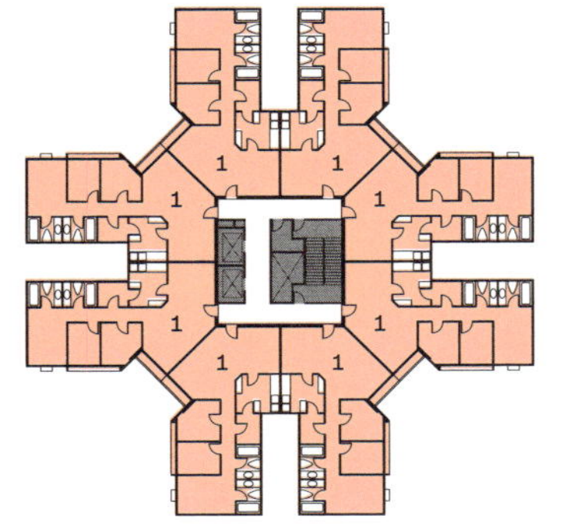

Phase 1 →/ **Block 1–5**

B¹

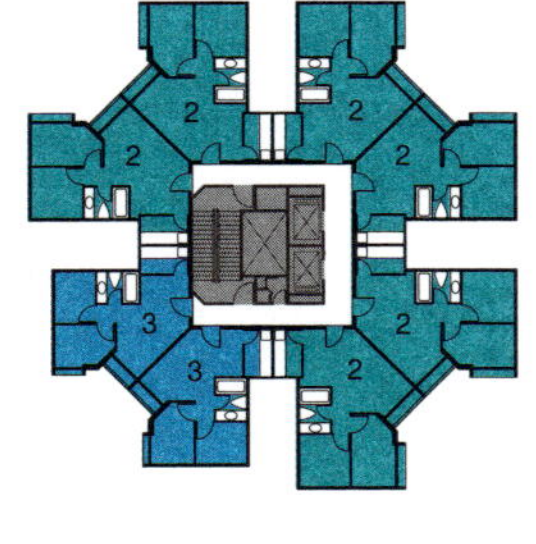

Phase 2 →/ **Block 6**

B²

2 Copies

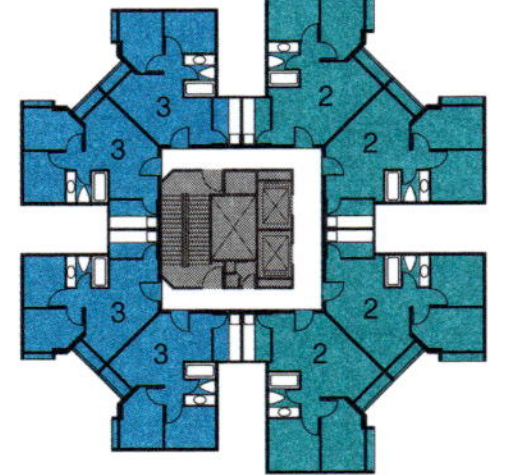

Phase 2 →/ **Block 7–9**

B³

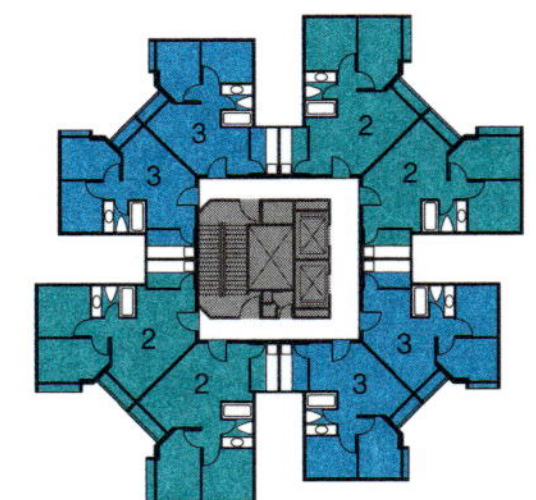

Phase 2 →/ **Block 10**

B⁴

1 Copy

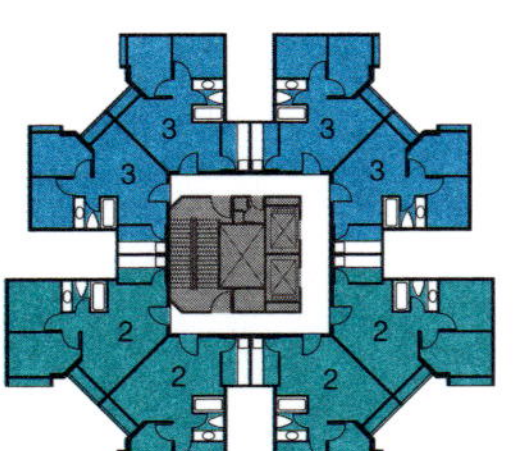

Phase 2 →/ **Block 11, 13**

B⁵

1 Copy

Phase 2 →/ **Block 12, 16**

B⁶

3 Copies

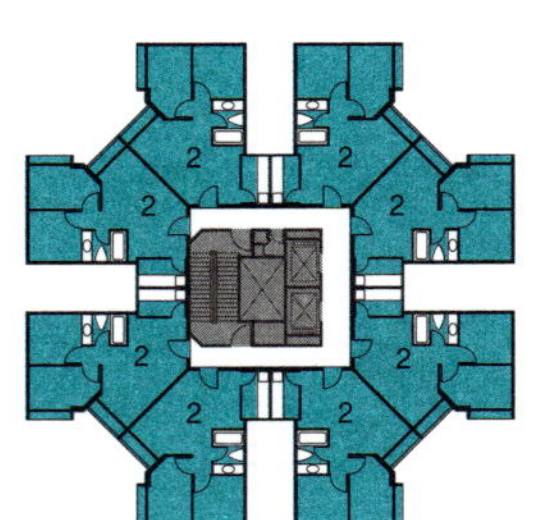

Phase 2 →/ **Block 14, 15, 17, 18**

B⁷

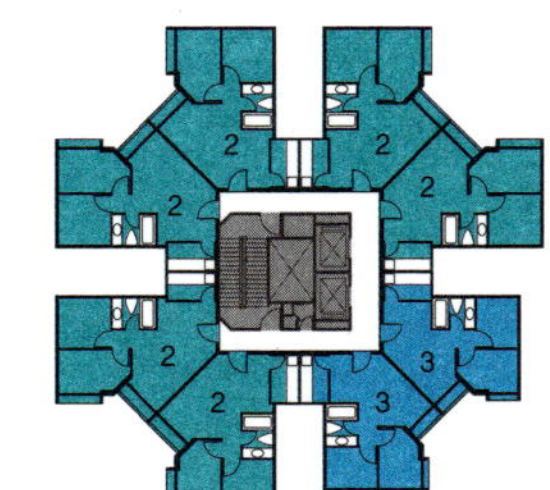

Phase 2 →/ **Block 19**

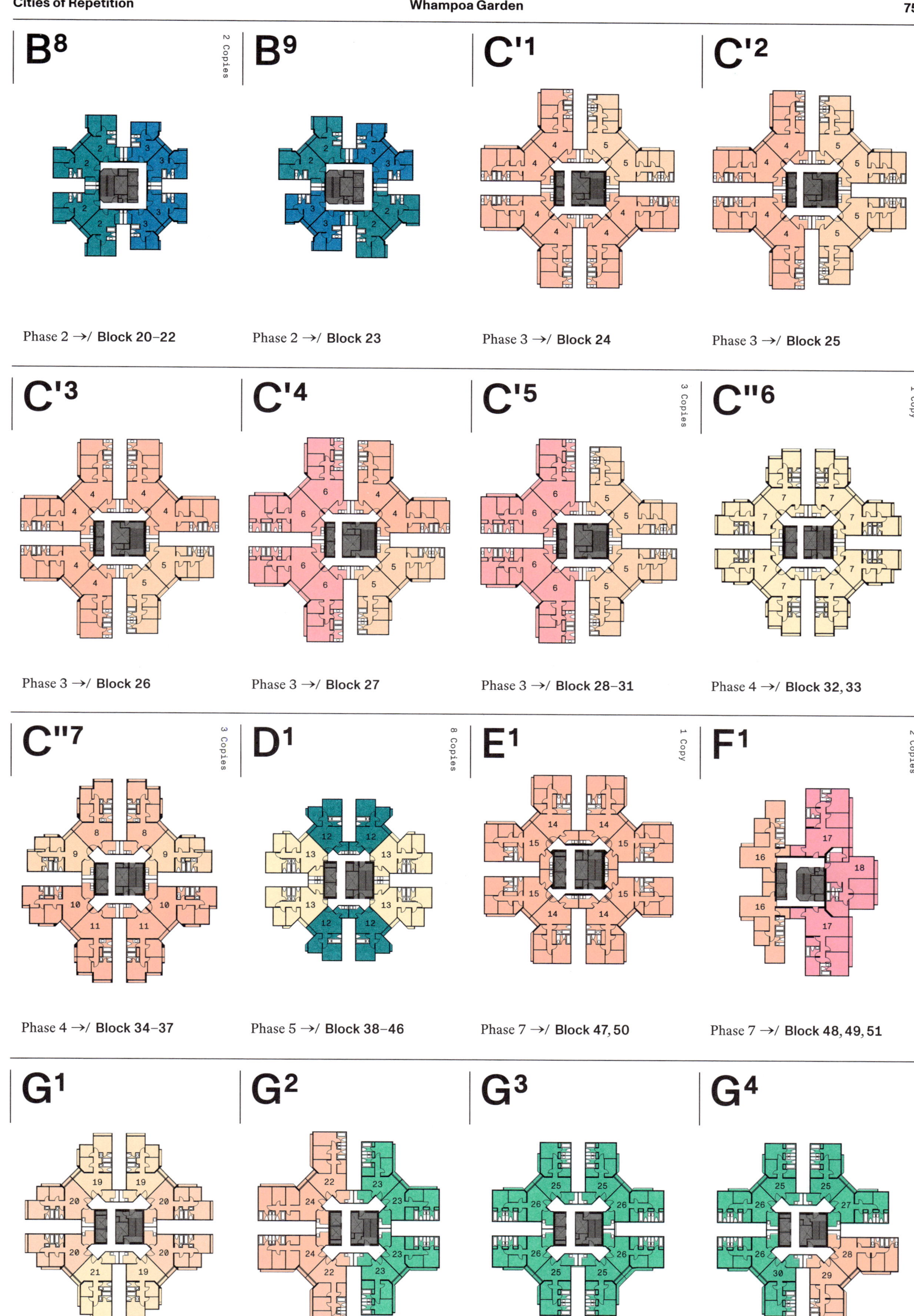
B8
2 Copies
Phase 2 →/ Block 20–22
B9
Phase 2 →/ Block 23
C'1
Phase 3 →/ Block 24
C'2
Phase 3 →/ Block 25
C'3
Phase 3 →/ Block 26
C'4
Phase 3 →/ Block 27
C'5
3 Copies
Phase 3 →/ Block 28–31
C''6
1 Copy
Phase 4 →/ Block 32, 33
C''7
3 Copies
Phase 4 →/ Block 34–37
D1
8 Copies
Phase 5 →/ Block 38–46
E1
1 Copy
Phase 7 →/ Block 47, 50
F1
2 Copies
Phase 7 →/ Block 48, 49, 51
G1
Phase 9 →/ Block 52
G2
Phase 9 →/ Block 53
G3
Phase 9 →/ Block 54
G4
Phase 9 →/ Block 55

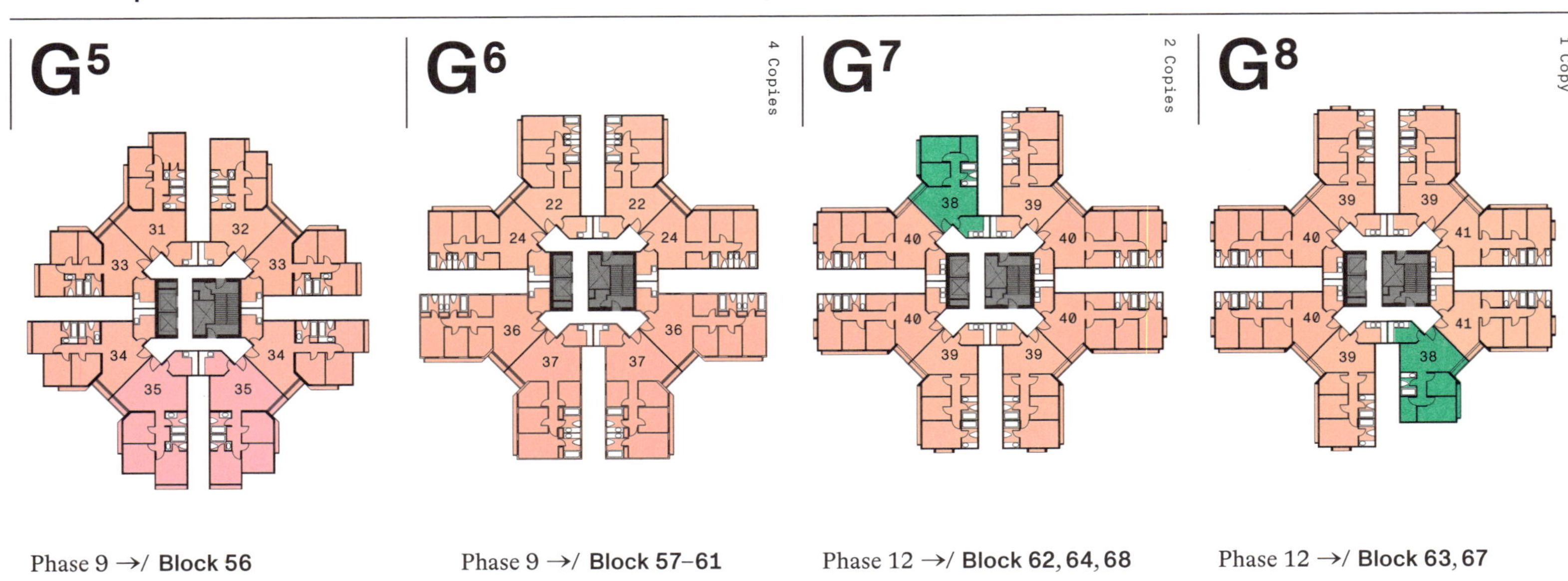

Phase 9 →/ **Block 56**

Phase 9 →/ **Block 57–61**

Phase 12 →/ **Block 62, 64, 68**

Phase 12 →/ **Block 63, 67**

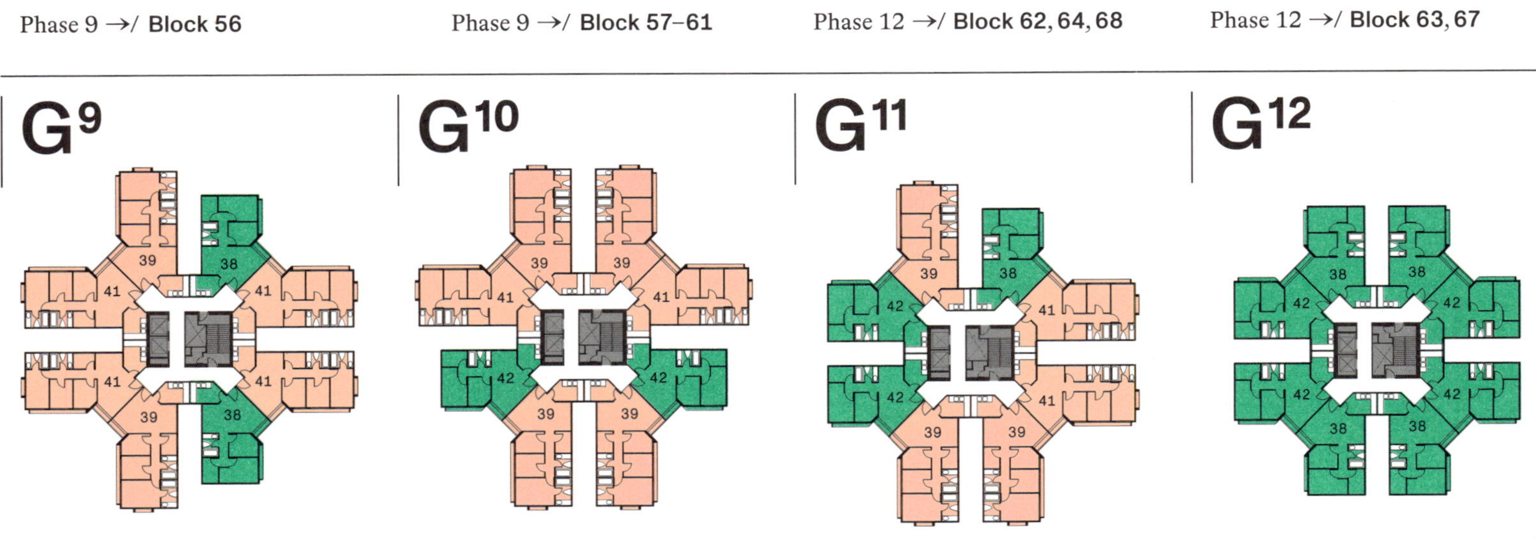

Phase 12 →/ **Block 65**

Phase 12 →/ **Block 66**

Phase 12 →/ **Block 69**

Phase 12 →/ **Block 70**

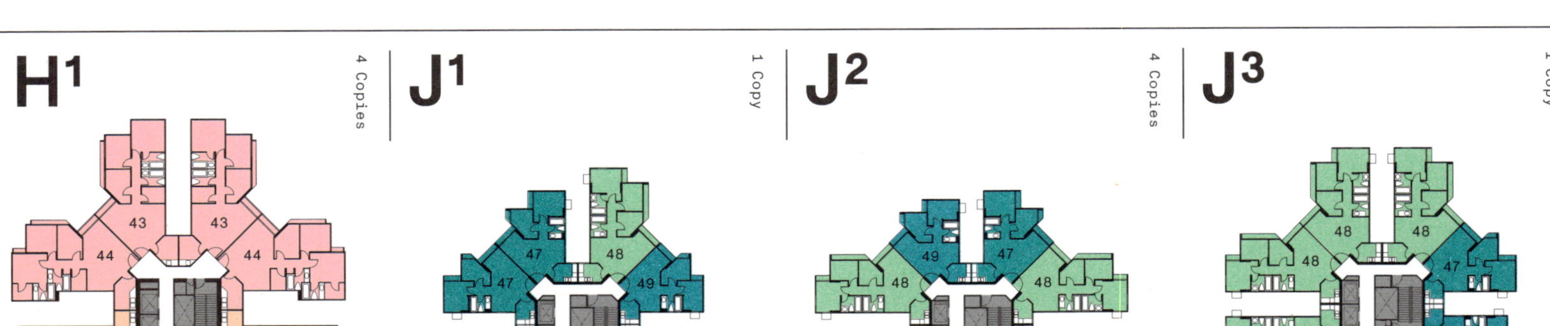

Phase 10 →/ **Block 71–75**

Phase 11 →/ **Block 76, 84**

Phase 11 →/ **Block 77, 78, 85–87**

Phase 11 →/ **Block 79, 88**

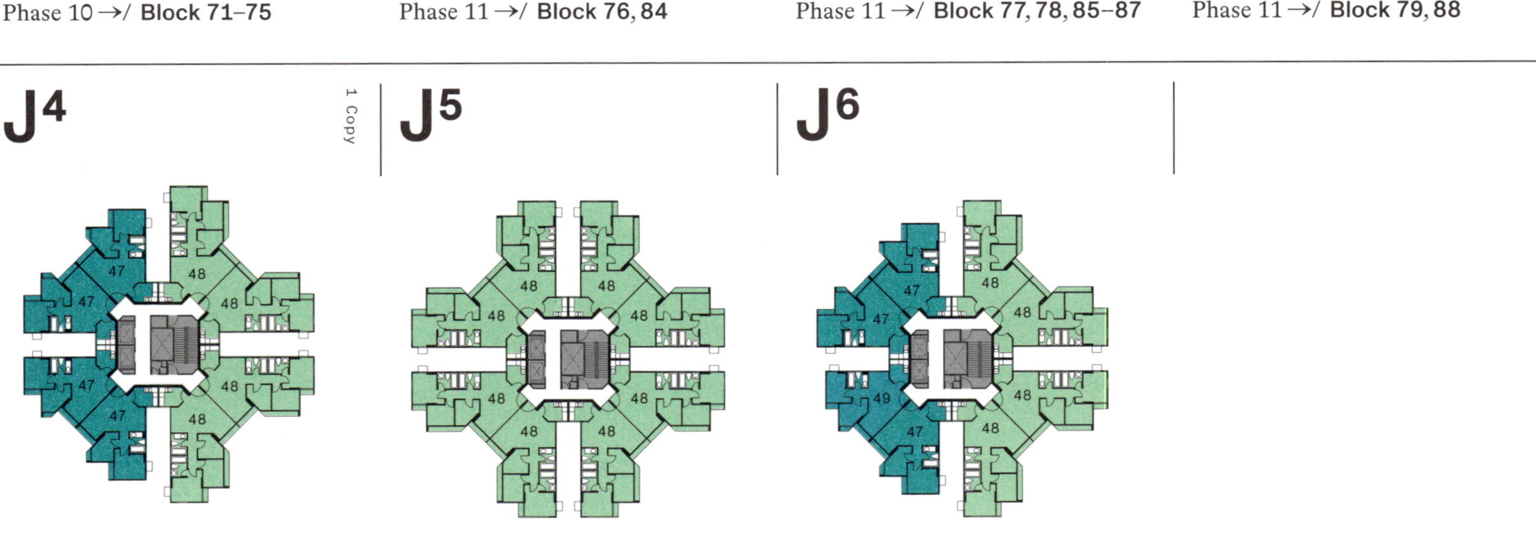

Phase 11 →/ **Block 80, 82**

Phase 11 →/ **Block 81**

Phase 11 →/ **Block 83**

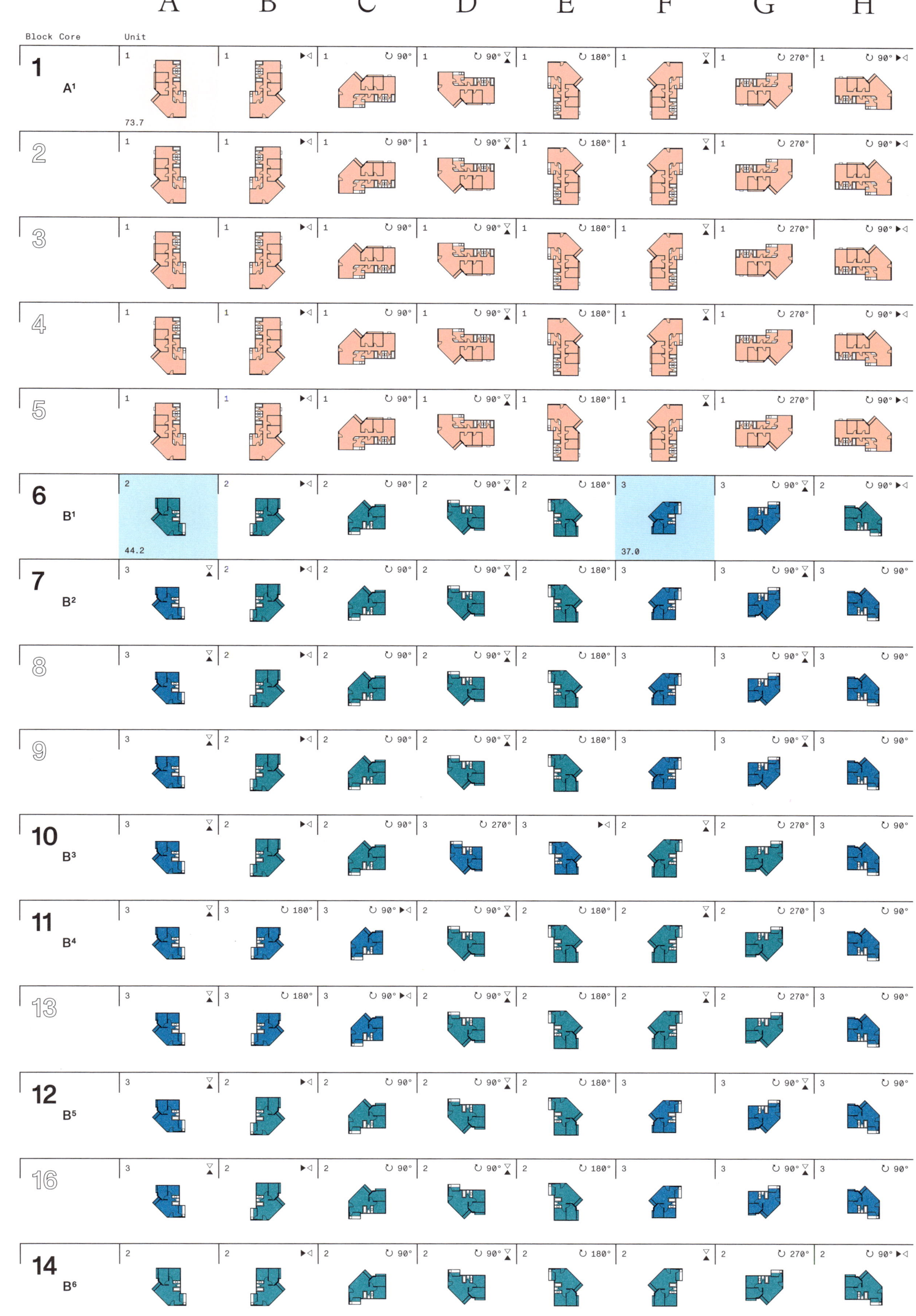
A
B
C
D
E
F
G
H
Block Core
Unit
1 A¹
73.7
2
3
4
5
6 B¹
44.2
37.0
7 B²
8
9
10 B³
11 B⁴
13
12 B⁵
16
14 B⁶

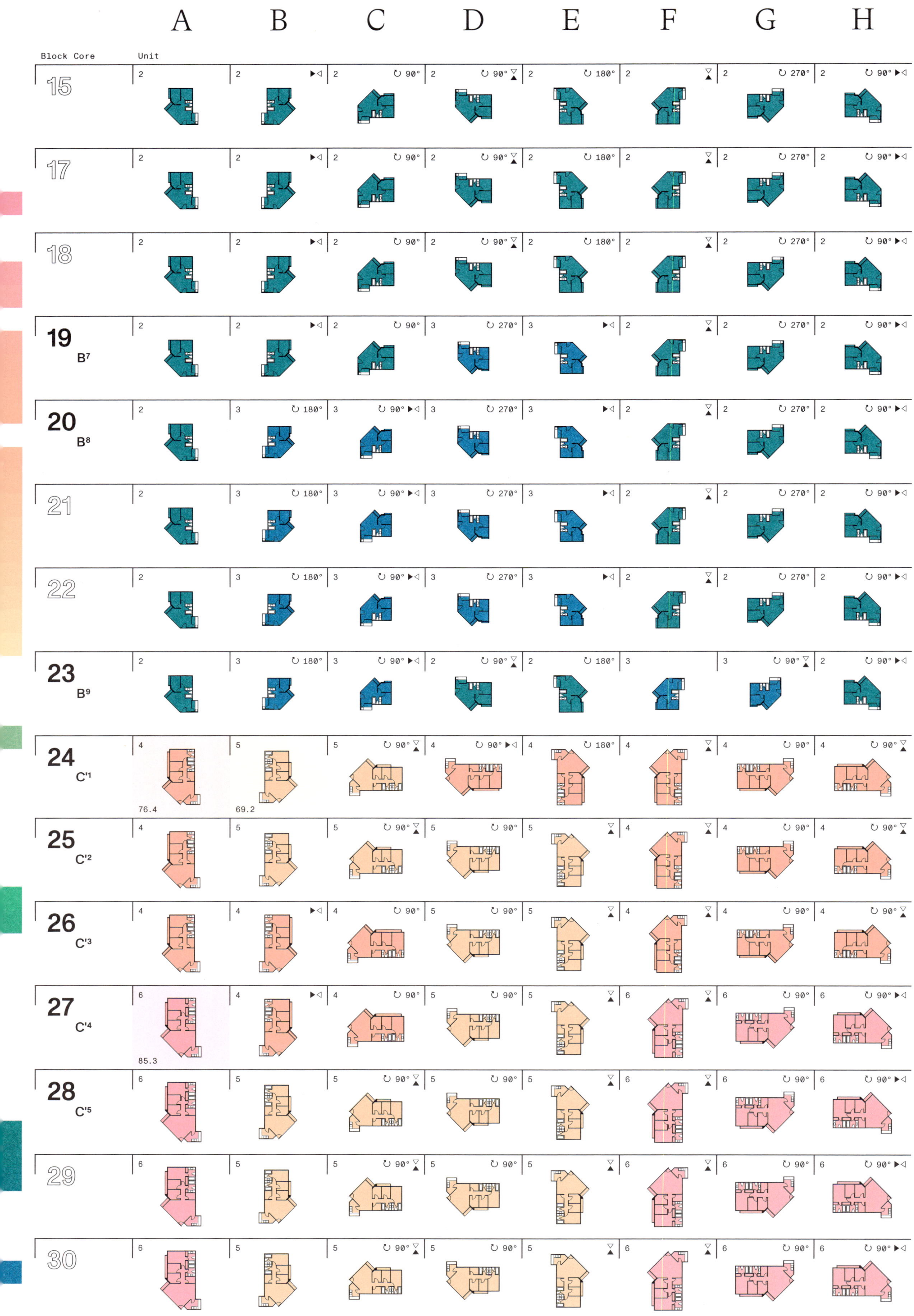

Block Core	A	B	C	D	E	F	G	H
	Unit							
15	2	2 ▶◁	2 ↻ 90°	2 ↻ 90° ▽▲	2 ↻ 180°	2 ▽▲	2 ↻ 270°	2 ↻ 90° ▶◁
17	2	2 ▶◁	2 ↻ 90°	2 ↻ 90° ▽▲	2 ↻ 180°	2 ▽▲	2 ↻ 270°	2 ↻ 90° ▶◁
18	2	2 ▶◁	2 ↻ 90°	2 ↻ 90° ▽▲	2 ↻ 180°	2 ▽▲	2 ↻ 270°	2 ↻ 90° ▶◁
19 B^{7}	2	2 ▶◁	2 ↻ 90°	3 ↻ 270°	3 ▶◁	2 ▽▲	2 ↻ 270°	2 ↻ 90° ▶◁
20 B^{8}	2	3 ↻ 180°	3 ↻ 90° ▶◁	3 ↻ 270°	3 ▶◁	2 ▽▲	2 ↻ 270°	2 ↻ 90° ▶◁
21	2	3 ↻ 180°	3 ↻ 90° ▶◁	3 ↻ 270°	3 ▶◁	2 ▽▲	2 ↻ 270°	2 ↻ 90° ▶◁
22	2	3 ↻ 180°	3 ↻ 90° ▶◁	3 ↻ 270°	3 ▶◁	2 ▽▲	2 ↻ 270°	2 ↻ 90° ▶◁
23 B^{9}	2	3 ↻ 180°	3 ↻ 90° ▶◁	2 ↻ 90° ▽▲	2 ↻ 180°	3	3 ↻ 90° ▽▲	2 ↻ 90° ▶◁
24 C'^{1}	4 76.4	5 69.2	5 ↻ 90° ▽▲	4 ↻ 90° ▶◁	4 ↻ 180°	4 ▽▲	4 ↻ 90°	4 ↻ 90° ▽▲
25 C'^{2}	4	5	5 ↻ 90° ▽▲	5 ↻ 90°	5 ▽▲	4 ▽▲	4 ↻ 90°	4 ↻ 90° ▽▲
26 C'^{3}	4	4 ▶◁	4 ↻ 90°	5 ↻ 90°	5 ▽▲	4 ▽▲	4 ↻ 90°	4 ↻ 90° ▽▲
27 C'^{4}	6 85.3	4 ▶◁	4 ↻ 90°	5 ↻ 90°	5 ▽▲	6 ▽▲	6 ↻ 90°	6 ↻ 90° ▶◁
28 C'^{5}	6	5	5 ↻ 90° ▽▲	5 ↻ 90°	5 ▽▲	6 ▽▲	6 ↻ 90°	6 ↻ 90° ▶◁
29	6	5	5 ↻ 90° ▽▲	5 ↻ 90°	5 ▽▲	6 ▽▲	6 ↻ 90°	6 ↻ 90° ▶◁
30	6	5	5 ↻ 90° ▽▲	5 ↻ 90°	5 ▽▲	6 ▽▲	6 ↻ 90°	6 ↻ 90° ▶◁

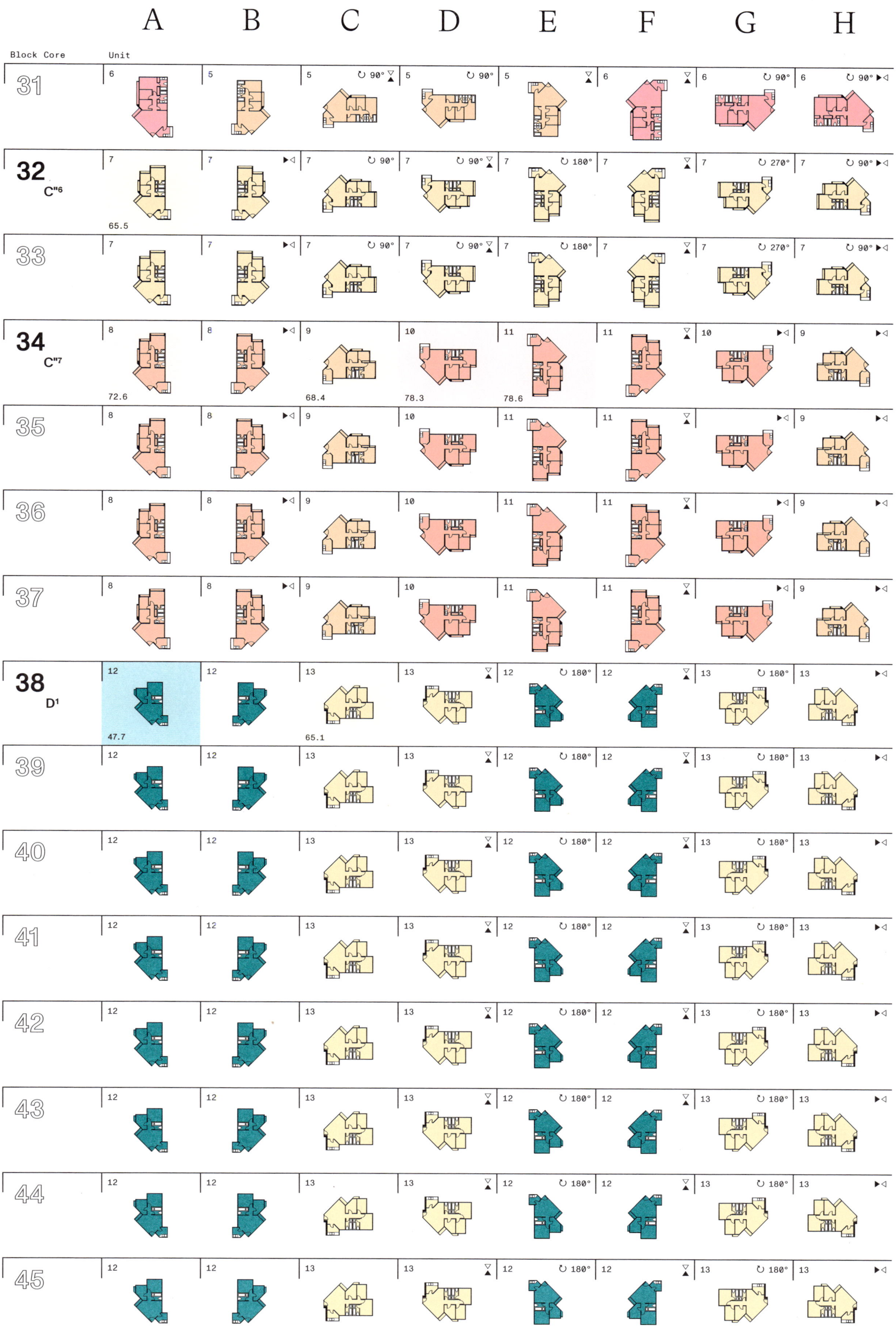

Block Core	A (Unit)	B	C	D	E	F	G	H
31	6	5	5 ↻ 90° ⧗	5 ↻ 90°	5 ⧗	6 ⧗	6 ↻ 90°	6 ↻ 90° ▶◁
32 C''^{6}	7 65.5	7 ▶◁	7 ↻ 90°	7 ↻ 90° ⧗	7 ↻ 180°	7 ⧗	7 ↻ 270°	7 ↻ 90° ▶◁
33	7	7 ▶◁	7 ↻ 90°	7 ↻ 90° ⧗	7 ↻ 180°	7 ⧗	7 ↻ 270°	7 ↻ 90° ▶◁
34 C''^{7}	8 72.6	8 ▶◁	9 68.4	10 78.3	11 78.6	11 ⧗	10 ▶◁	9 ▶◁
35	8	8 ▶◁	9	10	11	11 ⧗	▶◁	9 ▶◁
36	8	8 ▶◁	9	10	11	11 ⧗	▶◁	9 ▶◁
37	8	8 ▶◁	9	10	11	11 ⧗	▶◁	9 ▶◁
38 D^{1}	12 47.7	12	13 65.1	13 ⧗	12 ↻ 180°	12 ⧗	13 ↻ 180°	13 ▶◁
39	12	12	13	13 ⧗	12 ↻ 180°	12 ⧗	13 ↻ 180°	13 ▶◁
40	12	12	13	13 ⧗	12 ↻ 180°	12 ⧗	13 ↻ 180°	13 ▶◁
41	12	12	13	13 ⧗	12 ↻ 180°	12 ⧗	13 ↻ 180°	13 ▶◁
42	12	12	13	13 ⧗	12 ↻ 180°	12 ⧗	13 ↻ 180°	13 ▶◁
43	12	12	13	13 ⧗	12 ↻ 180°	12 ⧗	13 ↻ 180°	13 ▶◁
44	12	12	13	13 ⧗	12 ↻ 180°	12 ⧗	13 ↻ 180°	13 ▶◁
45	12	12	13	13 ⧗	12 ↻ 180°	12 ⧗	13 ↻ 180°	13 ▶◁

Block Core	A (Unit)	B	C	D	E	F	G	H
46	12	12	13	13 ⧗	12 ↻ 180°	12 ⧗	13 ↻ 180°	13 ▶◁
47 E^1	14 76.0	14 ▶◁	15 76.3	15 ⧗	14 ↻ 180°	14 ⧗	15 ↻ 180°	15 ▶◁
50	14	14 ▶◁	15	15 ⧗	14 ↻ 180°	14 ⧗	15 ↻ 180°	15 ▶◁
48 F^1	16 72.6	17 94.0	18 94.7	17 ⧗	16 ⧗			
49	16	17	18	17 ⧗	16 ⧗			
51	16	17	18	17 ⧗	16 ⧗			
52 G^1	19 67.8	19 ⧗	20 70.1	20 ⧗	19 ▶◁	21 66.7	20 ↻ 180°	20 ▶◁
53 G^2	22 72.5	23 59.7	23 ↻ 90° ⧗	23 ↻ 90°	23 ⧗	22 ⧗	24 73.6	24 ⧗
54 G^3	25 58.1	25 ▶◁	26 59.1	26 ⧗	25 ⧗	25 ↻ 180°	26 ↻ 180°	26 ▶◁
55 G^4	25	25 ▶◁	27 59.4	28 73.1	29 72.5	30 58.9	26 ↻ 180°	26 ▶◁
56 G^5	31 76.0	32 73.0	33 77.3	34 82.3	35 88.9	35	34	33
57 G^6	22	22 ▶◁	24 ↻ 180°	36 77.9	37 79.5	37 ▶◁	36 ▶◁	24 ⧗
58	22	22 ▶◁	24 ↻ 180°	36	37	37 ▶◁	36 ▶◁	24 ⧗
59	22	22 ▶◁	24 ↻ 180°	36	37	37 ▶◁	36 ▶◁	24 ⧗
60	22	22 ▶◁	24 ↻ 180°	36	37	37 ▶◁	36 ▶◁	24 ⧗

A B C D E F G H
Block Core
Unit
61
22 22 24 180° 36 37 37 36 24
62 G7
38 39 40 40 39 39 180° 40 180° 40
58.4 73.7 77.4
64
38 39 40 40 39 39 180° 40 180° 40
68
38 39 40 40 39 39 180° 40 180° 40
63 G8
39 39 41 41 38 180° 39 180° 40 180° 40
72.7
67
39 39 41 41 38 180° 39 180° 40 180° 40
65 G9
39 38 41 41 38 180° 39 180° 41 180° 41
66 G10
39 39 41 42 39 39 180° 42 41
58.7
69 G11
39 38 41 41 39 39 180° 42 42 180°
70 G12
38 38 42 42 38 180° 38 42 42 180°
71 H1
43 43 44 45 46 46 45 44
86.5 86.5 71.1 68.9
72
43 43 44 45 46 46 45 44
73
43 43 44 45 46 46 45 44
74
43 43 44 45 46 46 45 44
75
43 43 44 45 46 46 45 44

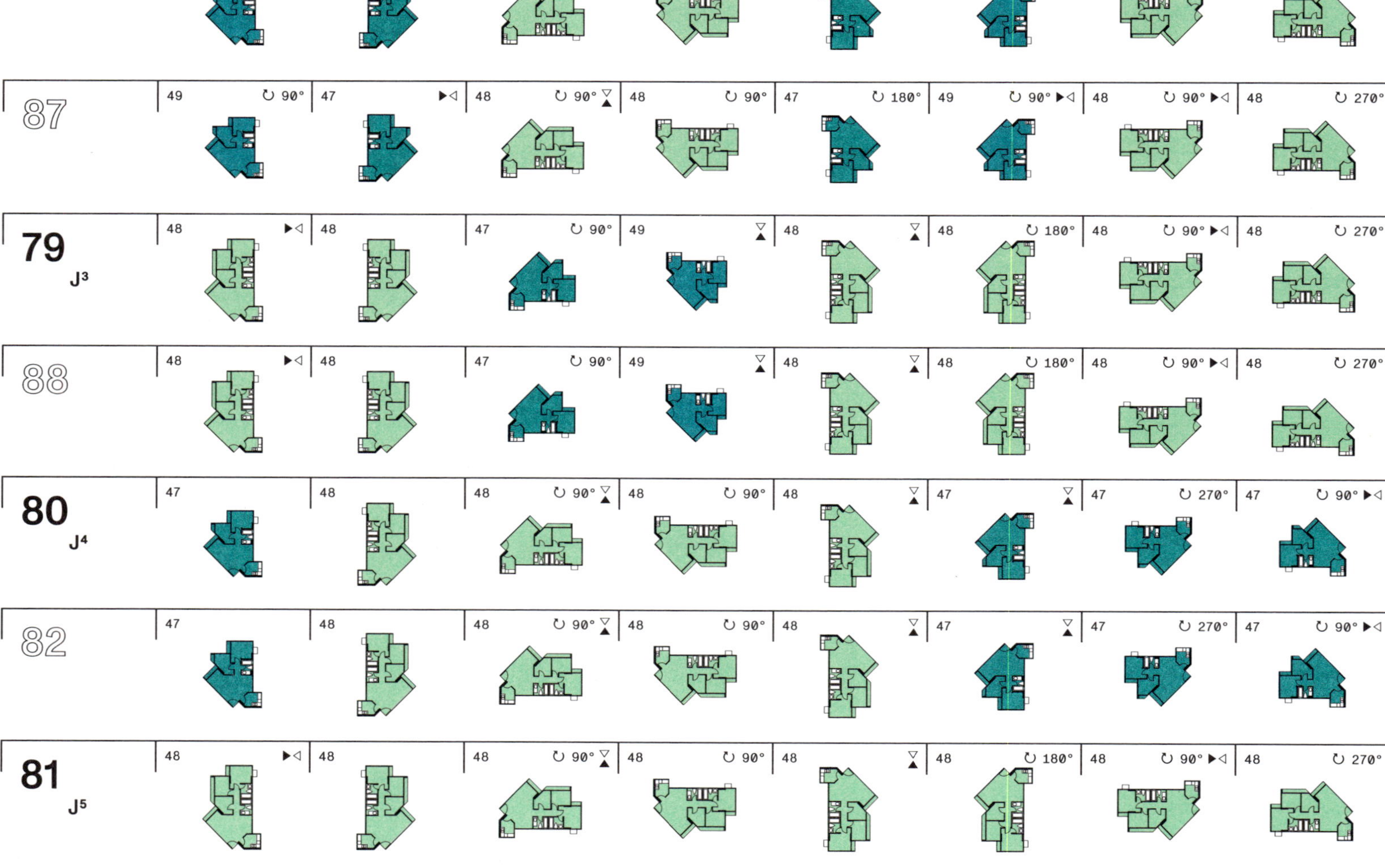

Block Core	Unit A	B	C	D	E	F	G	H
76 J^1	47 49.4	48 63.0	49 45.5	47 ↻ 90° ⧗	48 ⧗	47 ⧗	47 ↻ 270°	47 ↻ 90° ▶◁
84	47	48	49	47 ↻ 90° ⧗	48 ⧗	47 ⧗	47 ↻ 270°	47 ↻ 90° ▶◁
77 J^2	49 ↻ 90°	47 ▶◁	48 ↻ 90° ⧗	48 ↻ 90°	47 ↻ 180°	49 ↻ 90° ▶◁	48 ↻ 90° ▶◁	48 ↻ 270°
78	49 ↻ 90°	47 ▶◁	48 ↻ 90° ⧗	48 ↻ 90°	47 ↻ 180°	49 ↻ 90° ▶◁	48 ↻ 90° ▶◁	48 ↻ 270°
85	49 ↻ 90°	47 ▶◁	48 ↻ 90° ⧗	48 ↻ 90°	47 ↻ 180°	49 ↻ 90° ▶◁	48 ↻ 90° ▶◁	48 ↻ 270°
86	49 ↻ 90°	47 ▶◁	48 ↻ 90° ⧗	48 ↻ 90°	47 ↻ 180°	49 ↻ 90° ▶◁	48 ↻ 90° ▶◁	48 ↻ 270°
87	49 ↻ 90°	47 ▶◁	48 ↻ 90° ⧗	48 ↻ 90°	47 ↻ 180°	49 ↻ 90° ▶◁	48 ↻ 90° ▶◁	48 ↻ 270°
79 J^3	48 ▶◁	48	47 ↻ 90°	49 ⧗	48 ⧗	48 ↻ 180°	48 ↻ 90° ▶◁	48 ↻ 270°
88	48 ▶◁	48	47 ↻ 90°	49 ⧗	48 ⧗	48 ↻ 180°	48 ↻ 90° ▶◁	48 ↻ 270°
80 J^4	47	48	48 ↻ 90° ⧗	48 ↻ 90°	48 ⧗	47 ⧗	47 ↻ 270°	47 ↻ 90° ▶◁
82	47	48	48 ↻ 90° ⧗	48 ↻ 90°	48 ⧗	47 ⧗	47 ↻ 270°	47 ↻ 90° ▶◁
81 J^5	48 ▶◁	48	48 ↻ 90° ⧗	48 ↻ 90°	48 ⧗	48 ↻ 180°	48 ↻ 90° ▶◁	48 ↻ 270°
83 J^6	47	48	48 ↻ 90° ⧗	48 ↻ 90°	48 ⧗	47 ⧗	49 ↻ 180°	47 ↻ 90° ▶◁

Belvedere Garden 1987–91

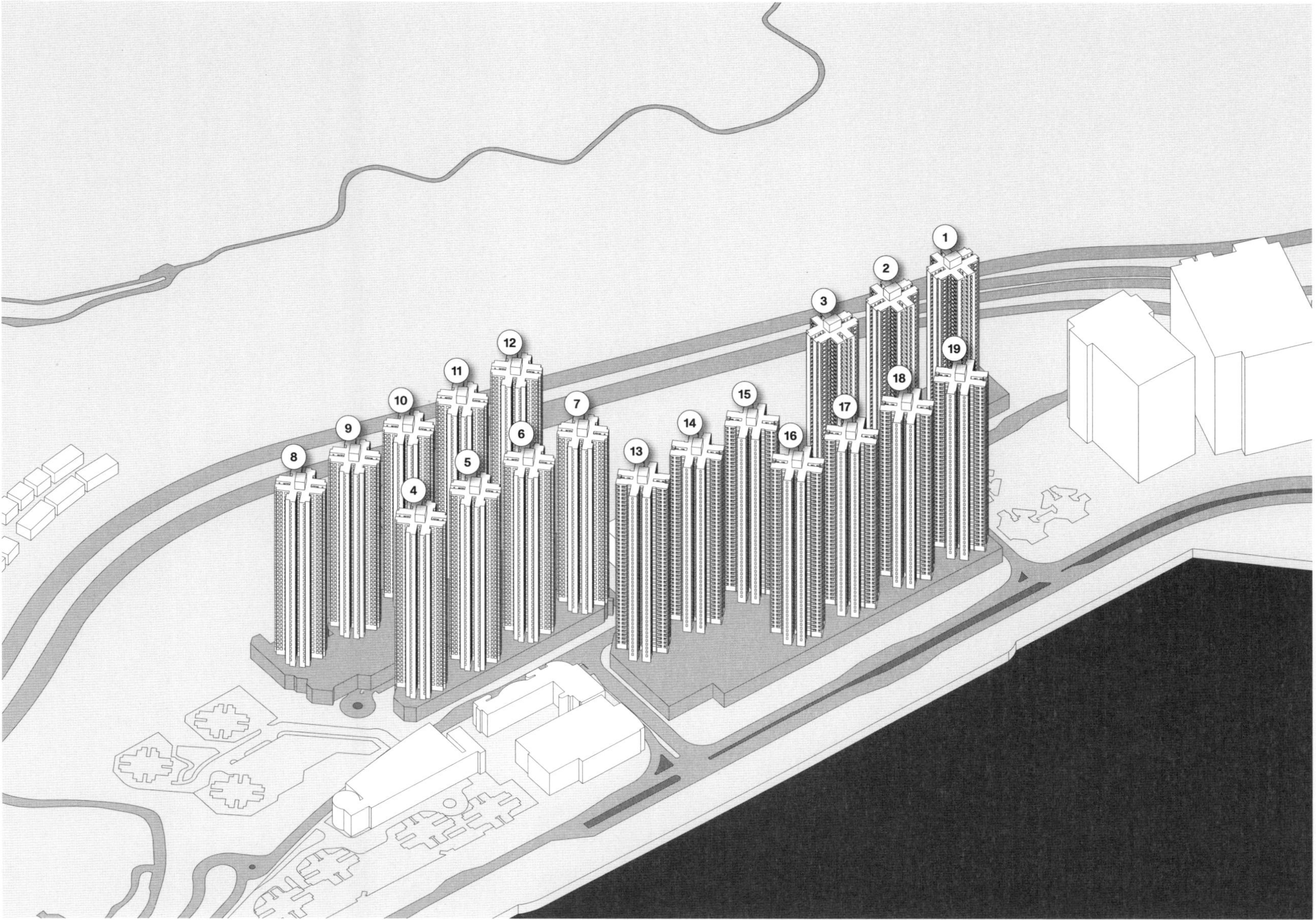

DEVELOPER:	Cheung Kong Holdings		
ARCHITECT:	Design 2 HK Ltd.		
LOCATION:	Tsuen Wan		
POPULATION:	20,423		
TOWERS:	19	UNIQUE:	3
APARTMENTS:	6,016	UNIT TYPES:	8
CORES:	3	PHASES:	3

Belvedere Garden is made of up nineteen densely packed towers organized in linear rows that climb up the hillside from the waterfront in the district of Tseun Wan. Tseun Wan was a satellite urban area developed in the 1960s under the Hong Kong British colonial government's New Town urbanization strategy. Present day Tseun Wan is built on land reclaimed from a shallow bay and wetlands and is home to a number of public and private housing estates. Belvedere Gardens is the largest private estate in Tseun Wan by population and the most densely populated estate of the ten estates surveyed in this book in terms of population to plot size. It has a population density of 4,300 people per hectare.

Towers are grouped into three clusters according to construction phase. Each phase in the estate has a separate podium with amenities below. The podium spaces are connected by elevated pedestrian walkways and together contain parking garages, recreational facilities, community services, schools, and shopping arcades.

Type A

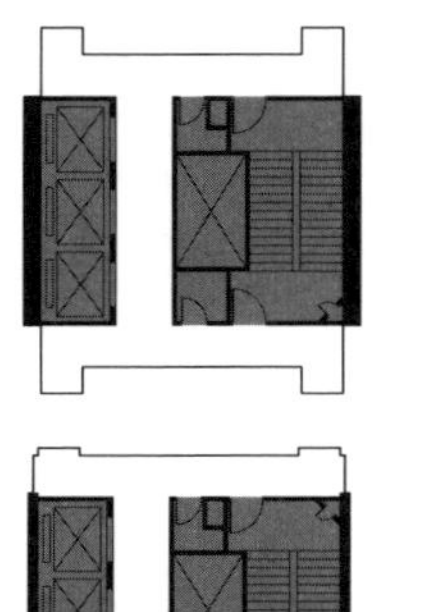

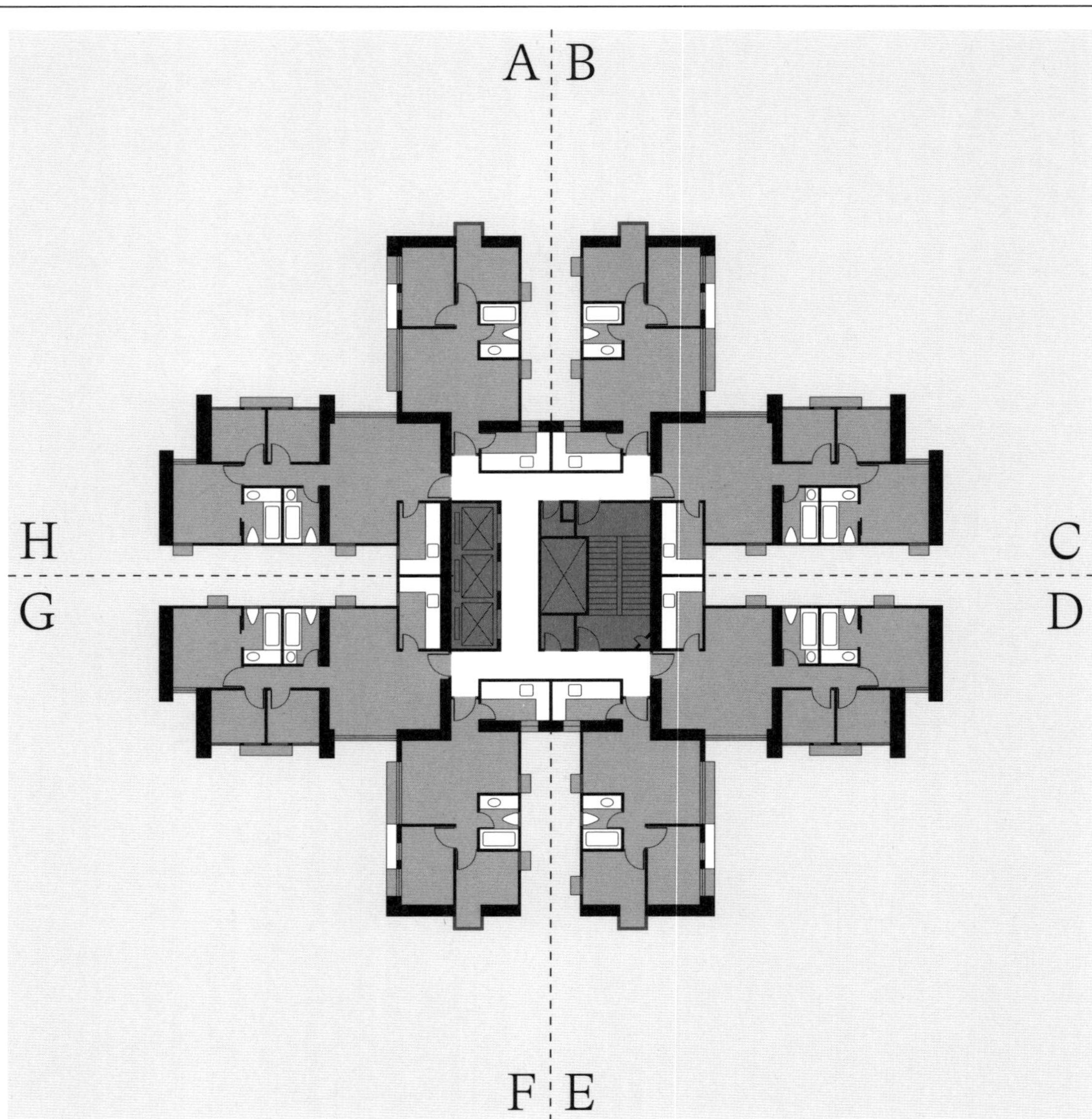

Type B

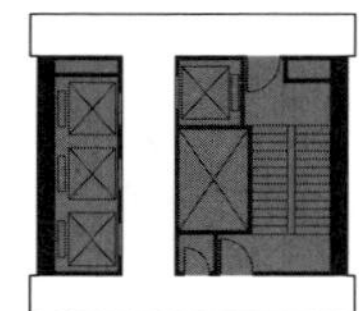

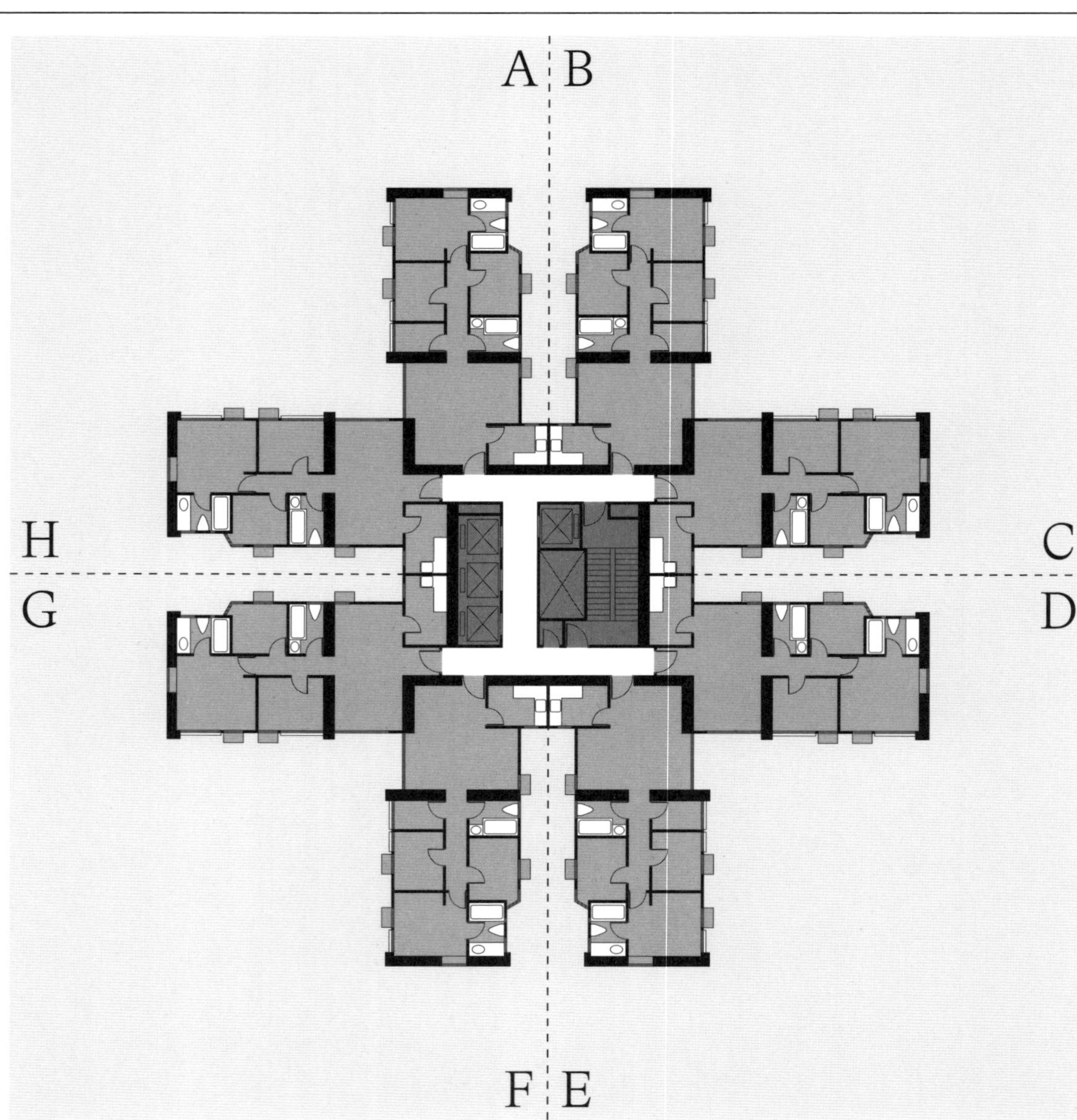

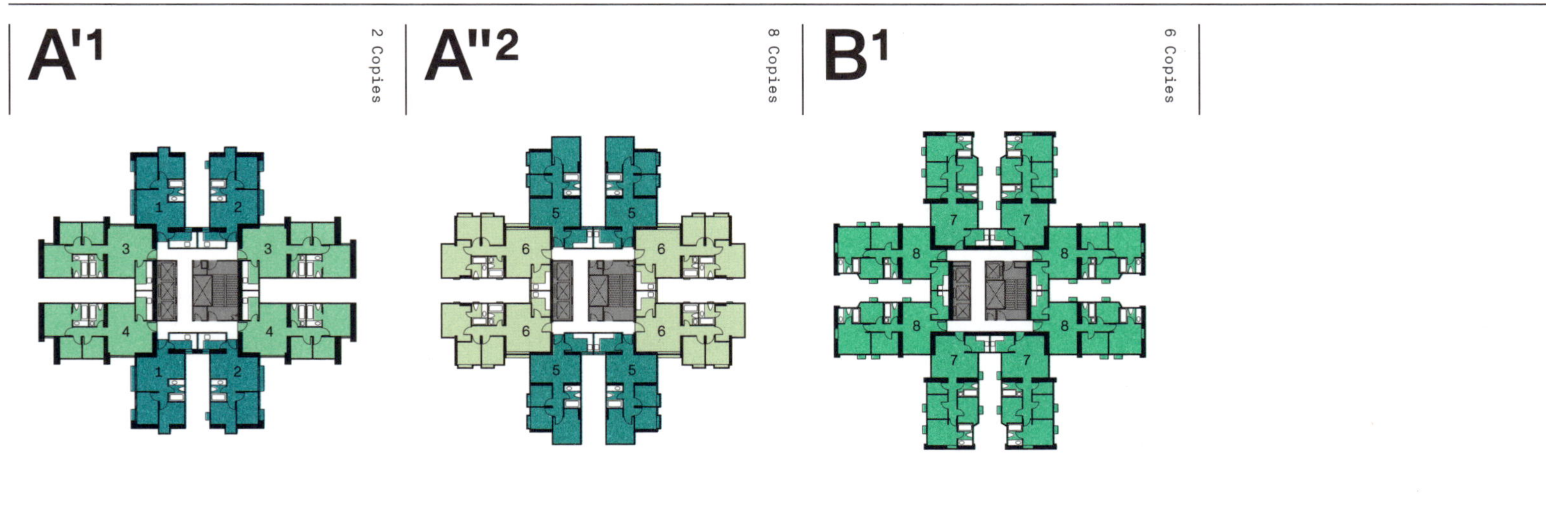

A'^1

2 Copies

Phase 1 →/ **Block 1–3**

A''^2

8 Copies

Phase 2 →/ **Block 4–12**

B^1

6 Copies

Phase 3 →/ **Block 13–19**

Block Core	Unit A	B	C	D	E	F	G	H
1 A'^1	1 48.2	2 48.5	3 62.7	4 62.7	2 ⧗	1 ⧗	4 ▶◁	3 ▶◁
2	1	2	3	4	2 ⧗	1 ⧗	4 ▶◁	3 ▶◁
3	1	2	3	4	2 ⧗	1 ⧗	4 ▶◁	3 ▶◁
4 A''^2	5 52.4	5 ▶◁	6 64.1	6 ⧗	5 ↻ 180°	5 ⧗	6 ↻ 180°	6 ▶◁
5	5	5 ▶◁	6	6 ⧗	5 ↻ 180°	5 ⧗	6 ↻ 180°	6 ▶◁
6	5	5 ▶◁	6	6 ⧗	5 ↻ 180°	5 ⧗	6 ↻ 180°	6 ▶◁
7	5	5 ▶◁	6	6 ⧗	5 ↻ 180°	5 ⧗	6 ↻ 180°	6 ▶◁
8	5	5 ▶◁	6	6 ⧗	5 ↻ 180°	5 ⧗	6 ↻ 180°	6 ▶◁
9	5	5 ▶◁	6	6 ⧗	5 ↻ 180°	5 ⧗	6 ↻ 180°	6 ▶◁
10	5	5 ▶◁	6	6 ⧗	5 ↻ 180°	5 ⧗	6 ↻ 180°	6 ▶◁
11	5	5 ▶◁	6	6 ⧗	5 ↻ 180°	5 ⧗	6 ↻ 180°	6 ▶◁

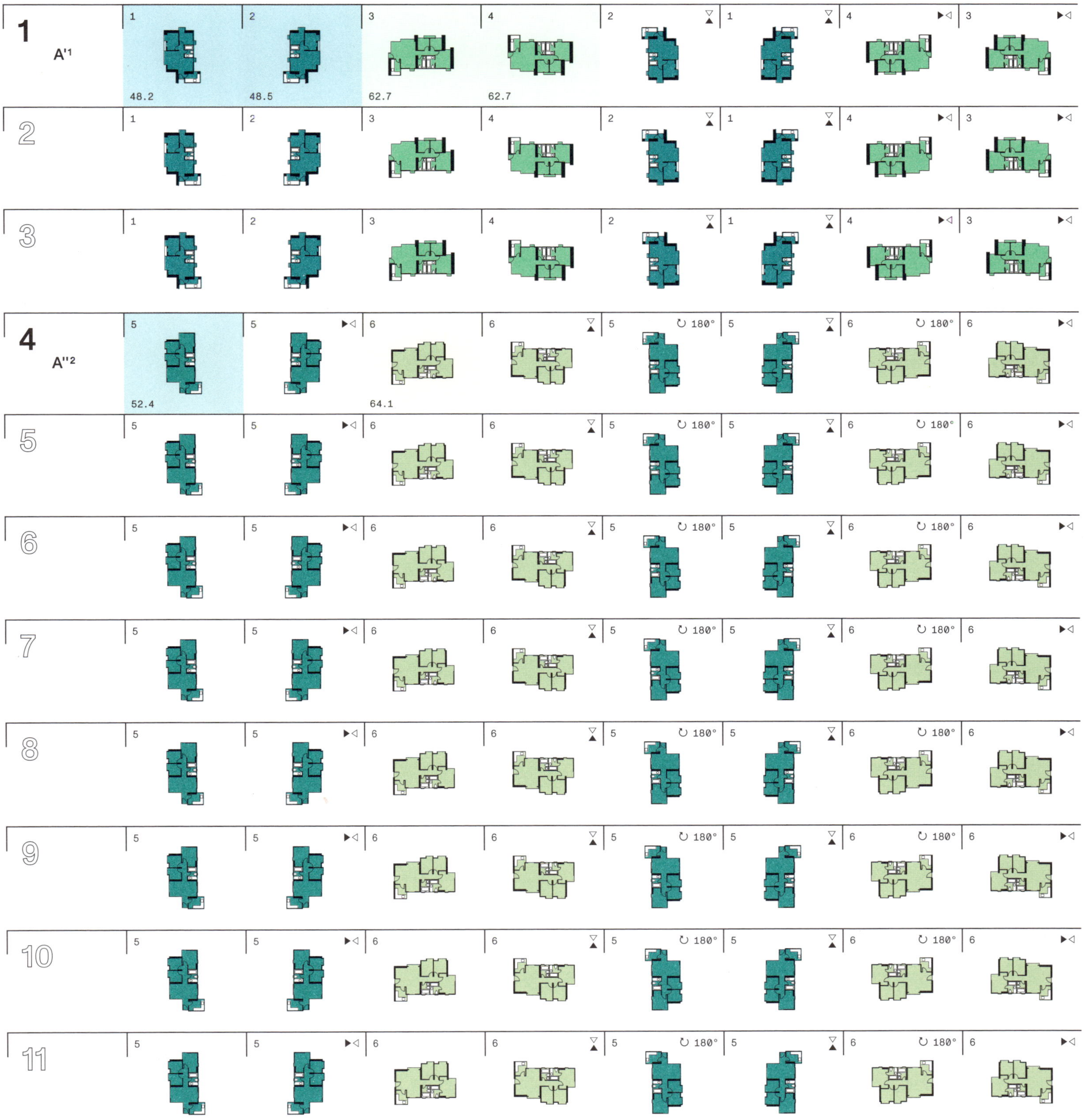

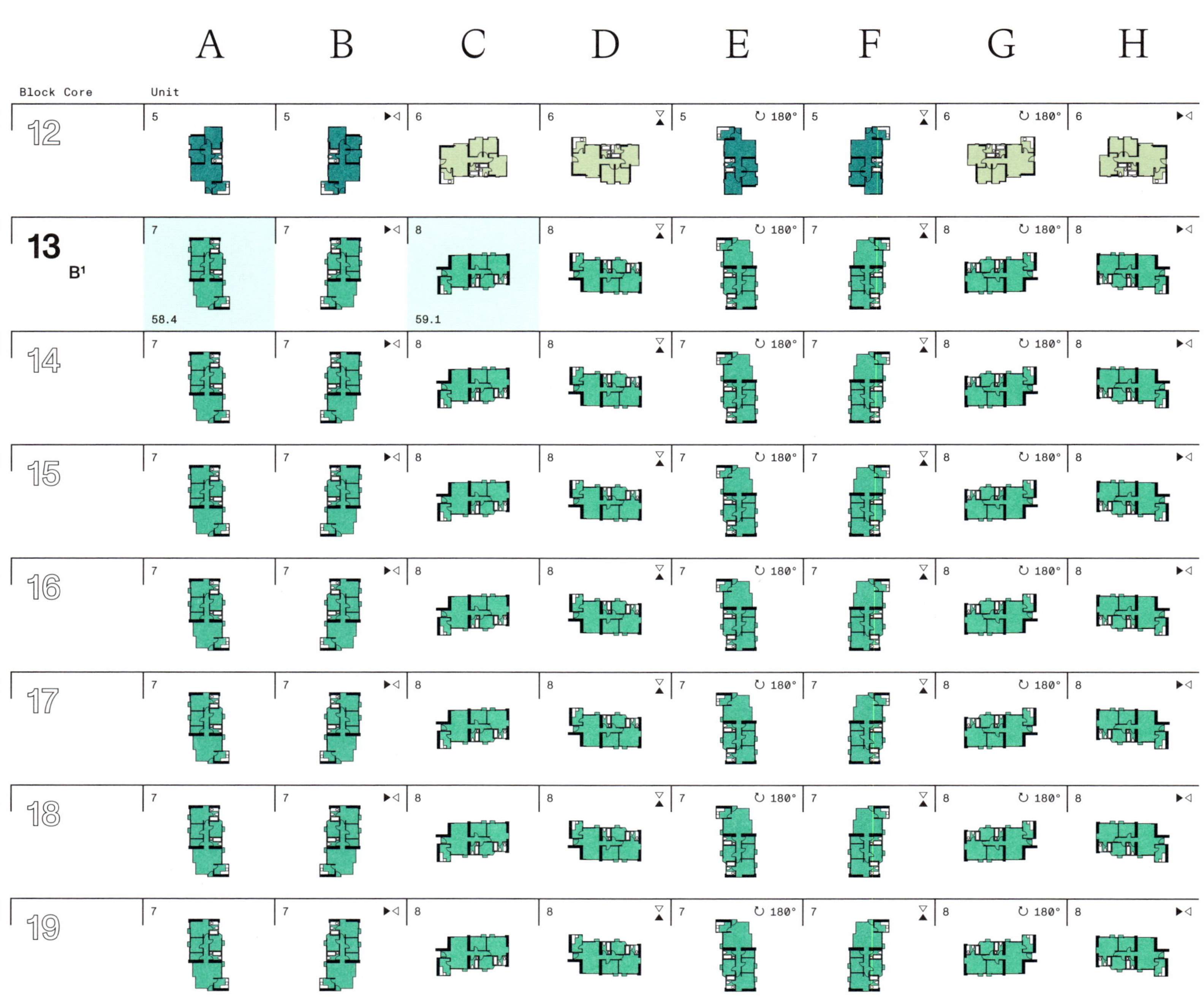
A
B
C
D
E
F
G
H
Block Core
Unit
12
13
B1
14
15
16
17
18
19
58.4
59.1
180°

Laguna City 1990–94

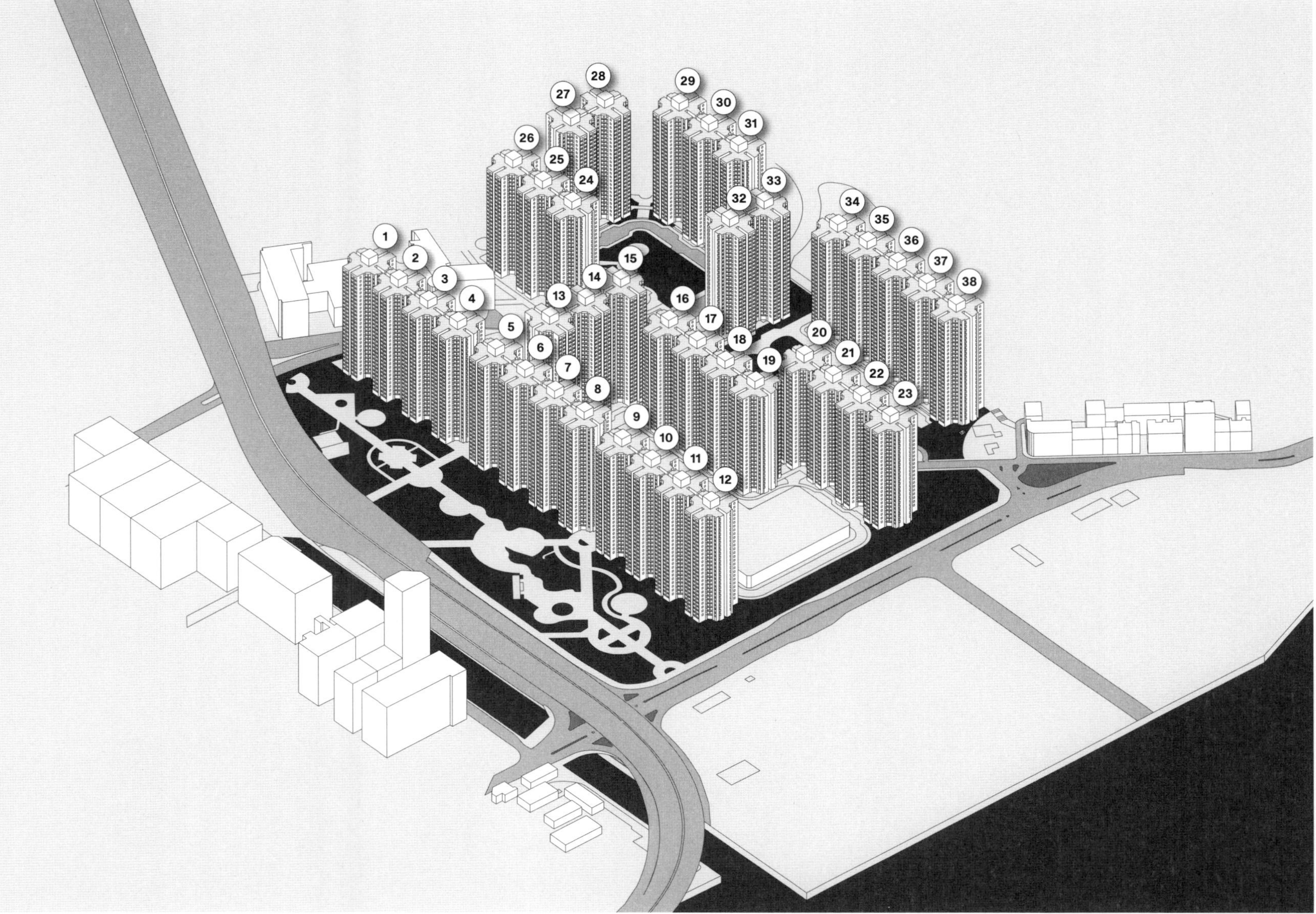

DEVELOPER:	Cheung Kong Holdings, Hutchinson Whampoa Property		
ARCHITECT:	DLN Architects & Engineers		
LOCATION:	Kwun Tong		
POPULATION:	23,354		
TOWERS:	38	UNIQUE:	4
APARTMENTS:	8,072	UNIT TYPES:	4
CORES:	1	PHASES:	4

Laguna City is located on a site adjacent to the Kwun Tong neighborhood of Kowloon and close to the Kai Tak Airport runway in Kowloon. The Kai Tak airport was in use until 1997 and now has been converted into a public park and a cruise ship terminal.

Like Mei Foo and South Horizons, the Laguna City housing estate was built on a former industrial site used for oil storage tanks. The site was purchased from the Shell Oil Company in the late 1980s by Cheung Kong Holdings. Cheung Kong developed the site into residential estate with Hutchison Whampoa Property Limited in the early 1990s.

Laguna City is an autonomous enclave as it is only indirectly connected to the Lam Tin MTR station by a pedestrian walkway and is effectively cut off from Kwun Tong's dense urban fabric by a vehicular overpass. Tower blocks are organized in rows that fit into an overall gridded pattern and are grouped by construction phase. Towers sit atop podiums that house parking and shopping centers. Recreational facilities including playing fields, tennis courts, and swimming pools are situated between the rows of towers.

Type A

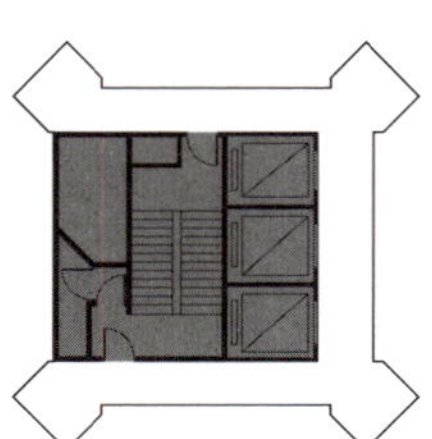

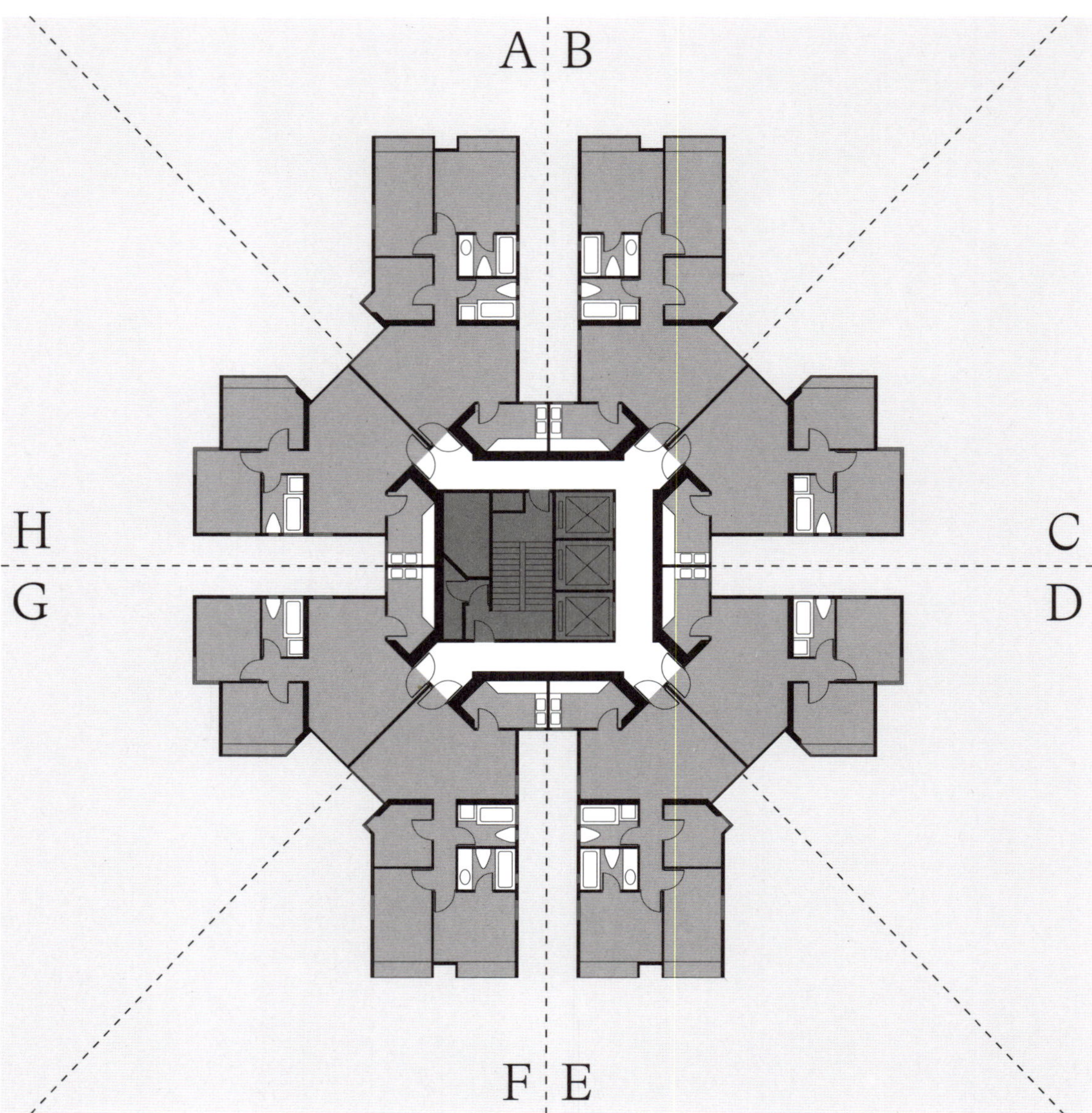

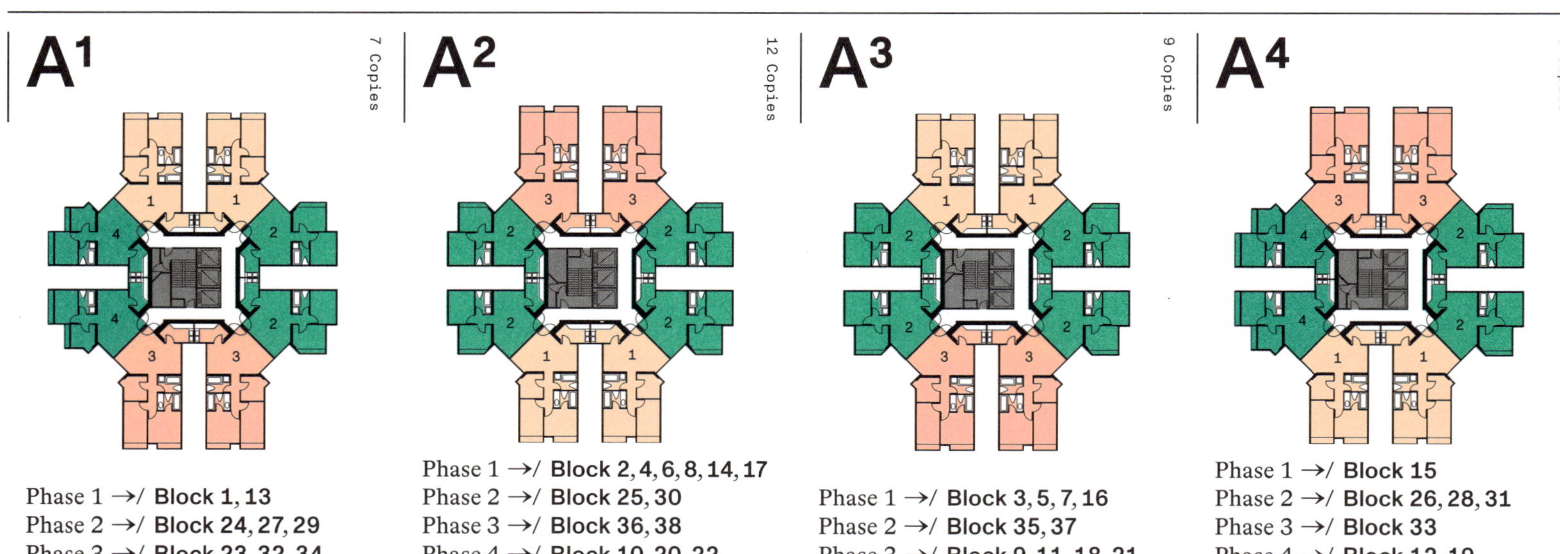

A^1 (7 Copies)

Phase 1 →/ **Block 1, 13**
Phase 2 →/ **Block 24, 27, 29**
Phase 3 →/ **Block 23, 32, 34**

A^2 (12 Copies)

Phase 1 →/ **Block 2, 4, 6, 8, 14, 17**
Phase 2 →/ **Block 25, 30**
Phase 3 →/ **Block 36, 38**
Phase 4 →/ **Block 10, 20, 22**

A^3 (9 Copies)

Phase 1 →/ **Block 3, 5, 7, 16**
Phase 2 →/ **Block 35, 37**
Phase 3 →/ **Block 9, 11, 18, 21**

A^4 (6 Copies)

Phase 1 →/ **Block 15**
Phase 2 →/ **Block 26, 28, 31**
Phase 3 →/ **Block 33**
Phase 4 →/ **Block 12, 19**

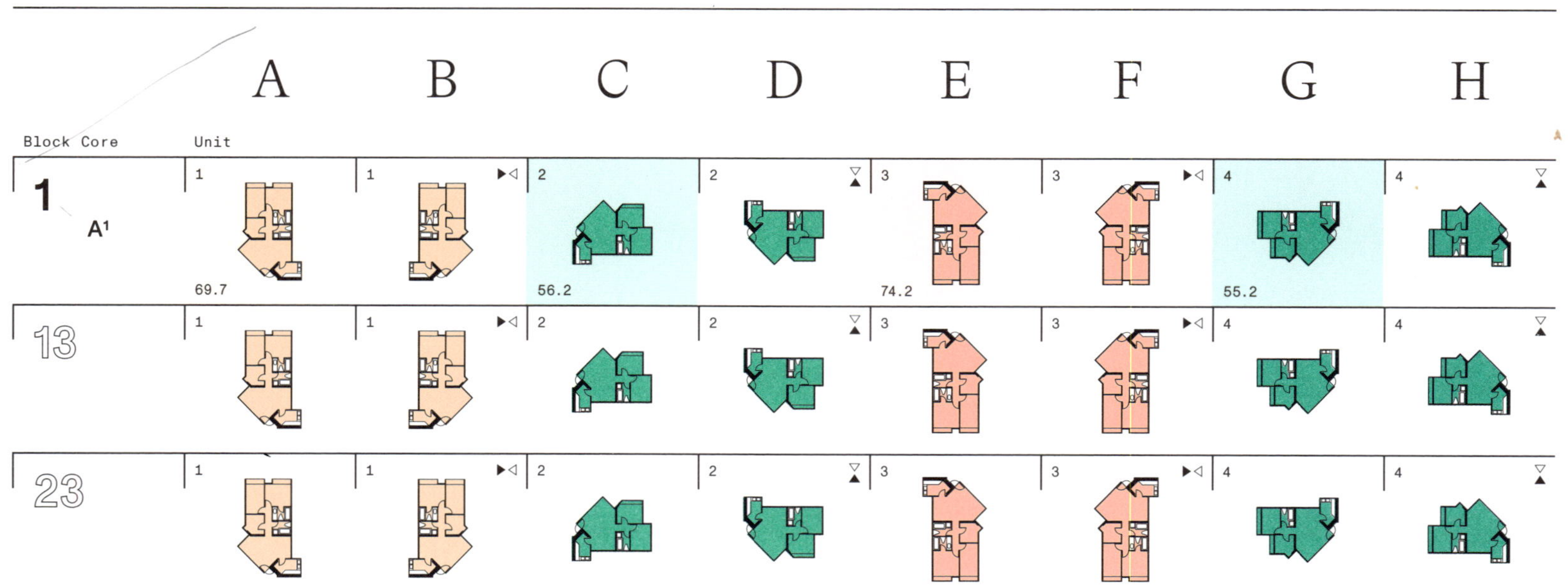

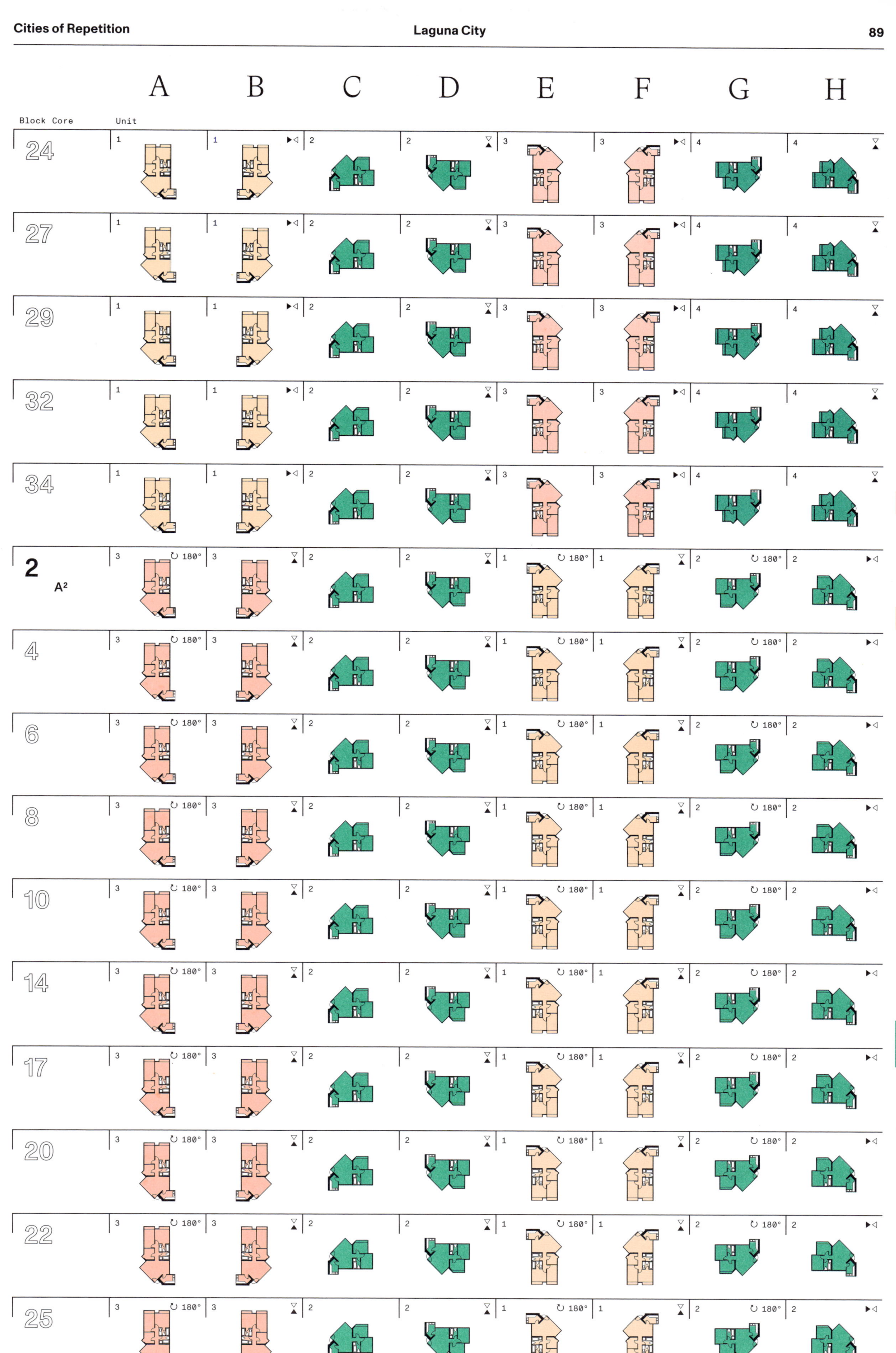
A B C D E F G H
Block Core
Unit
24 1 1 2 2 3 3 4 4
27 1 1 2 2 3 3 4 4
29 1 1 2 2 3 3 4 4
32 1 1 2 2 3 3 4 4
34 1 1 2 2 3 3 4 4
2 A² 3 180° 3 2 2 1 180° 1 2 180° 2
4 3 180° 3 2 2 1 180° 1 2 180° 2
6 3 180° 3 2 2 1 180° 1 2 180° 2
8 3 180° 3 2 2 1 180° 1 2 180° 2
10 3 180° 3 2 2 1 180° 1 2 180° 2
14 3 180° 3 2 2 1 180° 1 2 180° 2
17 3 180° 3 2 2 1 180° 1 2 180° 2
20 3 180° 3 2 2 1 180° 1 2 180° 2
22 3 180° 3 2 2 1 180° 1 2 180° 2
25 3 180° 3 2 2 1 180° 1 2 180° 2

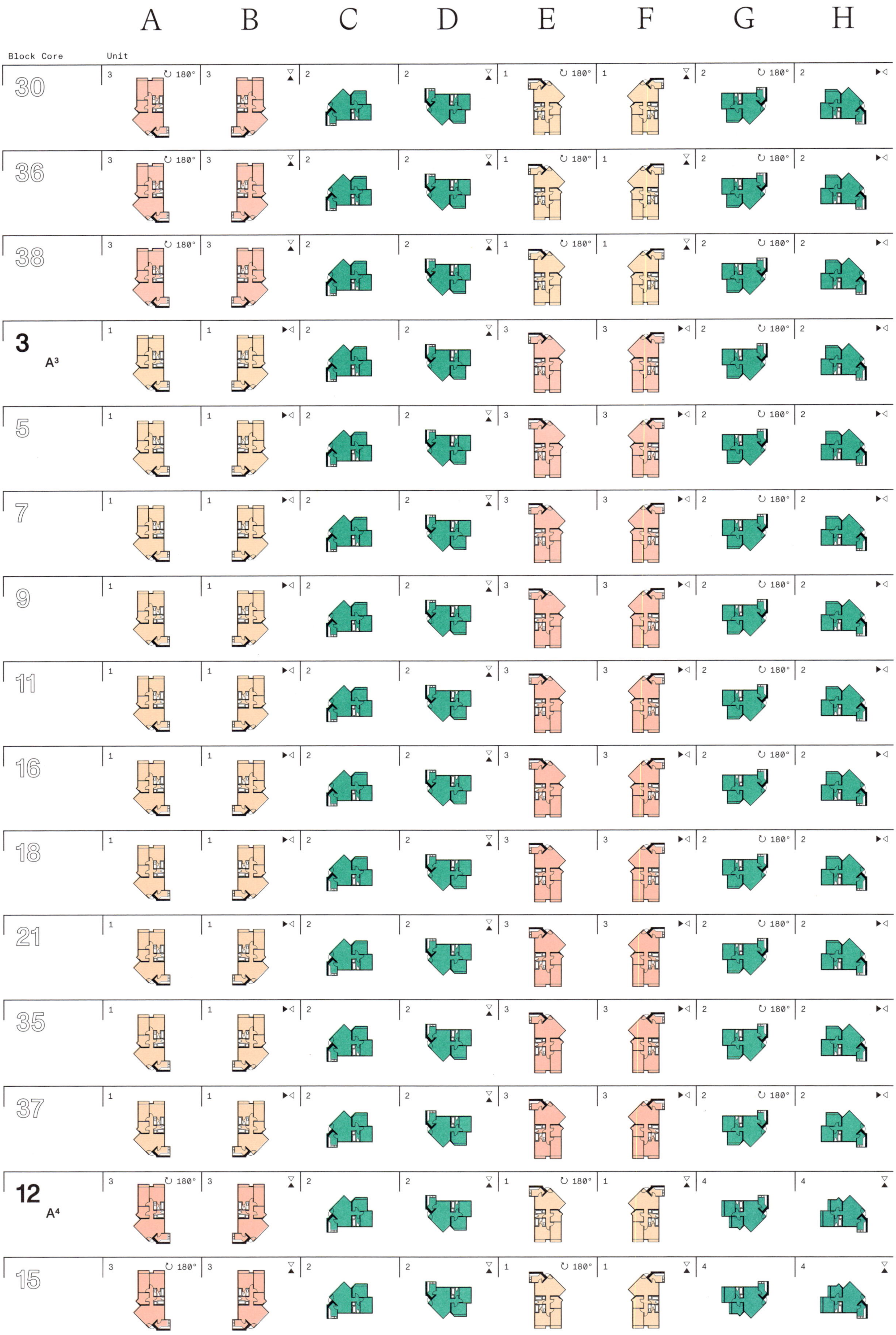

Block Core	Unit A	Unit B	Unit C	Unit D	Unit E	Unit F	Unit G	Unit H
30	3 ↻ 180°	3 ▽▲	2	2 ▽▲	1 ↻ 180°	1 ▽▲	2 ↻ 180°	2 ▶◁
36	3 ↻ 180°	3 ▽▲	2	2 ▽▲	1 ↻ 180°	1 ▽▲	2 ↻ 180°	2 ▶◁
38	3 ↻ 180°	3 ▽▲	2	2 ▽▲	1 ↻ 180°	1 ▽▲	2 ↻ 180°	2 ▶◁
3 A³	1	1 ▶◁	2	2 ▽▲	3	3 ▶◁	2 ↻ 180°	2 ▶◁
5	1	1 ▶◁	2	2 ▽▲	3	3 ▶◁	2 ↻ 180°	2 ▶◁
7	1	1 ▶◁	2	2 ▽▲	3	3 ▶◁	2 ↻ 180°	2 ▶◁
9	1	1 ▶◁	2	2 ▽▲	3	3 ▶◁	2 ↻ 180°	2 ▶◁
11	1	1 ▶◁	2	2 ▽▲	3	3 ▶◁	2 ↻ 180°	2 ▶◁
16	1	1 ▶◁	2	2 ▽▲	3	3 ▶◁	2 ↻ 180°	2 ▶◁
18	1	1 ▶◁	2	2 ▽▲	3	3 ▶◁	2 ↻ 180°	2 ▶◁
21	1	1 ▶◁	2	2 ▽▲	3	3 ▶◁	2 ↻ 180°	2 ▶◁
35	1	1 ▶◁	2	2 ▽▲	3	3 ▶◁	2 ↻ 180°	2 ▶◁
37	1	1 ▶◁	2	2 ▽▲	3	3 ▶◁	2 ↻ 180°	2 ▶◁
12 A⁴	3 ↻ 180°	3 ▽▲	2	2 ▽▲	1 ↻ 180°	1 ▽▲	4	4 ▽▲
15	3 ↻ 180°	3 ▽▲	2	2 ▽▲	1 ↻ 180°	1 ▽▲	4	4 ▽▲

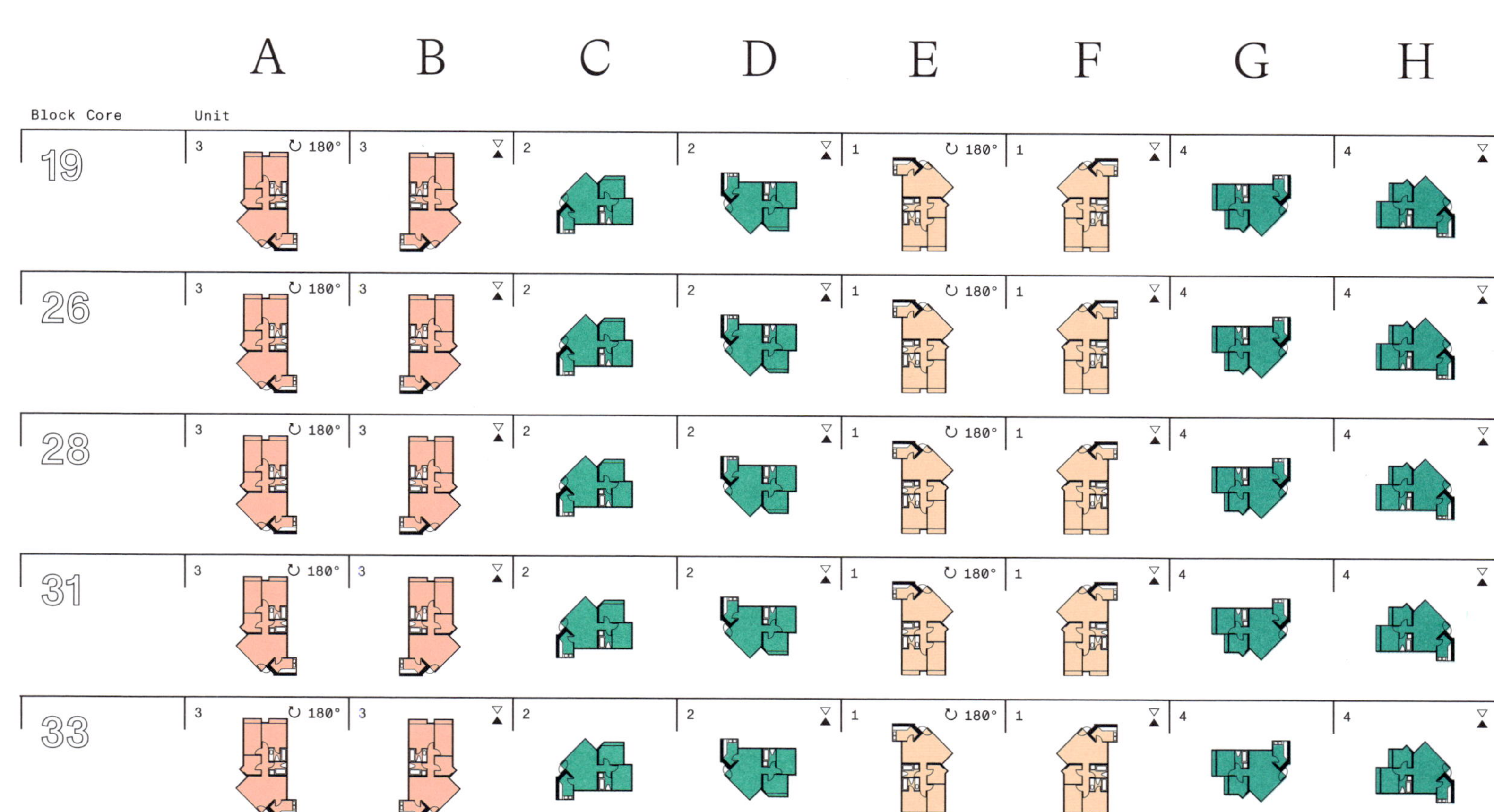
A
B
C
D
E
F
G
H
Block Core
Unit
19
26
28
31
33
3
180°
2
1
4

South Horizons 1991–95

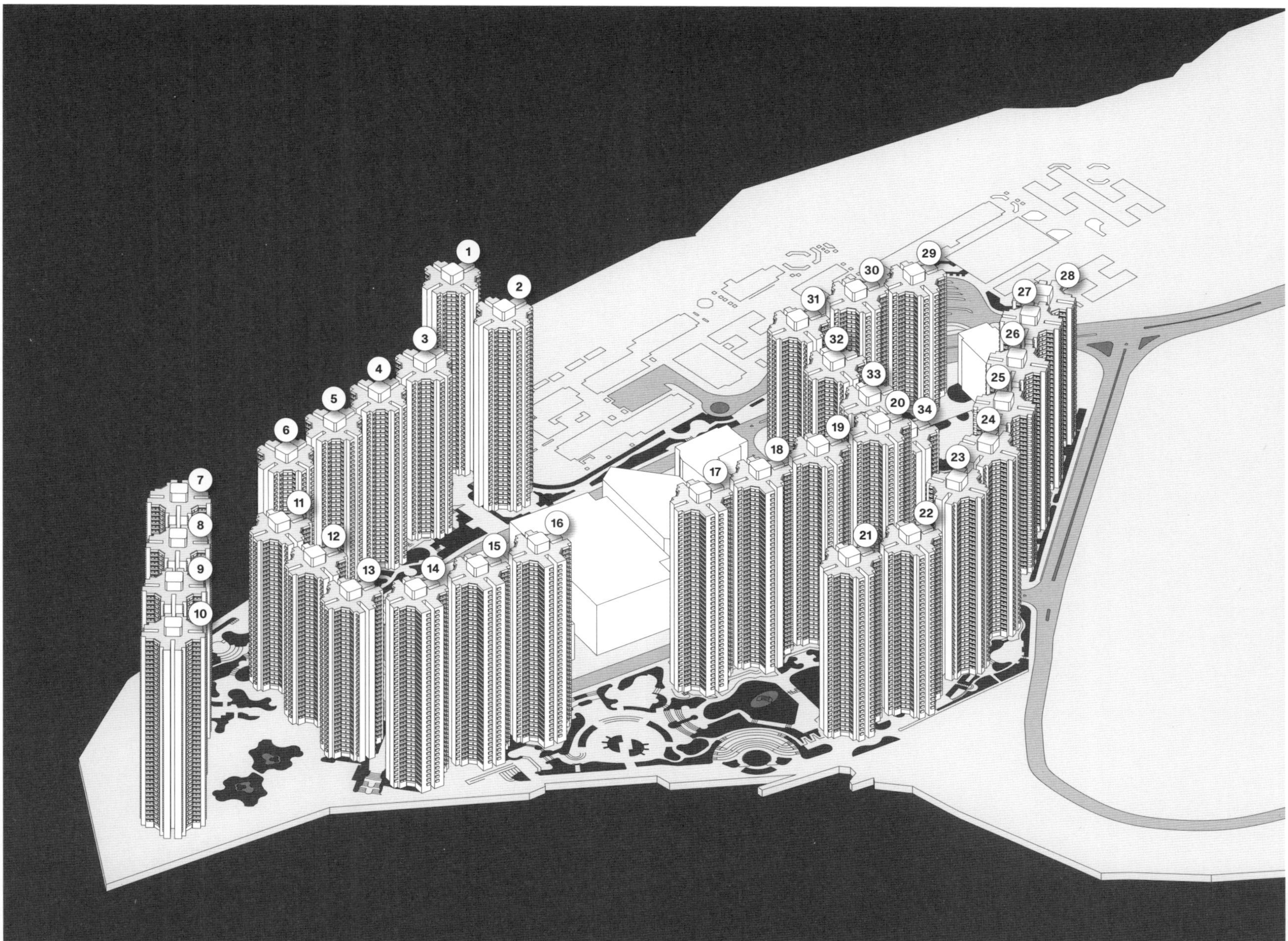

DEVELOPER:	Hutchison Whampoa Ltd.		
ARCHITECT:	Hsin Yieh Architects & Engineers Ltd.		
LOCATION:	Southern		
POPULATION:	31,496		
TOWERS:	34	UNIQUE:	16
APARTMENTS:	9,813	UNIT TYPES:	31
CORES:	2	PHASES:	4

South Horizons is located on one of the world's most densely populated islands, Ap Lei Chau. Ap Lei Chau lies on the south side of Hong Kong Island, close to the urban district of Aberdeen. Before it was developed into a housing estate, the western end of Ap Lei Chau was home to an oil depot owned by Shell Hong Kong Ltd. and a power station owned by the utility company, Hong Kong Electric. The power station was relocated to nearby Lamma Island in the late 1980s and South Horizons was constructed between 1991 and 1995. The thirty-four tower blocks of the estate were built in four phases. While the towers in the center of the estate are organized in a large rectangular grid, tower blocks on the periphery of the site are laid out in axial rows that follow the coastline and mountainous terrain of the island. The overall planning of the estate includes a publically accessible waterfront promenade and allows for many of the residential units to have open views of the South China Sea and Aberdeen Harbor.

Two large shopping centers and a bus depot occupies the center of the site. An MTR transit line was extended to Ap Lei Chau with the construction of the South Island Line and has a terminal station at South Horizons that opened in late 2016.

Type A

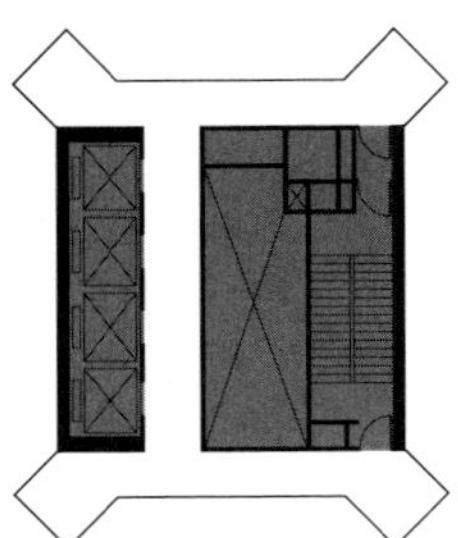

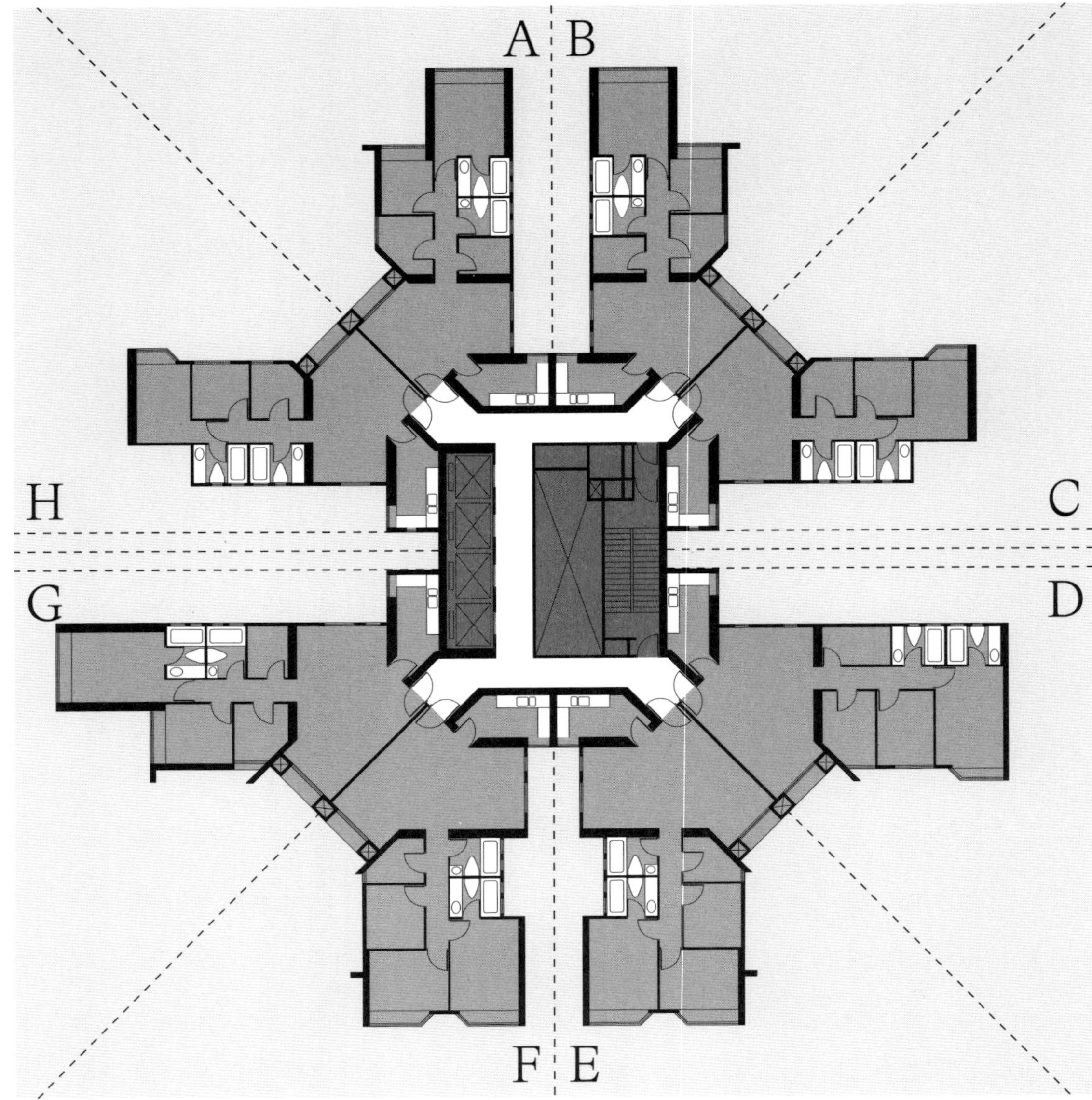

Type B

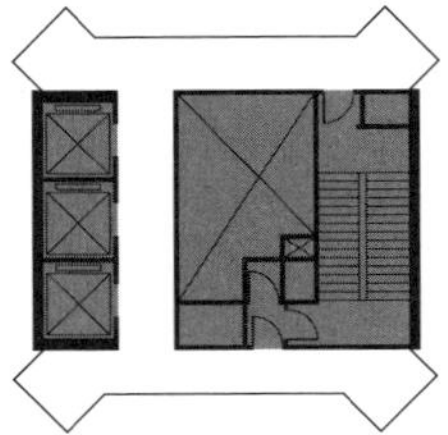

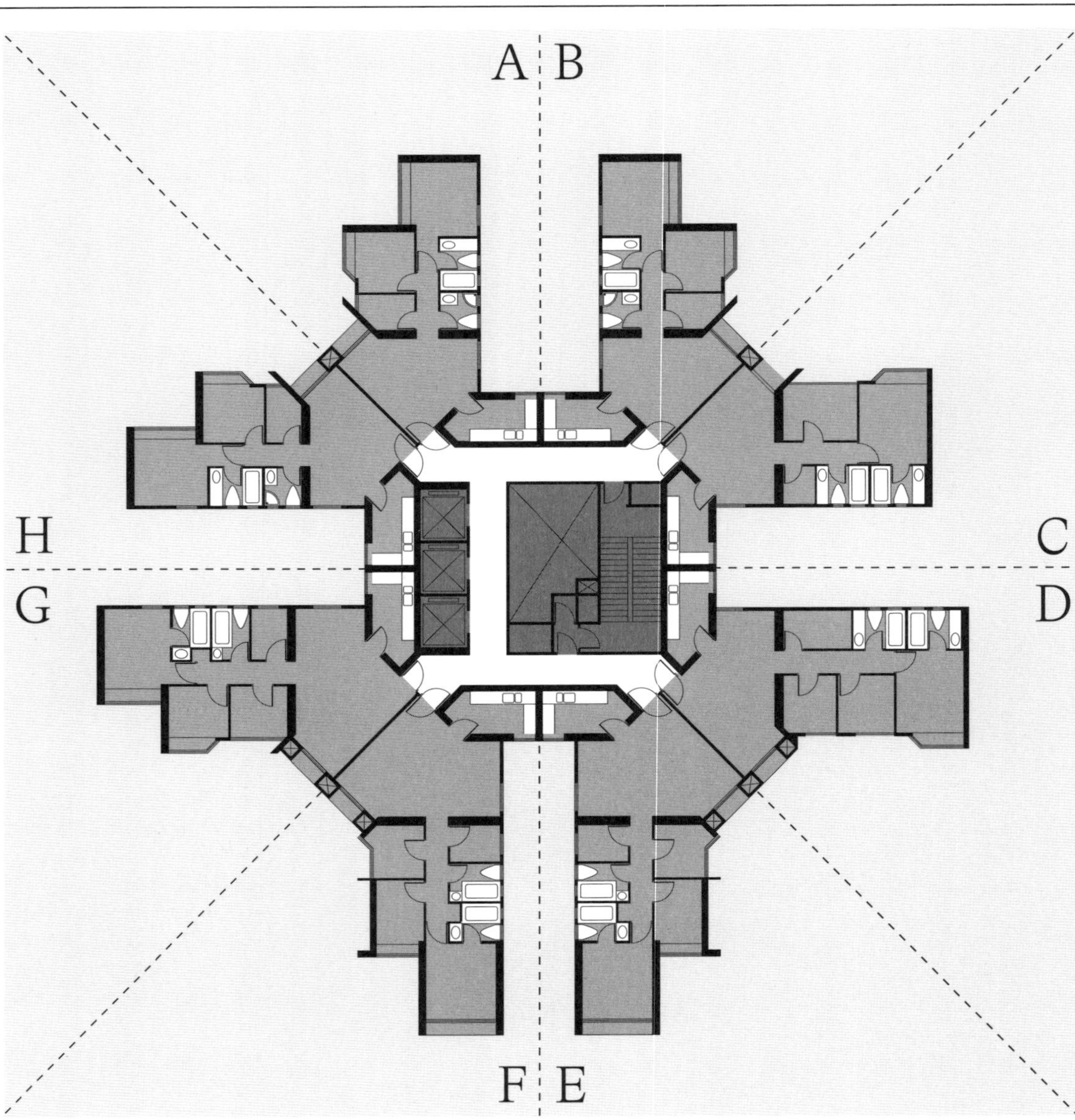

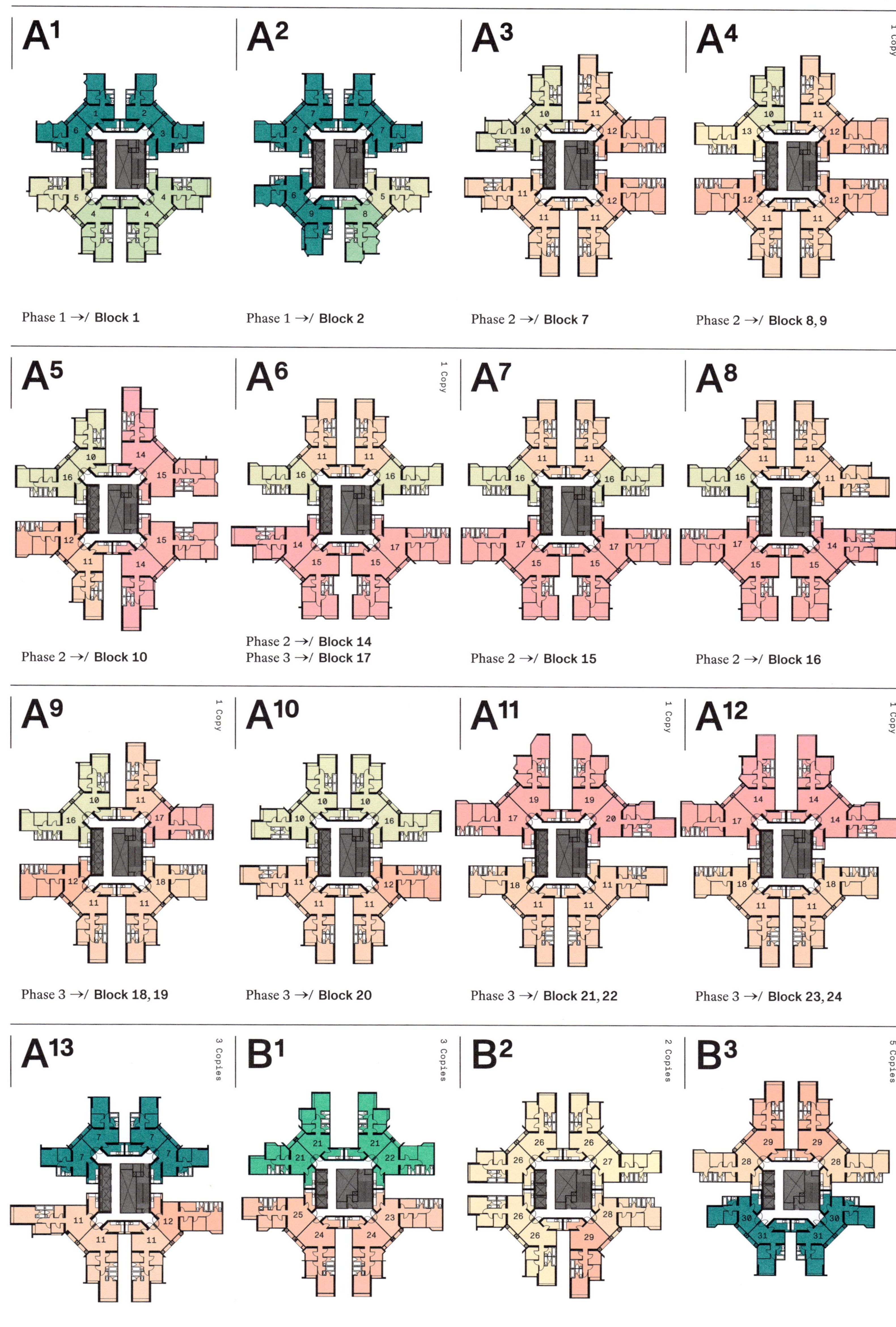
A1
Phase 1 →/ Block 1
A2
Phase 1 →/ Block 2
A3
Phase 2 →/ Block 7
A4
1 Copy
Phase 2 →/ Block 8, 9
A5
Phase 2 →/ Block 10
A6
1 Copy
Phase 2 →/ Block 14
Phase 3 →/ Block 17
A7
Phase 2 →/ Block 15
A8
Phase 2 →/ Block 16
A9
1 Copy
Phase 3 →/ Block 18, 19
A10
Phase 3 →/ Block 20
A11
1 Copy
Phase 3 →/ Block 21, 22
A12
1 Copy
Phase 3 →/ Block 23, 24
A13
3 Copies
Phase 4 →/ Block 25–28
B1
3 Copies
Phase 1 →/ Block 3–6
B2
2 Copies
Phase 2 →/ Block 11–13
B3
5 Copies
Phase 4 →/ Block 29–34

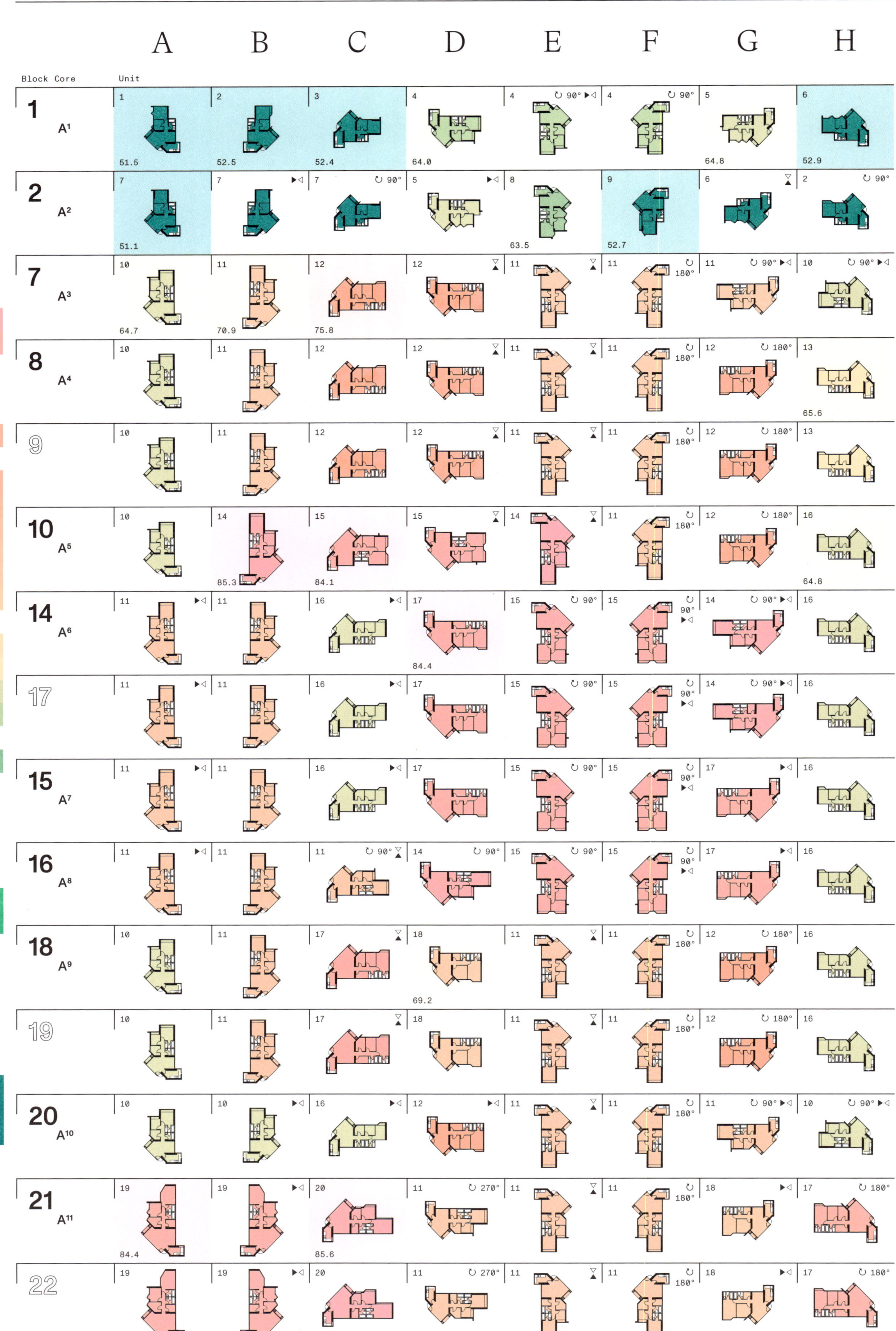
A B C D E F G H
Block Core
Unit
1 A¹ | 1 51.5 | 2 52.5 | 3 52.4 | 4 64.0 | 4 ↻ 90° ▶◁ | 4 ↻ 90° | 5 64.8 | 6 52.9
2 A² | 7 51.1 | 7 ▶◁ | 7 ↻ 90° | 5 ▶◁ | 8 63.5 | 9 52.7 | 6 | 2 ↻ 90°
7 A³ | 10 64.7 | 11 70.9 | 12 75.8 | 12 | 11 | 11 ↻ 180° | 11 ↻ 90° ▶◁ | 10 ↻ 90° ▶◁
8 A⁴ | 10 | 11 | 12 | 12 | 11 | 11 ↻ 180° | 12 ↻ 180° | 13 65.6
9 | 10 | 11 | 12 | 12 | 11 | 11 ↻ 180° | 12 ↻ 180° | 13
10 A⁵ | 10 | 14 85.3 | 15 84.1 | 15 | 14 | 11 ↻ 180° | 12 ↻ 180° | 16 64.8
14 A⁶ | 11 ▶◁ | 11 | 16 ▶◁ | 17 84.4 | 15 ↻ 90° | 15 ↻ 90° ▶◁ | 14 ↻ 90° ▶◁ | 16
17 | 11 ▶◁ | 11 | 16 ▶◁ | 17 | 15 ↻ 90° | 15 ↻ 90° ▶◁ | 14 ↻ 90° ▶◁ | 16
15 A⁷ | 11 ▶◁ | 11 | 16 ▶◁ | 17 | 15 ↻ 90° | 15 ↻ 90° ▶◁ | 17 ▶◁ | 16
16 A⁸ | 11 ▶◁ | 11 | 11 ↻ 90° | 14 ↻ 90° | 15 ↻ 90° | 15 ↻ 90° ▶◁ | 17 ▶◁ | 16
18 A⁹ | 10 | 11 | 17 | 18 69.2 | 11 | 11 ↻ 180° | 12 ↻ 180° | 16
19 | 10 | 11 | 17 | 18 | 11 | 11 ↻ 180° | 12 ↻ 180° | 16
20 A¹⁰ | 10 | 10 ▶◁ | 16 ▶◁ | 12 ▶◁ | 11 | 11 ↻ 180° | 11 ↻ 90° ▶◁ | 10 ↻ 90° ▶◁
21 A¹¹ | 19 84.4 | 19 ▶◁ | 20 85.6 | 11 ↻ 270° | 11 | 11 ↻ 180° | 18 ▶◁ | 17 ↻ 180°
22 | 19 | 19 ▶◁ | 20 | 11 ↻ 270° | 11 | 11 ↻ 180° | 18 ▶◁ | 17 ↻ 180°

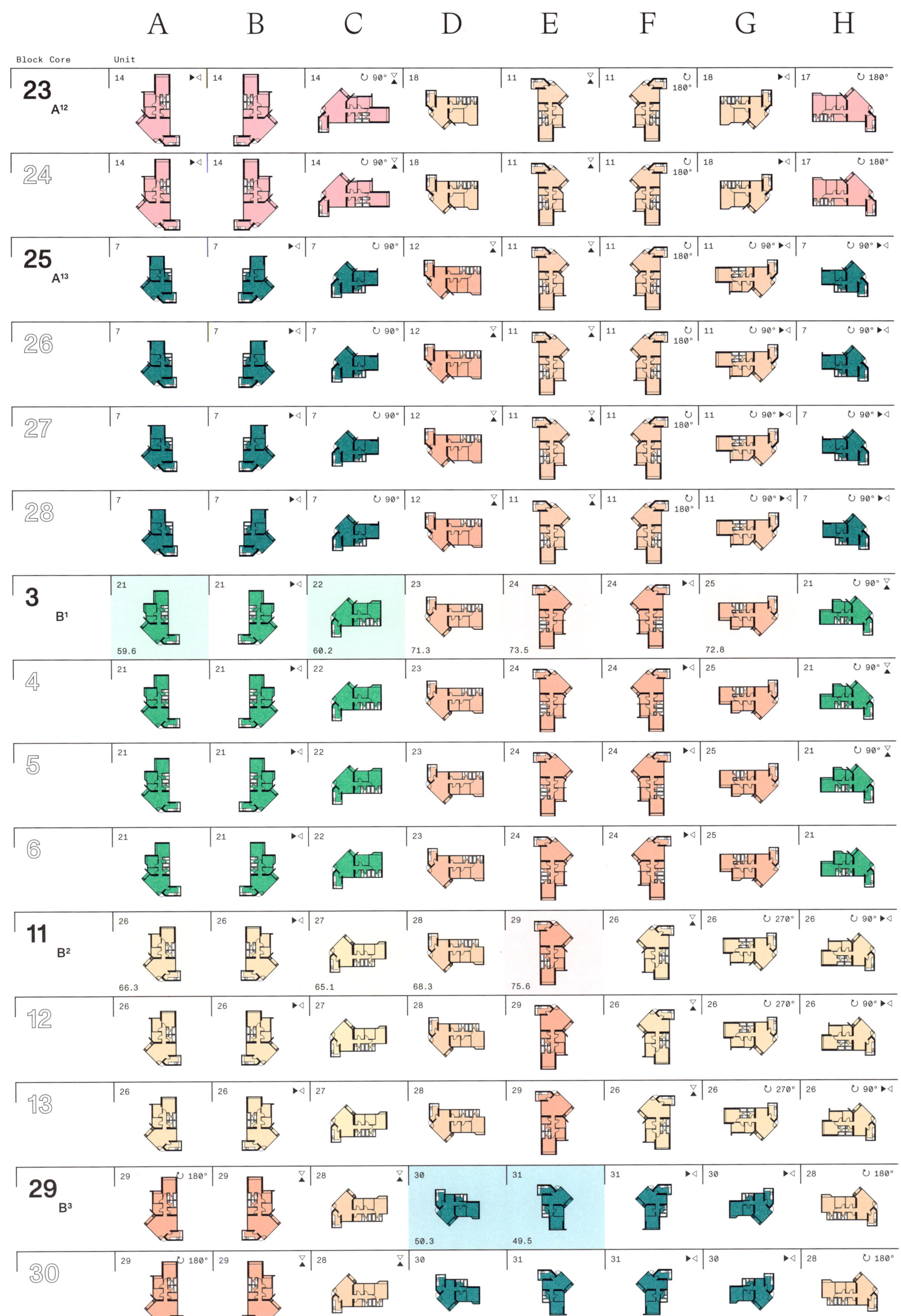
A
B
C
D
E
F
G
H
Block Core
Unit
23
A12
24
25
A13
26
27
28
3
B1
4
5
6
11
B2
12
13
29
B3
30
59.6
60.2
71.3
73.5
72.8
66.3
65.1
68.3
75.6
50.3
49.5

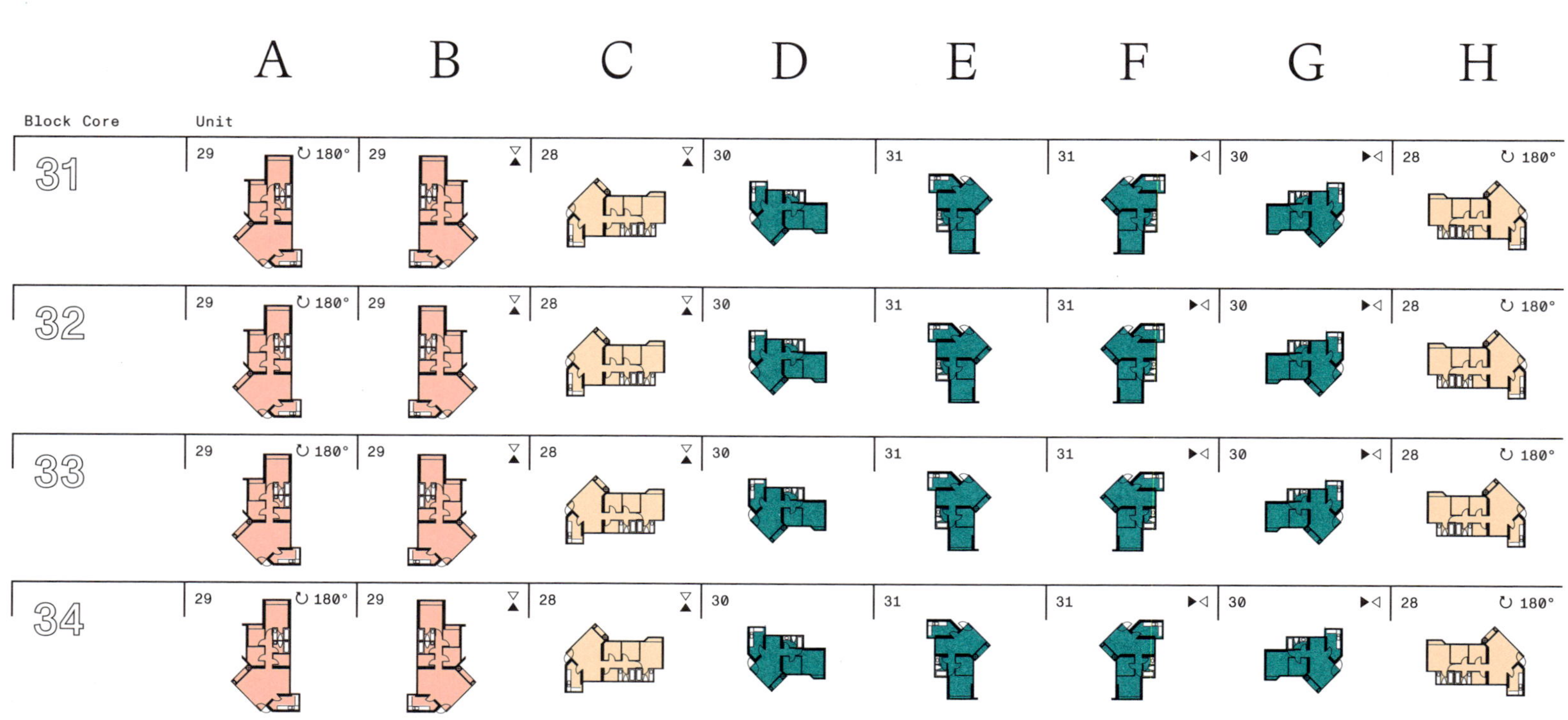
A
B
C
D
E
F
G
H
Block Core
Unit
31
32
33
34
29
28
30
31
180°

Kingswood Villas 1991–99

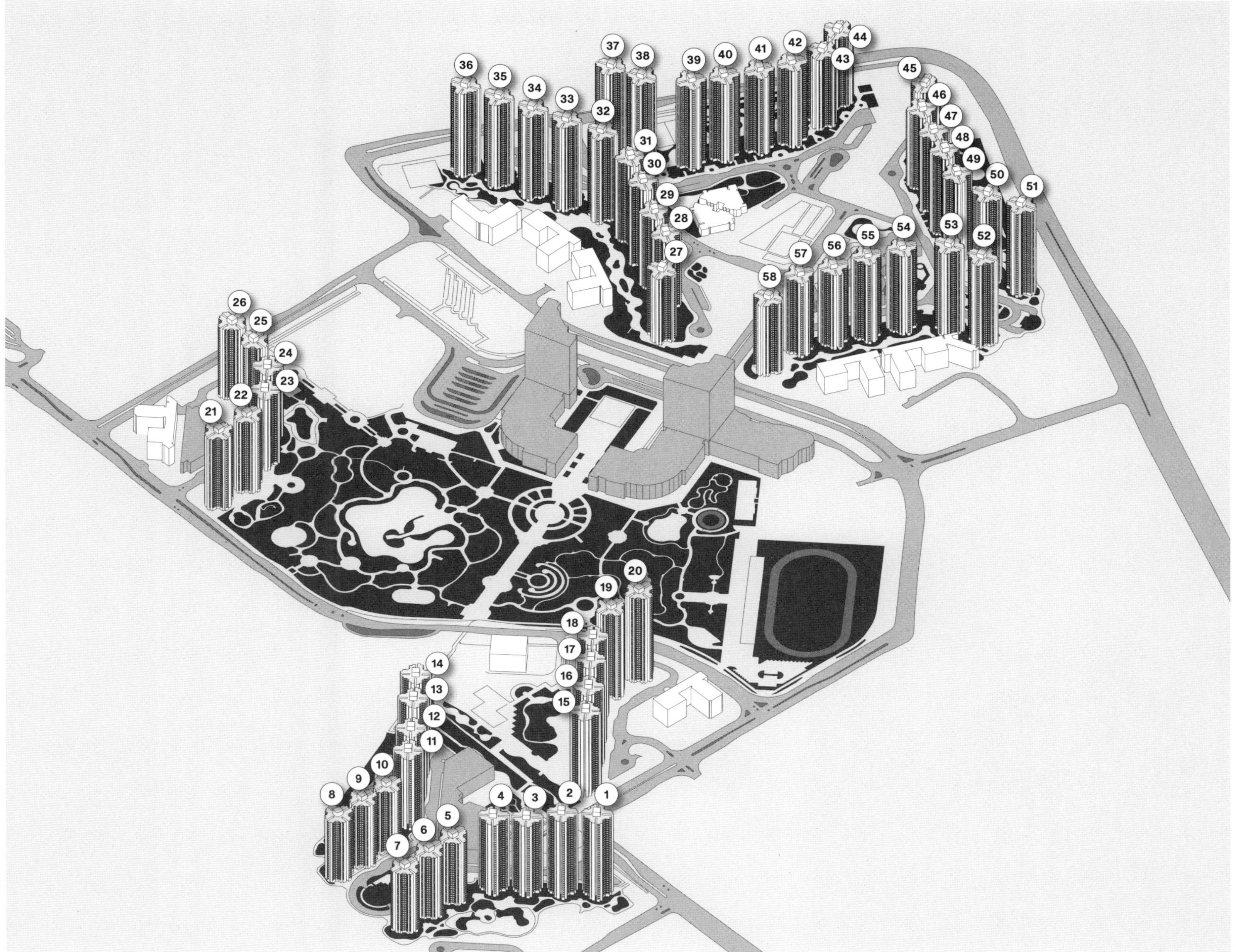

DEVELOPER:	Cheung Kong Holdings		
ARCHITECT:	DLN Architects & Engineers.		
LOCATION:	Yuen Long		
POPULATION:	39,361		
TOWERS:	58	UNIQUE:	2
APARTMENTS:	15,927	UNIT TYPES:	15
CORES:	1	PHASES:	6

Of the housing estates surveyed in this book, **Kingswood Villas** is the most recently constructed and the largest estate in terms of built area. Fifty-eight towers are spread across the site in a series of long, curvilinear arcs. These strips of towers are connected at the ground level by an atypical podium that is set back from the street edge. Communal amenities such as clubhouses and swimming pools are situated in the space behind each crescent of towers.

Kingswood Villas is also the most repetitive estate of those surveyed in terms of tower type. Of the fifty-eight towers on the estate, there are only two unique tower types that vary only slightly from each other in terms of floor plan.

Type A

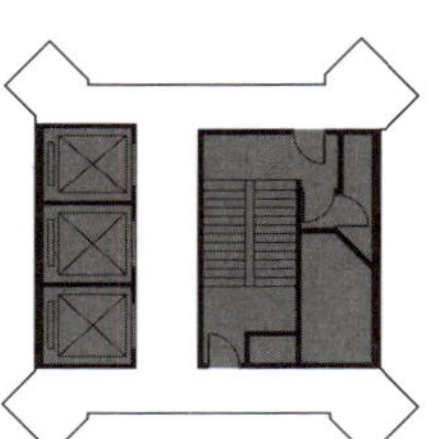

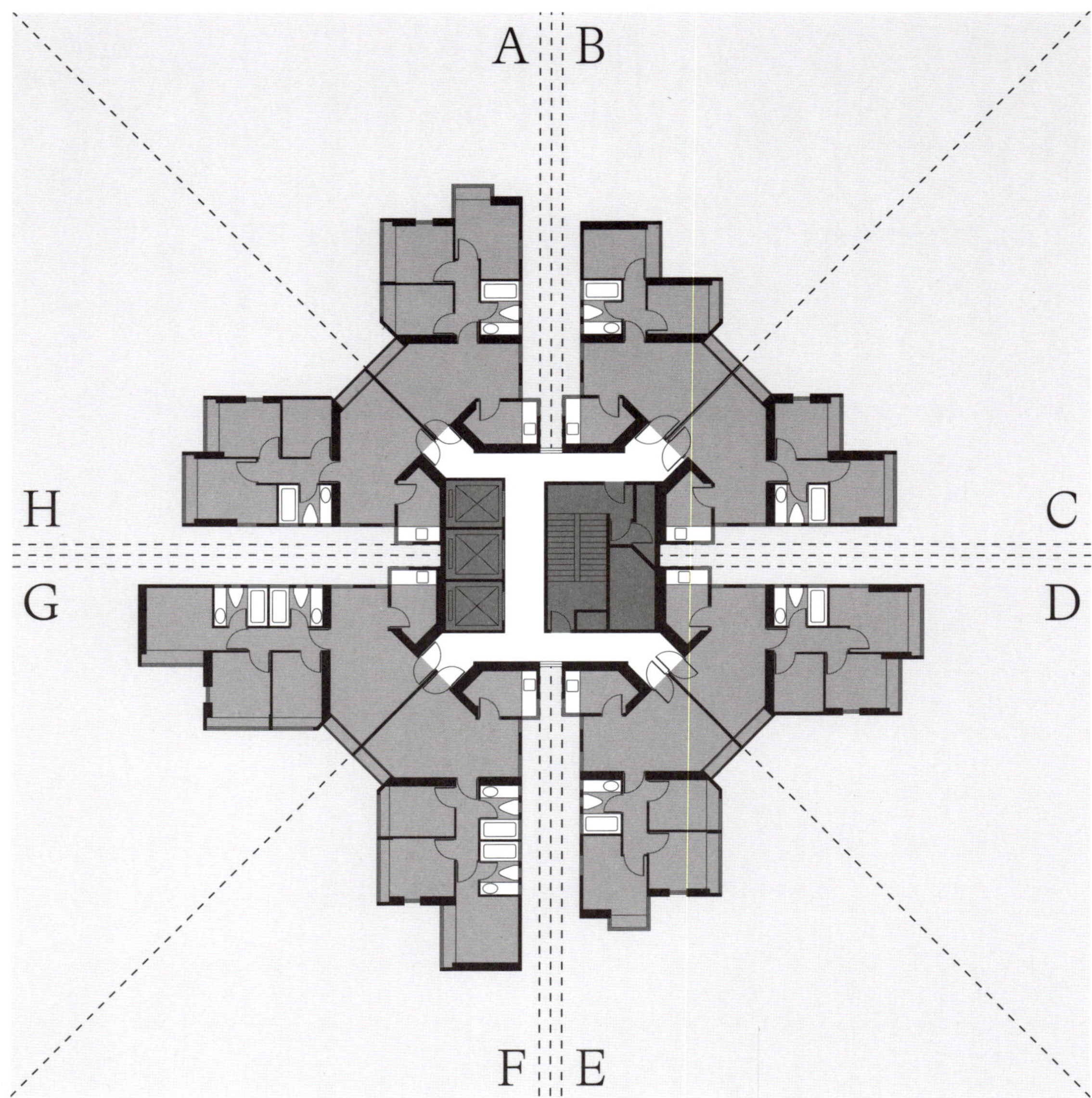

A¹

29 Copies

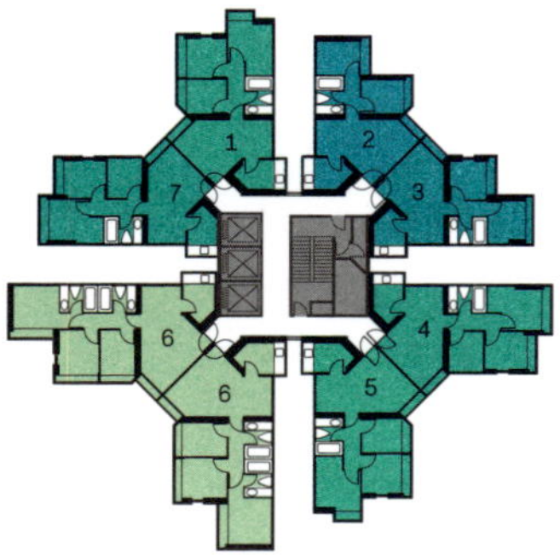

1
2
3
4
5
6
6
7

Phase 1 →/ **Block 1–7**
Phase 2 →/ **Block 8–14**
Phase 3 →/ **Block 15–20**
Phase 4 →/ **Block 21–26**
Phase 5 →/ **Block 27–30**

A²

27 Copies

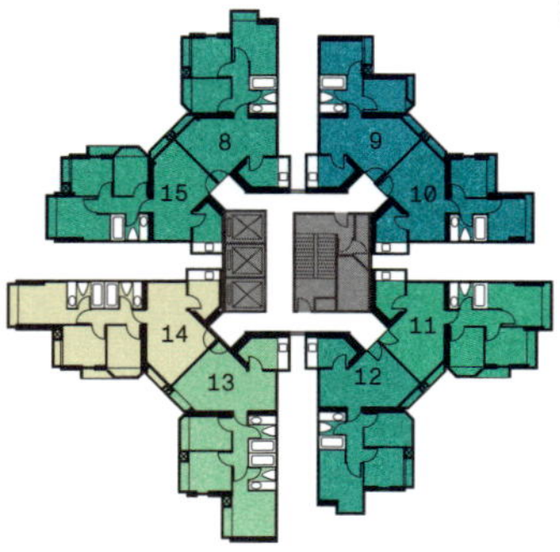

8
9
10
11
12
13
14
15

Phase 5 →/ **Block 31–36**
Phase 6 →/ **Block 37–44**
Phase 7 →/ **Block 45–58**

Block Core	Unit A	B	C	D	E	F	G	H
1 A¹	1 54.2	2 45.0	3 46.3	4 54.3	5 55.1	6 63.8	6 ↻ 90°	7 53.3
2	1	2	3	4	5	6	6 ↻ 90°	7
3	1	2	3	4	5	6	6 ↻ 90°	7

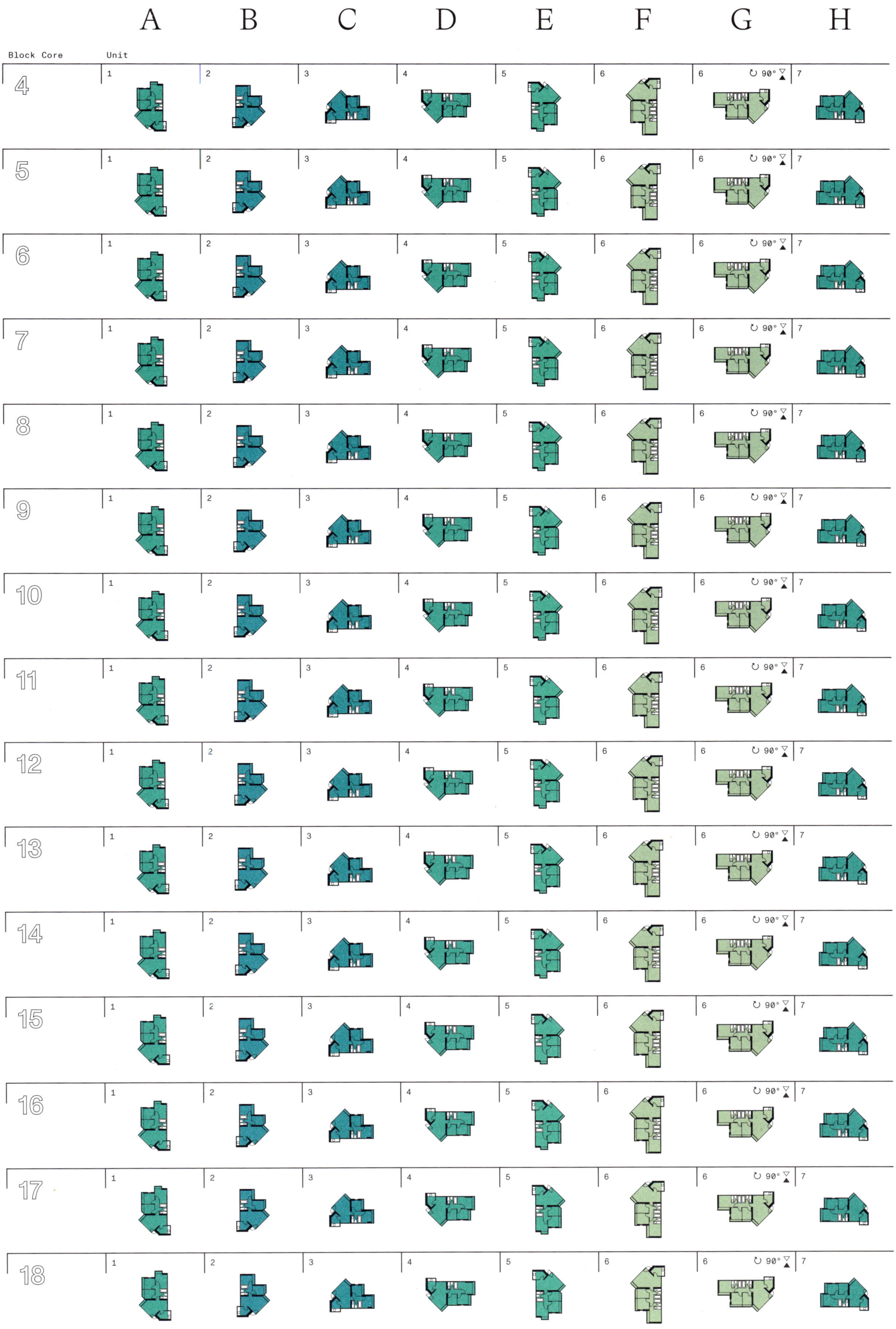

Block Core	A	B	C	D	E	F	G	H
	Unit							
4	1	2	3	4	5	6	6 ↻ 90°	7
5	1	2	3	4	5	6	6 ↻ 90°	7
6	1	2	3	4	5	6	6 ↻ 90°	7
7	1	2	3	4	5	6	6 ↻ 90°	7
8	1	2	3	4	5	6	6 ↻ 90°	7
9	1	2	3	4	5	6	6 ↻ 90°	7
10	1	2	3	4	5	6	6 ↻ 90°	7
11	1	2	3	4	5	6	6 ↻ 90°	7
12	1	2	3	4	5	6	6 ↻ 90°	7
13	1	2	3	4	5	6	6 ↻ 90°	7
14	1	2	3	4	5	6	6 ↻ 90°	7
15	1	2	3	4	5	6	6 ↻ 90°	7
16	1	2	3	4	5	6	6 ↻ 90°	7
17	1	2	3	4	5	6	6 ↻ 90°	7
18	1	2	3	4	5	6	6 ↻ 90°	7

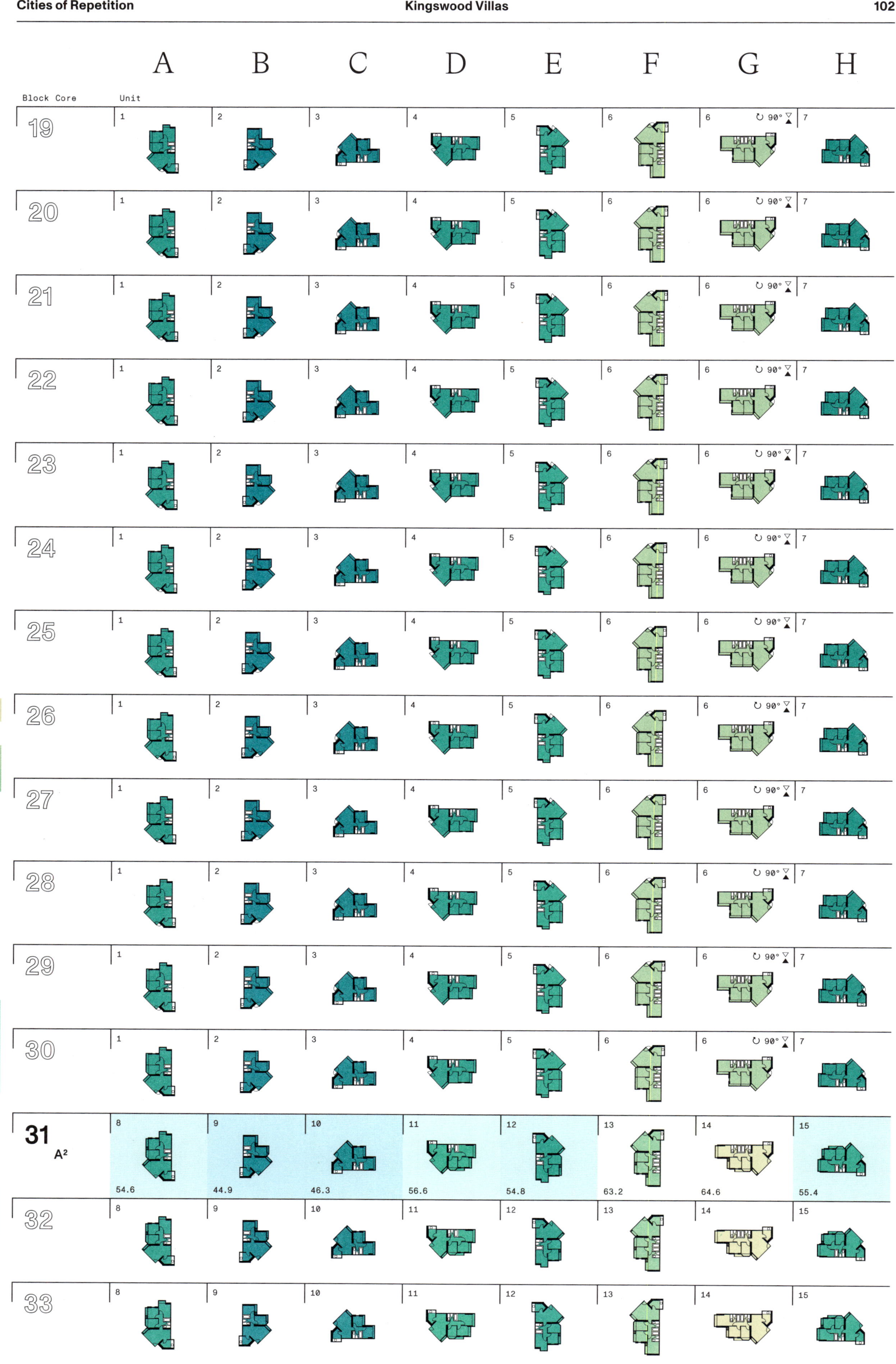
A B C D E F G H
Block Core
Unit
19
20
21
22
23
24
25
26
27
28
29
30
31 A²
32
33
90°
54.6
44.9
46.3
56.6
54.8
63.2
64.6
55.4

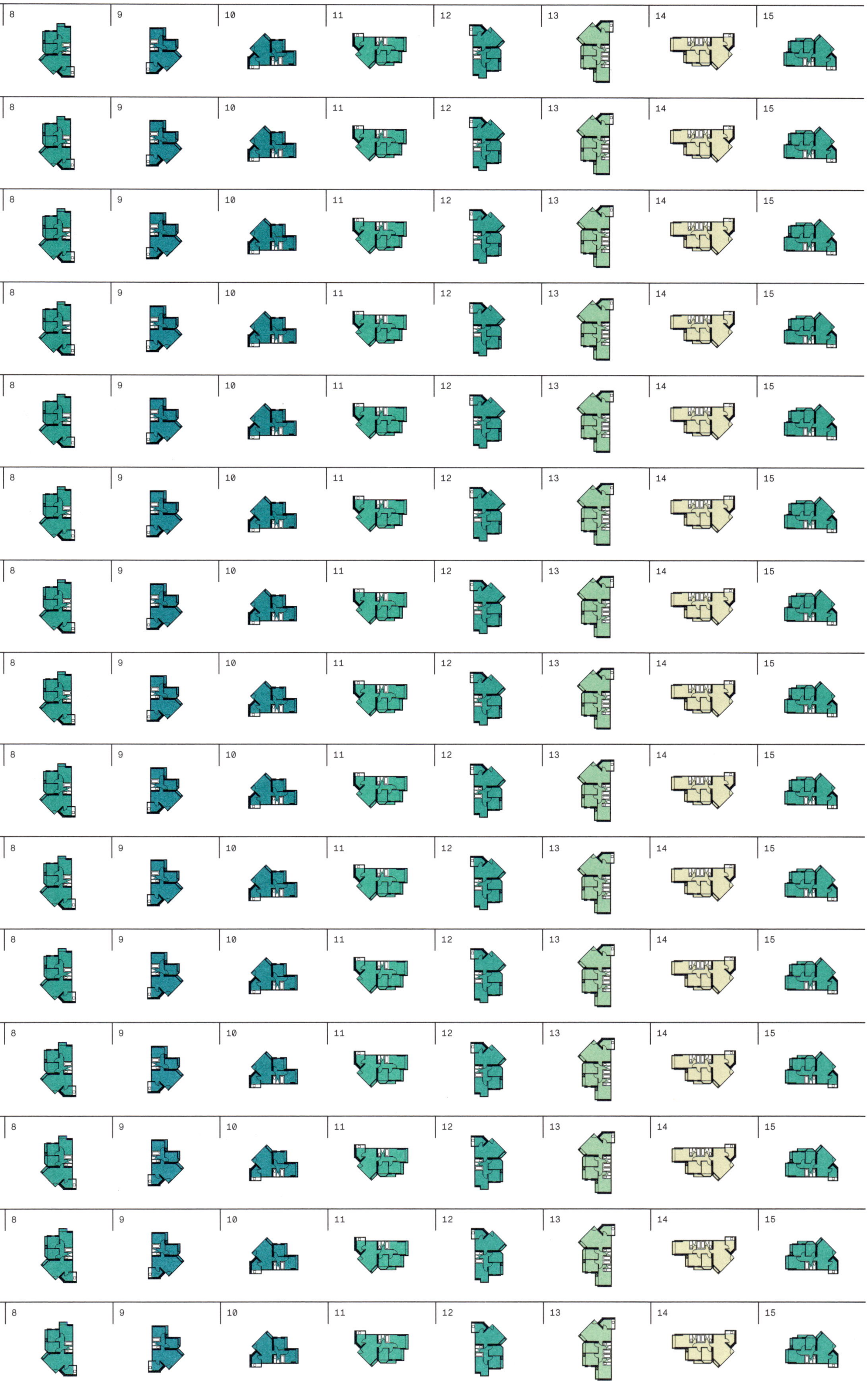

Block Core	A	B	C	D	E	F	G	H
	Unit							
34	8	9	10	11	12	13	14	15
35	8	9	10	11	12	13	14	15
36	8	9	10	11	12	13	14	15
37	8	9	10	11	12	13	14	15
38	8	9	10	11	12	13	14	15
39	8	9	10	11	12	13	14	15
40	8	9	10	11	12	13	14	15
41	8	9	10	11	12	13	14	15
42	8	9	10	11	12	13	14	15
43	8	9	10	11	12	13	14	15
44	8	9	10	11	12	13	14	15
45	8	9	10	11	12	13	14	15
46	8	9	10	11	12	13	14	15
47	8	9	10	11	12	13	14	15
48	8	9	10	11	12	13	14	15

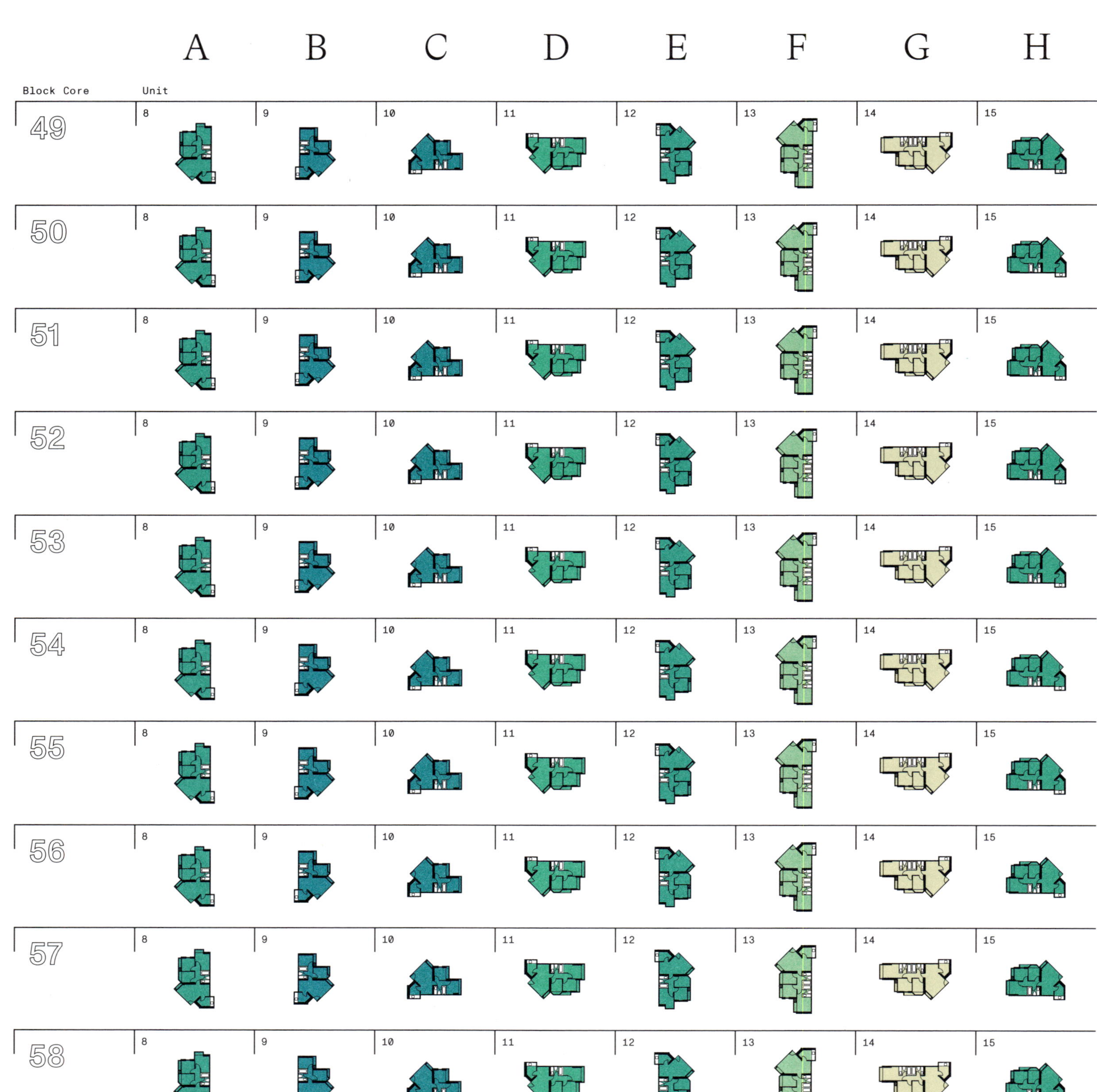
A B C D E F G H
Block Core
Unit
8 9 10 11 12 13 14 15
49
50
51
52
53
54
55
56
57
58

Analysis
Quantifying Repetition

Overview

This section of the book brings together graphic information and comparative data from the previous sections to better understand the enormous scale, extreme population densities, and relative degrees of standardization and repetition exhibited by these ten remarkable housing estates. While the drawings in the 10 Estates section reveal the organization, homogeneity, and diversity of towers and units through extensive color-coding, this section provides a deeper comparative understanding through graphical and statistical analysis across these highly repetitive micro-cities.

As the development period of the ten cities of repetition spans more than three decades, the purpose of this analysis is to reveal different design tendencies and evolutionary development over time. The two-dimensional architectural drawings and three-dimensional digital models created for the study provided a data set for quantitative and qualitative examination. Data was captured by measuring the drawings to calculate various parameters to give an understanding of the differences. Several diagrams highlight the design strategies and represent calculable aspects of physical space, architectural form, degrees of standardization, and trajectories of homogenization over time. The analysis helps to demonstrate the extreme impact of building code on the built environment of Hong Kong.

Estates are presented chronologically so that graphics can be read as a bar chart to reveal trends or trajectories. Supplementary information, including demographic and economic data from Hong Kong, is provided at the end of this section to draw correlations between socio-economic developments, architectural forms, and urban design strategies. While the differences between the estates are sometimes subtle, the analysis reveals a discernable trajectory of tower designs becoming taller, more repetitive, and more standardized.

Toward Congruency

Each of the estates consists of dozens of towers. At first glance, they not only look very alike from the outside but also, as the next pages of diagrams show, follow very similar organizational principles. From the beginning of this project, the critical interest in researching these estates was to understand how deeply the concept of standardization actually shaped built form and how repetition materialized. It was clear from the beginning of the study that little evolution happened in the concept of the cruciform plan type over time. Comparative diagrams show that any experimentation or invention within the design and planning of these hundreds of towers was very subtle.

The 529 towers that make up the ten estates consist of 198 unique blocks and 331 copies. The figure-ground matrix [fig. 1, p. 6], which reads from top to bottom and left to right through all estates, visualizes this ratio. The diagram clearly shows the repetitive nature of these shapes – but also how the variation develops in the tower outline. While core typology has been established as one aspect to differentiate, other attributes to distinguish between the various towers include outlines, dimensions, and size.

As stated in the Hong Kong Housing chapter, the perimeter of the typical tower had to adapt to building code requirements. Kitchens and bathrooms had to be naturally ventilated, and layout planning became more streamlined and followed principles of efficiency. Hence, deep incisions, or "re-entrants," into outside walls were introduced in order to make the layout work. Moreover, the introduction of the projected bay window—which is so typical for Hong Kong's housing landscape—extended the perimeter of the outline even further. For a period of time starting in the 1980s, projecting (bay) windows were exempt from GFA calculations and became a standard in the design of housing towers.

Efficiency Factor = Overall Floor Plan Area / Common Area

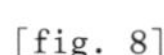
[fig. 8]

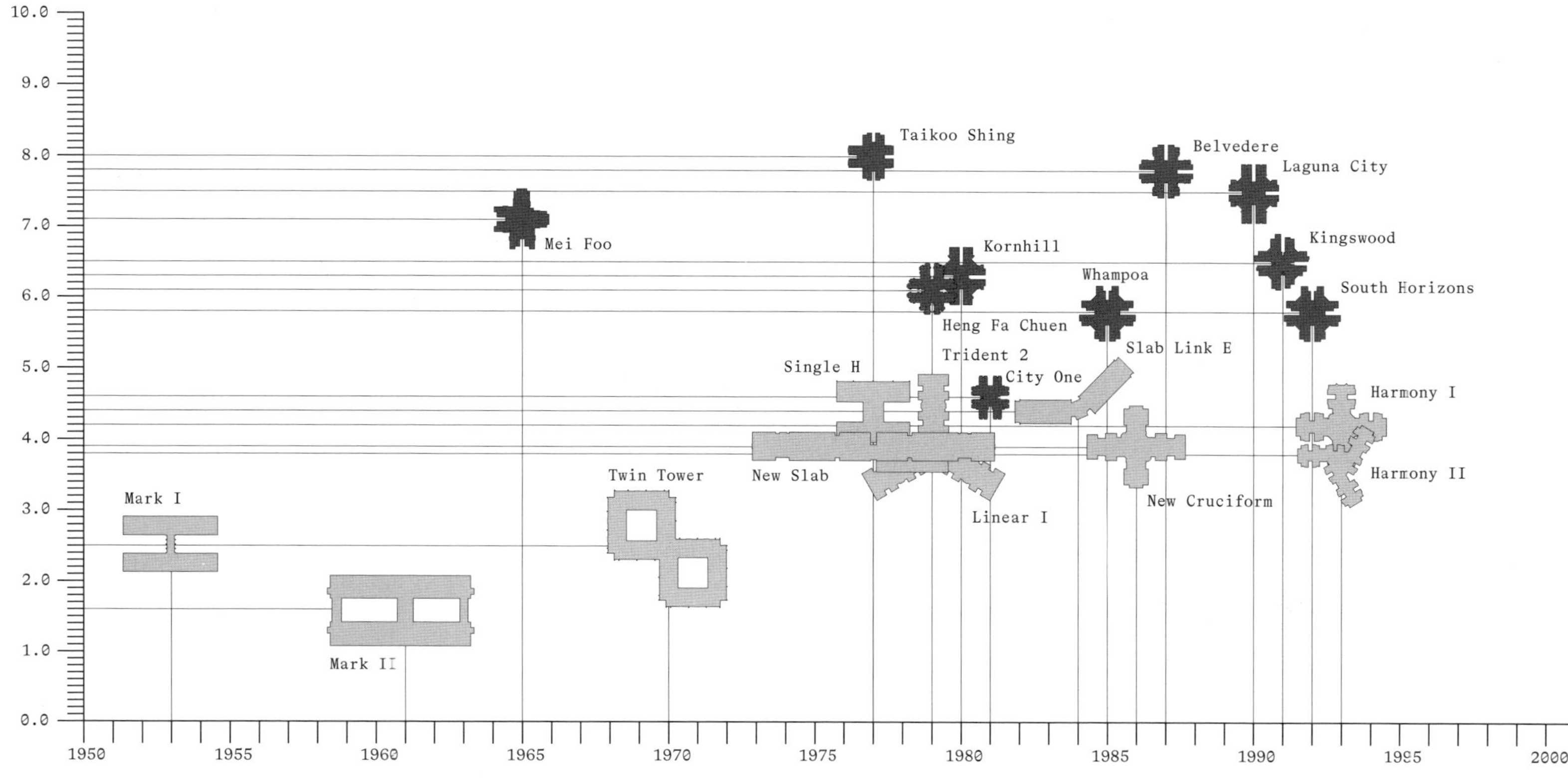

Figure 9 shows the relationship between the tower area and its respective perimeter length. The diagram clearly shows the impact of the building code, with towers of the early Mei Foo estate positioned on the left side (shorter perimeter) and towers of South Horizons and Kingswood Villas set on the right side (more extended perimeter). So, the evolutionary process within private housing in Hong Kong from the 1960s to 1990s was not so much about designing ever-so-new organizational typologies driven by aspects of space but more about subtle changes in the building code that impacted the outline of the standard cruciform plan.

Repetition as a strategy has been used in all estates. However, copies are not evenly found throughout the individual towers. Figure 10 represents a comparison between towers and their area and the respective number of copies in each estate. Although the graph does not show an absolute recognizable trend through time, it is apparent that the more recently built estates exhibit more repetition in general, as exemplified by Belvedere Garden, Laguna City, and Kingswood Villas. It is interesting, though, to note that beyond the most repetitious examples, the other estates tend to show a range of tower area variation. However, there doesn't seem to be a universal answer as to why some of the estate towers have more copies than others. Varying site constraints or opportunities likely affected where and when standardization was applied. In Taikoo Shing for example, the most repetitive tower is the block type that faces the waterfront and has the largest units. One might expect this particular real estate condition with valuable views to have a wide range of tower types with a multitude of unit layouts. In the case of Hong Kong, it's the opposite.

Density and Scale

The diagrams on pages 108–109 present a comparative study of the estates with regard to built area, population density, and tower repetition. As mentioned in the introduction, the ten estates in this project were chosen for analysis due to their population size and extraordinary number of towers. Each estate represents a community of residents housed in highly repetitive high-rise tower blocks with overall populations ranging from approximately 19,000 to 40,000 inhabitants per estate. Every one of inhabitants of the more than 529 towers in this study lives in a variation of a cruciform-shaped floor plan.

While each of the estates has different site conditions and constraints, there is a notable trajectory toward more densely developed plots with less space per resident. Some of this tower density is directly related to tower height. Taller building heights, especially among the estates built in the 1990s, allow more residents to live on a smaller footprint. Laguna City, for example, has a built-area to plot-area ratio of 8.79. That means with more than 621,025 square meters of built area for 70,620 square meters of site area, the developers have created more than eight times the amount of residential space compared with the amount of land.

In terms of built space per resident, Belvedere Garden is the densest estate with only 20.5 square meters per inhabitant. By comparison, Mei Foo Sun Chuen, which developed in decades with less expensive land prices, provides residents with 26.3 square meters per inhabitant.

The diagrams also show the degrees of tower standardization across time. Again, the estates built in recent years exhibit more extreme ratios of unique tower types to copies of those tower types. Kingswood Villas, is the most repetitive with only two original tower blocks and 56 duplicates.

Advancements in structural engineering and construction technologies allowed steel-reinforced concrete buildings worldwide to grow steadily taller decade by decade in the latter half of the twentieth century. Towers in Hong Kong followed this trend. Tower height grew steadily in the estates studied in this project from a standard height of 21 stories, in the Mei Foo and Heng Fa Chuen estates, to a maximum of 42 stories in the tallest towers in the Belvedere Garden and South Horizons estates. There are notable exceptions to this trajectory of increasing tower height through the decades. For instance, tower height for the Whampoa Gardens estate was restricted to 16 stories due to its proximity to the then-active Kai Tak Airport.

Increasing tower height throughout these decades, combined with the nature of extreme repetition of tower floor plans, contributes to even higher instances of repetitive units. Since all the floor plans in each of the tower studies are simply extruded vertically, taller towers create greater ratios of unit copies. Tower heights for the ten estates studied are presented numerically (by number of stories). More recently, built estates such as Belvedere Garden represent extremes of unit standardization with only eight unit types utilized for over six thousand apartment units. That presents a standardization ratio of 752 copies for every distinct unit type. (8 unit types per 6,016 total units = 1:752.) Laguna City is even more repetitive in its unit types with only four unit types for over eight thousand apartment units. When calculated, the ratio of distinct units to copies is one to 2,018 (4 unit types per 8,072 total units = 1:2,018).

Standardization and Deviation

One way of graphically demonstrating degrees of standardization across the ten largest estates is through a series of drawings which superimpose outlines of plans. The two-page spread of drawings, entitled Standardization/Deviation (pages 110–111), provides a graphic comparison of tower diversity in each estate through six different drawing types. Row A presents a set of regulating lines for each of the estates. These diagrams reveal axes of unit organization around a central core; they also show divisions between units and lines of mirroring and reflection. Underlying geometries reveal more complex or simplistic strategies of variation from estate to estate through architectural cutting, pasting, rotating, and mirroring.

By layering the dozens of tower block plans over each other in Row B, different degrees of visual sharpness or fuzziness are revealed, providing an understanding of gradations of relative standardization or variability. Estates built from the 1960s through the 1980s reveal fuzzier, superimposed outlines when compared with the most recent estates. Of the estates developed in the 1990s, Belvedere Garden (19 Towers), Laguna City (38 Towers), and Kingswood Villas (58 Towers) all exhibit strikingly thin perimeter lines and therefore nearly identical tower plans used across the estates.

[fig. 9]

Superimposition

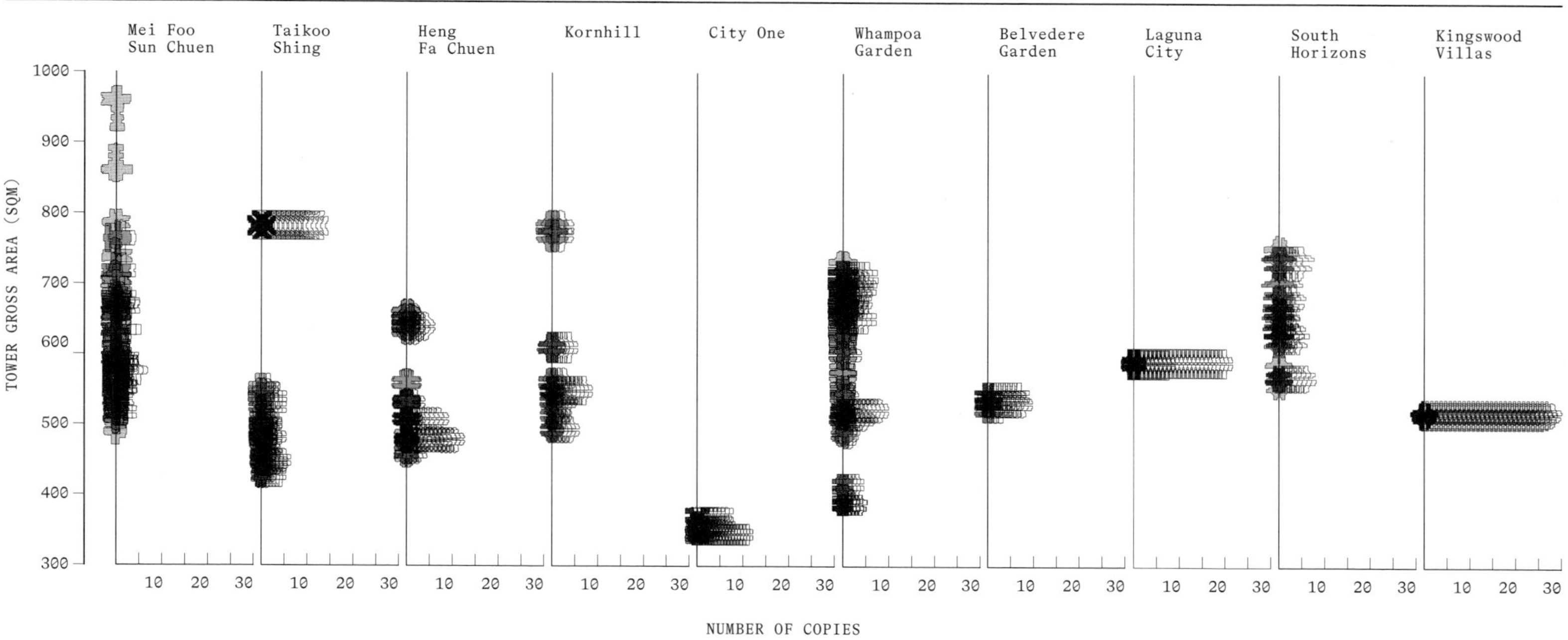

[fig. 10]

Row D reveals a composite drawing of each estate. It represents the outermost outline of the superimposed plans on Rows B and C creating, in effect, a Rorschach inkblot of each entire estate. This composite perimeter drawing is juxtaposed against the smallest tower block in each estate in Row E. The highlighted area between the outline of the smallest tower and the composite perimeter shows a measure of the variation in tower forms. This is greatest in the six earliest estates and almost negligible in three of the four most recently developed estates, again showing a trend toward greater standardization. The diagrams on the bottom of the page in Row F compare the composite perimeter form with the most dominant tower plan in each of the ten estates. Statistically speaking, the diagrams in Rows B and C are graphical representations of the "sum" of the towers in each estate; Rows D and E represent the "range" of tower types; and the dominant tower types shown in Row F represent the "mode" in each estate. The "mode" is defined as the number, or in this case, instance in a data set that occurs most frequently. All of the ten estates surveyed exhibit remarkably high levels of plan efficiency, especially when compared with Hong Kong's public housing types within the same time period [fig. 8].

Dominant Tower Types

The spread of analytical diagrams on pages 112–113 compares the most dominant tower-type floor plans from each of the ten estates studied. The six rows of diagrams show a range of organizational strategies across the estates and their increasing architectural efficiency over time. Row B measures the relationship of tower floor plates to the vertical circulation core. More compact cores can minimize circulation and common space and yield higher percentages of salable area for developers.

Rows C and D compare plan organization of service areas, such as kitchens and bathrooms with living rooms and bedroom spaces. Tower and unit plans reveal similar organizational and planning strategies between estates based on code requirements and planning efficiency. Service-oriented spaces like kitchens and bathrooms have window requirements for ventilation but are most often located closer to the tower vertical circulation core. Most buildings in Hong Kong feature plumbing supply and drainage pipes on the outside of building envelopes. Positioning rooms with plumbing on the inner recesses, or "re-entrants," of towers is advantageous to visually conceal externally mounted pipes. Smaller windows for air ventilation in service-oriented rooms are not given priority for views. Living rooms, dining areas, and bedroom spaces are typically given priority in floor plans with better light and views. Living spaces and dining area spaces in the estates studied are typically combined in one room. Window sizes in kitchens and rooms for habitation are regulated in the Hong Kong building code to ensure minimum standards for natural light: "the aggregate superficial area of glass in the window or windows is not less than one-tenth of the area of the floor of the room."

In a cruciform building plan, these more communal spaces are always positioned closer to the tower vertical circulation core to accommodate proximity to the kitchen and entry door. Bedrooms are typically placed on the outer periphery of tower plans in a more private zone of the apartment, away from the tower core. In more recently built estates, diagrammatic floor plans reveal kitchen and bathroom spaces that are more similar from unit to unit and more thoroughly integrated into central building cores.

Unit Types

The trajectory of increasing standardization of unit types and layouts follows the trajectory of tower repetition in the ten estates. The cruciform-type towers measured in this project are, with a few exceptions, almost always built with eight units surrounding a central circulation core. When these typical tower floor plans are extruded vertically, there is no vertical variation in unit type for the entire height of the tower block. Naturally, it follows that estates with greater tower standardization will have increased unit standardization. This is shown clearly on pages 114–115 in the comparison of unit outlines across the ten estates. Fuzzier outlines of superimposed unit plans represent more spatial diversity across the apartments for earlier developed estates; cleaner lines denote higher degrees of sameness.

Unit analysis also reveals trends in the space of unit-layout geometry over time. While the geometry of interior room spaces in Mei Foo Sun Chuen is consistently rectilinear, there is a shift to a diagonal geometry in the living room spaces of Taikoo Shing's Type B tower blocks. These towers, with diagonal walls in the living room spaces, were designed and built at the latter stage of Taikoo Shing's development, primarily along periphery of the estate, near the water's edge. The diagonal angle of the living room redirects the view away from adjacent towers in the grid and instead into the spaces between towers or out into the harbor. Diagonal living rooms are found in the majority of the estates after Taikoo Shing, including Heng Fa Chuen, Kornhill, Whampoa Garden, Laguna City, South Horizons, and Kingswood Villas. Beyond re-directing views away from neighboring towers, the diagonal chamfering of the living room area in floor plans is convenient to increasing standardization and design efficiencies in that it makes the flipping and mirroring of unit types easier. While the diagonally shaped living room space can make design and tower layout more efficient in planning, it makes interior spaces harder to furnish with standard furniture. As noted by Carolin Fong on page 118, after frustrating generations of Hong Kong residents with wedge-shaped living rooms, Hong Kong's developers and architects later reverted back to rectilinear room layouts in more recent decades based on user preference.

Socio-Economic Factors

Graphs of socio-economic data, [p. 116] charted across the latter half of the twentieth century, reveal factors that have increased pressure on the housing market during the timeframe in which these ten estates were designed and developed. Hong Kong's population grew steadily from the 1960s through the year 2000 with more rapid rates of growth in the late 1970s. Mirroring growth in population is Hong Kong's Gross Domestic Product, with a significant jump in the early 1990s. Residential property costs, as measured by the Hong Kong Rating and Valuation Department's Housing Price Index, also grew rapidly in the period recorded, from 1980 through 2000. Numbers dropped in the mid-2000s due to the outbreak of the SARS virus and the global economic turndown in 2007–08. Since that time, prices have continued to climb, mounting additional pressure on developers' bottom lines. Viewed in comparison with the remarkable measurements of scale, density, and repetition, with regard to these ten privately developed estates, Hong Kong's continued economic development and population growth over the last decades has had a profound effect on the shape and form of the built environment.

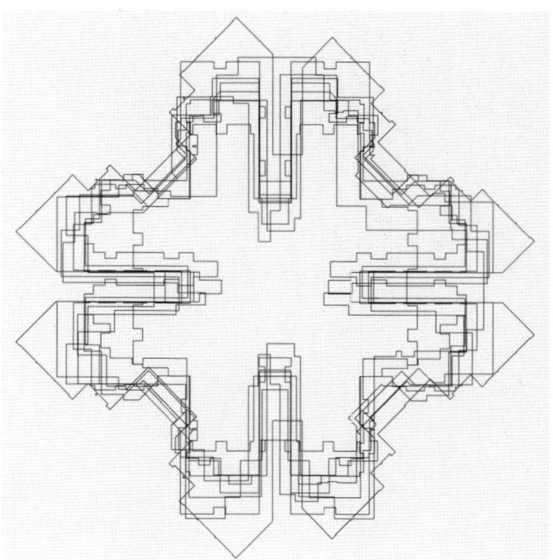

[fig. 11] The diagram shows the superimposition of the single most dominant tower block outline from each of the ten largest private estates.

Density and Scale

Mei Foo Sun Chuen
1965–1978

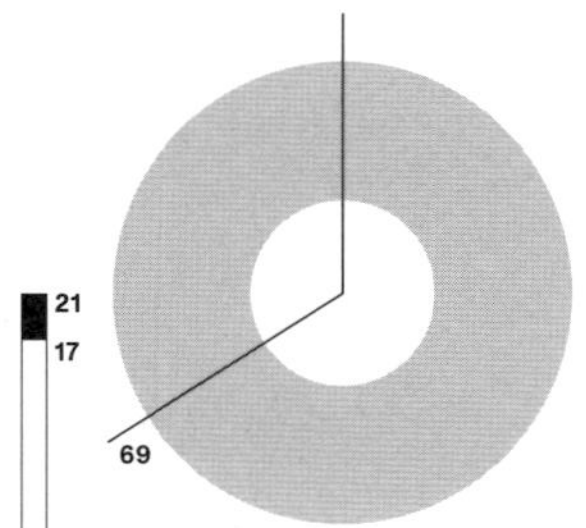

Tower Uniqueness Ratio: 69 %
Floor Area Ratio: 6.23
Tower Stories Range: 17–21

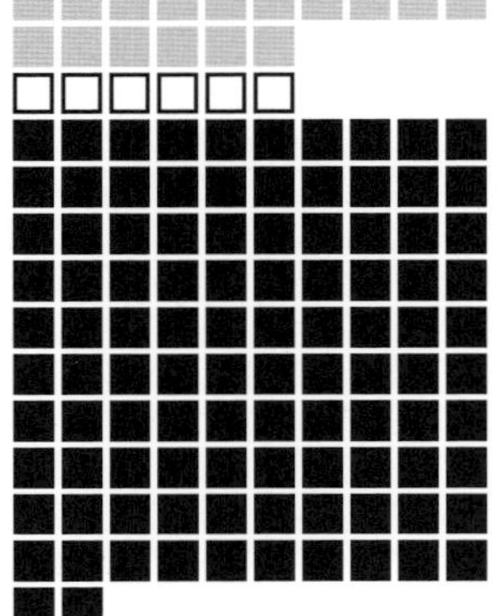

Site Area: 164,456 m²
Footprint Area: 56,229 m²
Built Area: 1,024,253 m²
Floor Area Ratio: 6.23
Plot Site Coverage: 34.2 %

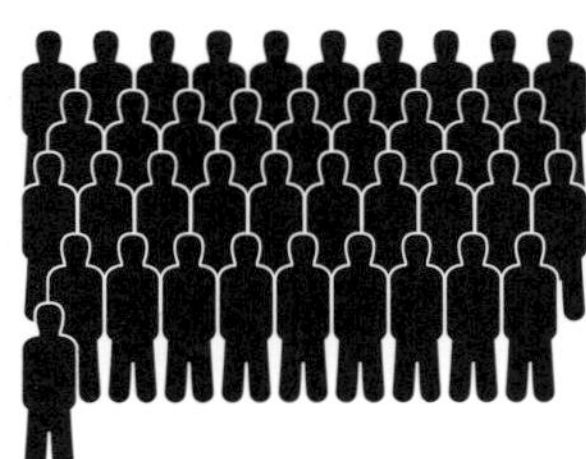

Population: 38,974 Residents

Site Area Density:
4.22 m²/Res

Built Area Density:
26.28 m²/Res

99 Blocks
68 Unique
31 Copies
Tower Uniqueness Ratio: 69 %

Taikoo Shing
1977–1989

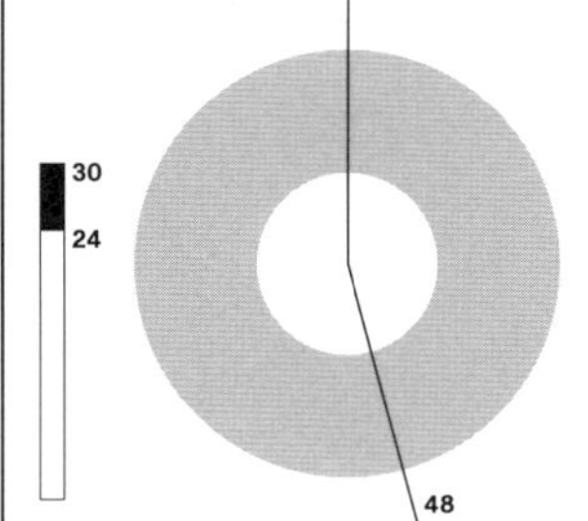

Tower Uniqueness Ratio: 48 %
Floor Area Ratio: 5.52
Tower Stories Range: 24–30

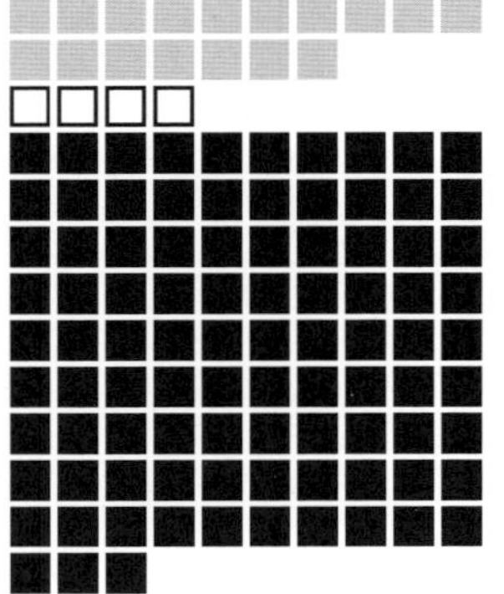

Site Area: 168,512 m²
Footprint Area: 35,606 m²
Built Area: 930,637 m²
Floor Area Ratio: 5.52
Plot Site Coverage: 21.1 %

Population: 36,796 Residents

Site Area Density:
4.58 m²/Res

Built Area Density:
25.29 m²/Res

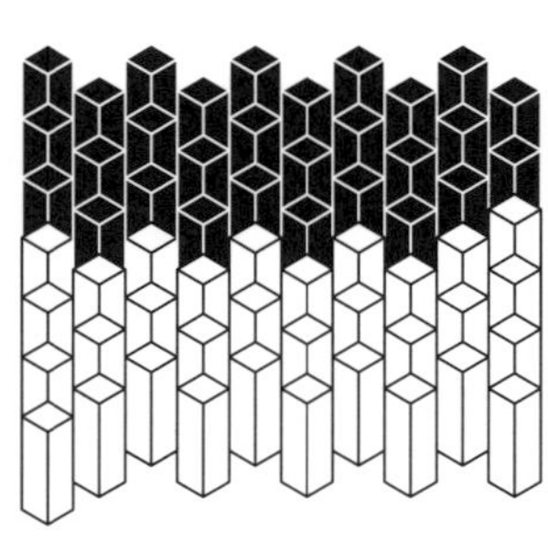

61 Blocks
29 Unique
32 Copies
Tower Uniqueness Ratio: 48 %

Heng Fa Chuen
1978–1989

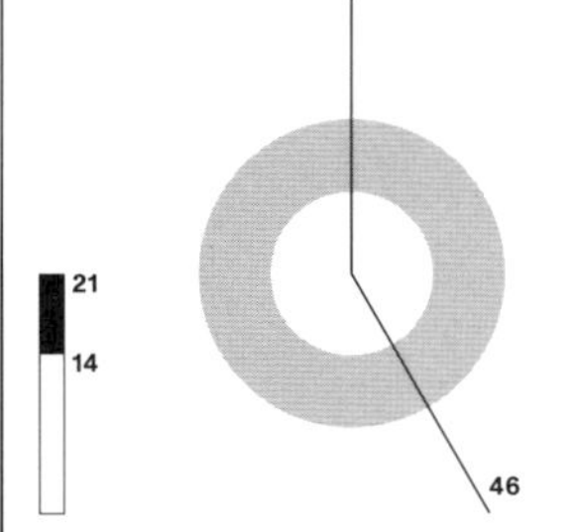

Tower Uniqueness Ratio: 46 %
Floor Area Ratio: 3.56
Tower Stories Range: 14–21

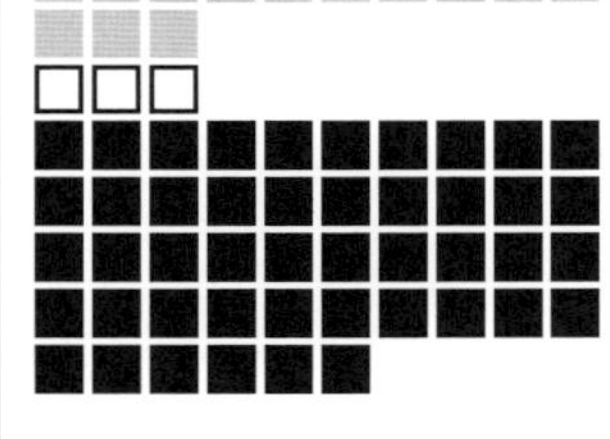

Site Area: 129,887 m²
Footprint Area: 26,248 m²
Built Area: 462,334 m²
Floor Area Ratio: 3.56
Plot Site Coverage: 20.2 %

Population: 18,921 Residents

Site Area Density:
6.86 m²/Res

Built Area Density:
24.45 m²/Res

48 Blocks
22 Unique
26 Copies
Tower Uniqueness Ratio: 46 %

Kornhill
1980–1987

31
19
28

Tower Uniqueness Ratio: 28 %
Floor Area Ratio: 7.10
Tower Stories Range: 19–31

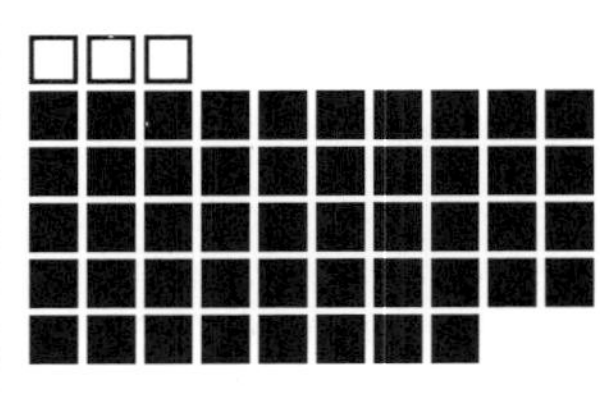

Site Area: 67,393 m²
Footprint Area: 17,274 m²
Built Area: 478,343 m²
Floor Area Ratio: 7.10
Plot Site Coverage: 25.6 %

Population: 18,450 Residents

Site Area Density:
3.65 m²/Res

Built Area Density:
25.93 m²/Res

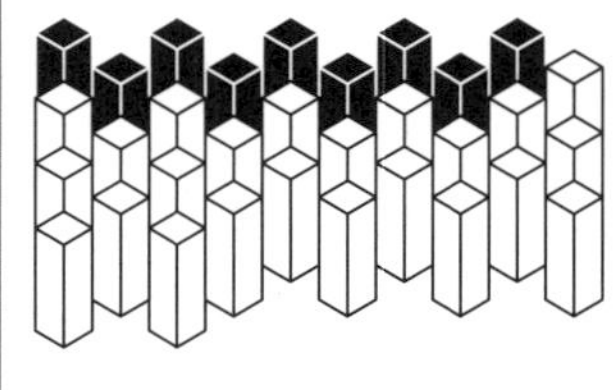

32 Blocks
9 Unique
23 Copies
Tower Uniqueness Ratio: 19 %

City One
1980–1988

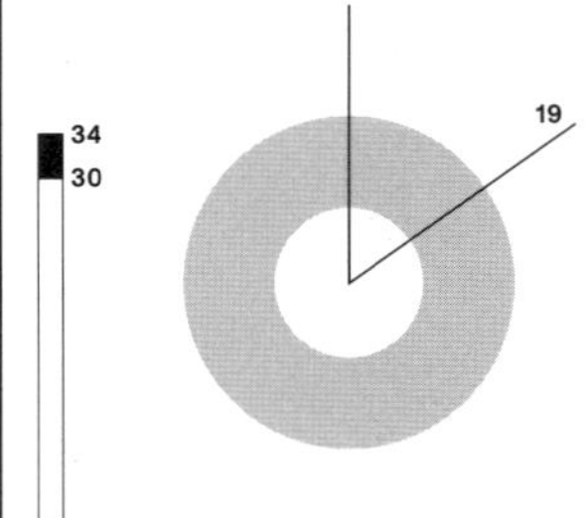

Tower Uniqueness Ratio: 19 %
Floor Area Ratio: 4.94
Tower Stories Range: 30–34

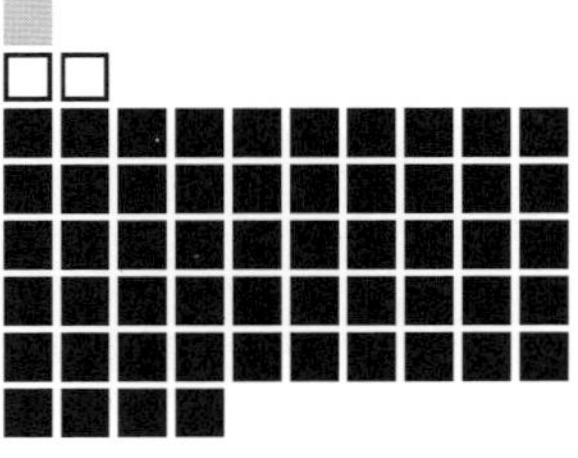

Site Area: 108,768 m²
Footprint Area: 17,546 m²
Built Area: 537,739 m²
Floor Area Ratio: 4.94
Plot Site Coverage: 16.1 %

Population: 24,758 Residents

Site Area Density:
4.39 m²/Res

Built Area Density:
21.72 m²/Res

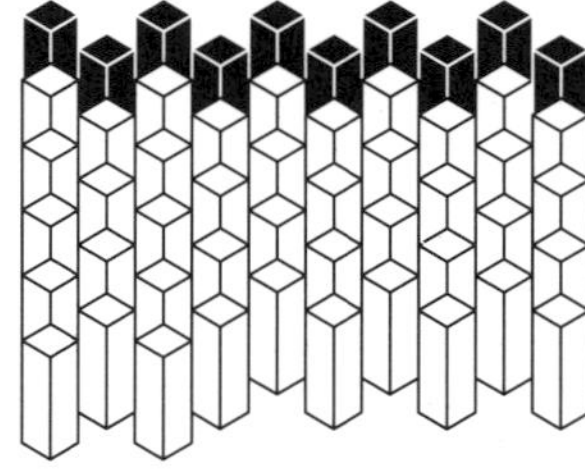

52 Blocks
10 Unique
42 Copies
Tower Uniqueness Ratio: 19 %

 Floor Area Ratio

 Tower Uniqueness

□ = 10,000 m²

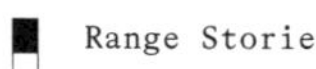 Range Stories

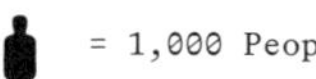 = 1,000 People

 = 1 m²

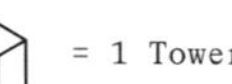 = 1 Tower

Whampoa Garden 1985–1991

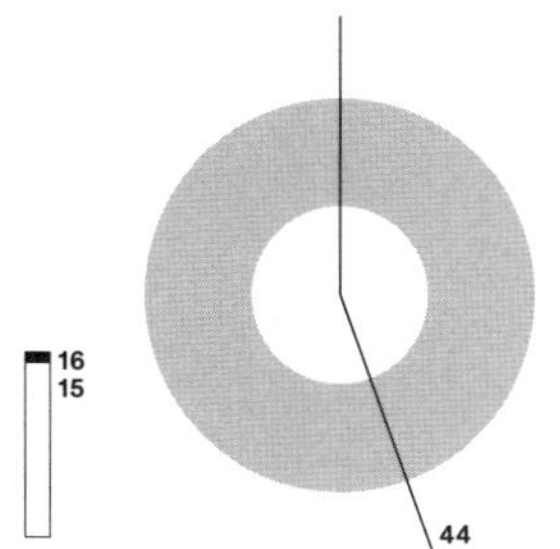

Tower Uniqueness Ratio: 44 %
Floor Area Ratio: 4.92
Tower Stories Range: 15–16

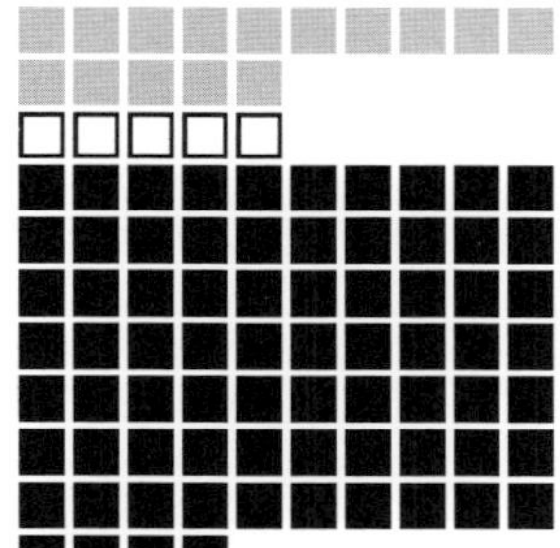

Site Area: 150,143 m²
Footprint Area: 48,941 m²
Built Area: 738,965 m²
Floor Area Ratio: 4.92
Plot Site Coverage: 32.6 %

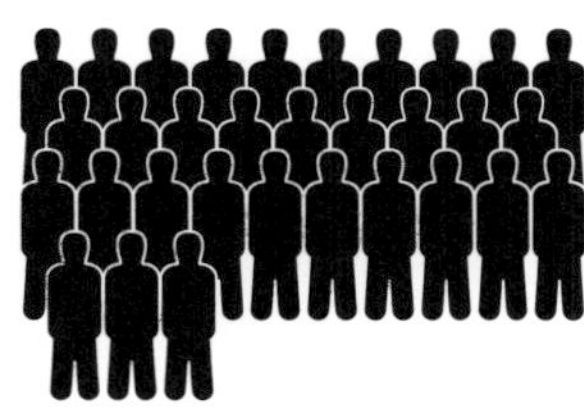

Population: 31,613 Residents

Site Area Density:
4.75 m²/Res

Built Area Density:
23.38 m²/Res

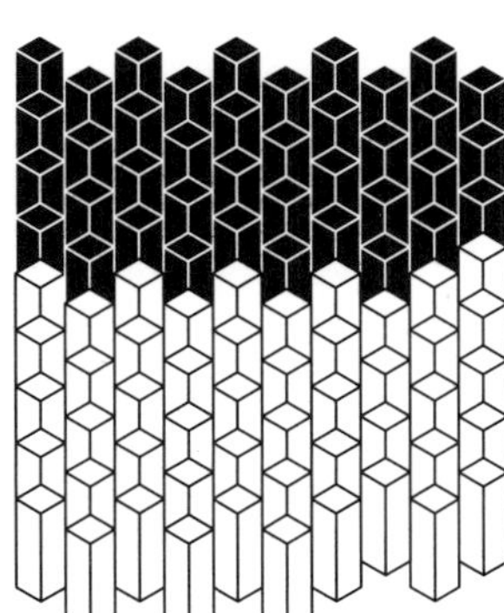

88 Blocks
39 Unique
49 Copies
Tower Uniqueness Ratio: 44 %

Belvedere Garden 1987–1991

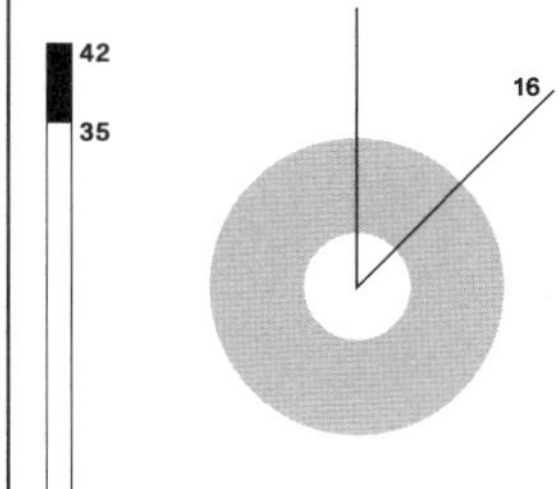

Tower Uniqueness Ratio: 16 %
Floor Area Ratio: 7.67
Tower Stories Range: 35–42

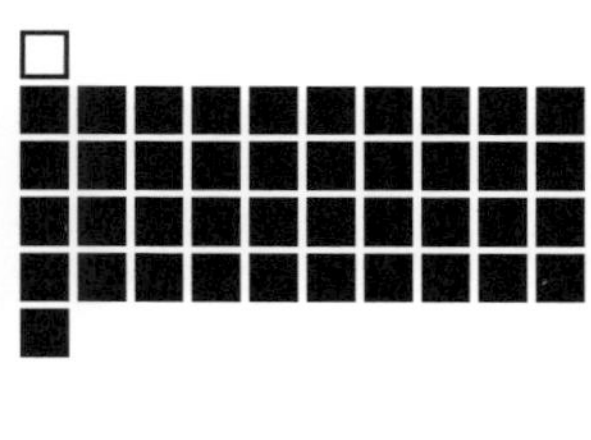

Site Area: 54,710 m²
Footprint Area: 9,920 m²
Built Area: 419,780 m²
Floor Area Ratio: 7.67
Plot Site Coverage: 18.1 %

Population: 20,423 Residents

Site Area Density:
2.68 m²/Res

Built Area Density:
20.55 m²/Res

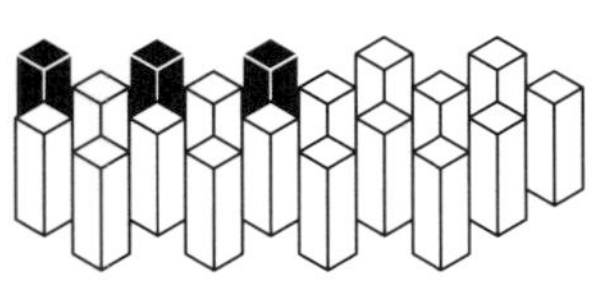

19 Blocks
3 Unique
16 Copies
Tower Uniqueness Ratio: 16 %

Laguna City 1991–1994

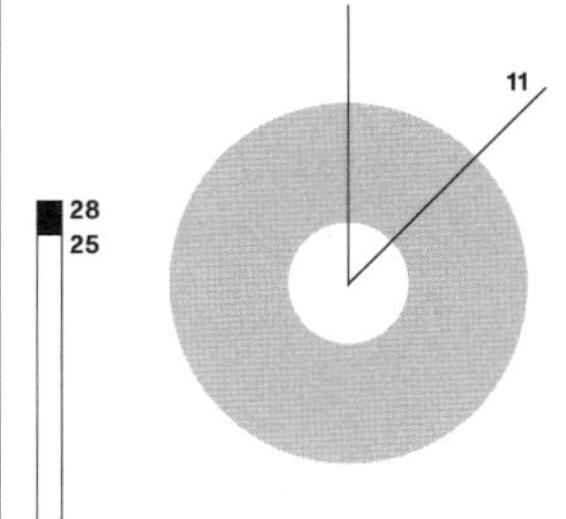

Tower Uniqueness Ratio: 11 %
Floor Area Ratio: 8.79
Tower Stories Range: 25–28

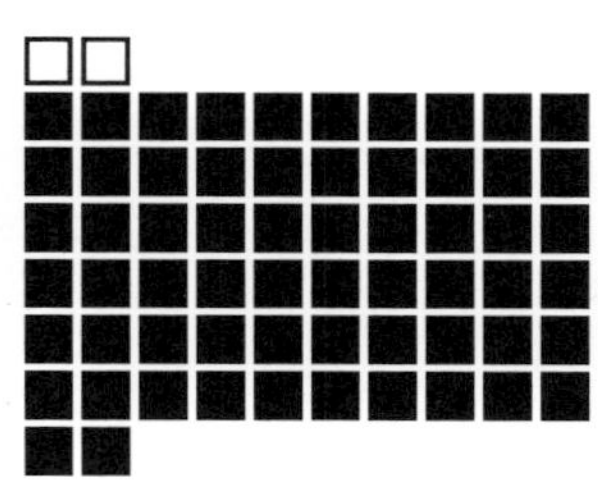

Site Area: 70,620 m²
Footprint Area: 23,388 m²
Built Area: 621,025 m²
Floor Area Ratio: 8.79
Plot Site Coverage: 33.1 %

Population: 23,354 Residents

Site Area Density:
3.02 m²/Res

Built Area Density:
26.59 m²/Res

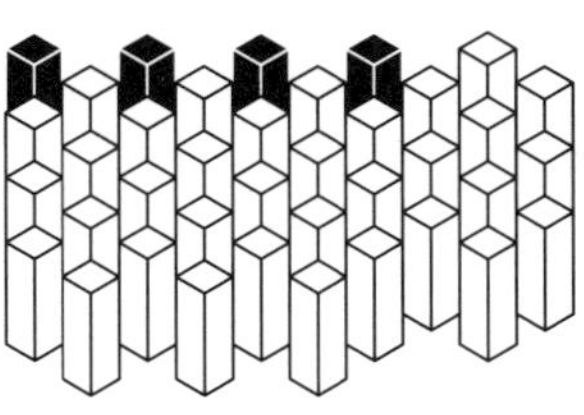

38 Blocks
4 Unique
34 Copies
Tower Uniqueness Ratio: 11 %

South Horizons 1991–1995

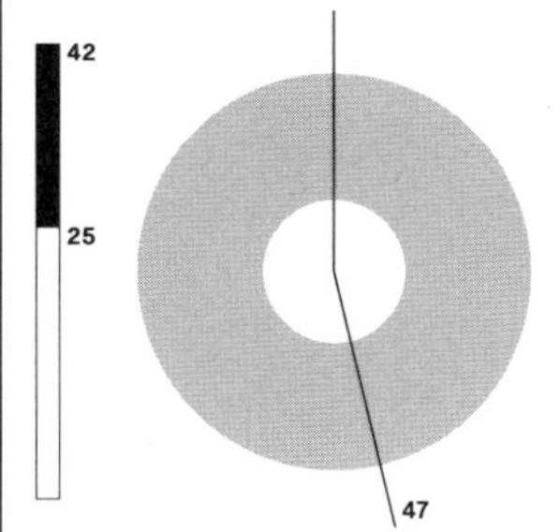

Tower Uniqueness Ratio: 47 %
Floor Area Ratio: 7.69
Tower Stories Range: 25–42

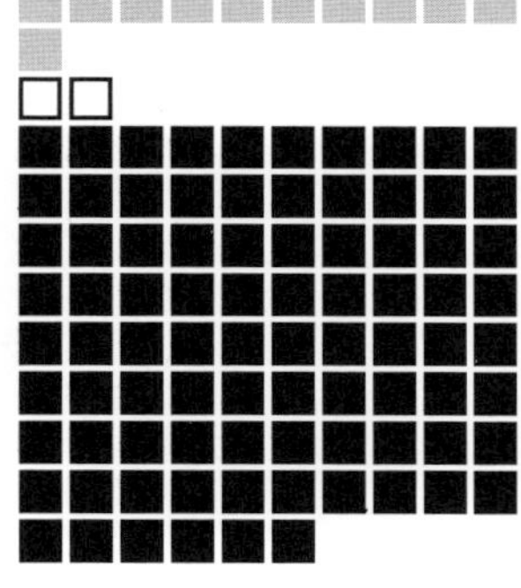

Site Area: 111,623 m²
Footprint Area: 23,867 m²
Built Area: 858,253 m²
Floor Area Ratio: 7.69
Plot Site Coverage: 21.4 %

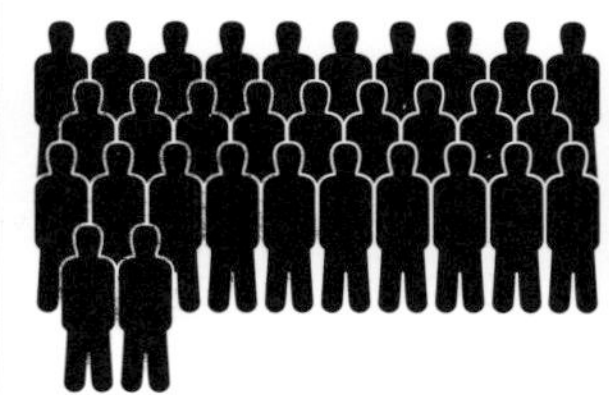

Population: 31,496 Residents

Site Area Density:
3.52 m²/Res

Built Area Density:
27.25 m²/Res

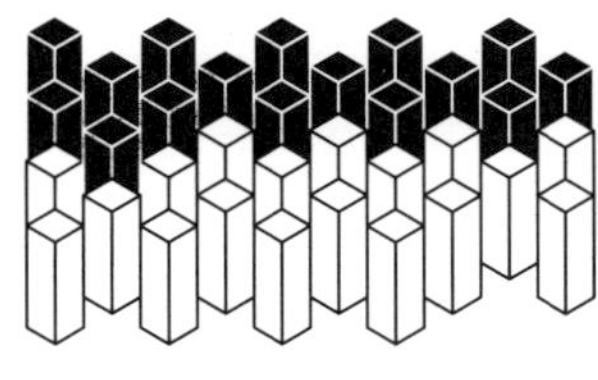

34 Blocks
16 Unique
18 Copies
Tower Uniqueness Ratio: 47 %

Kingswood Villas 1991–1999

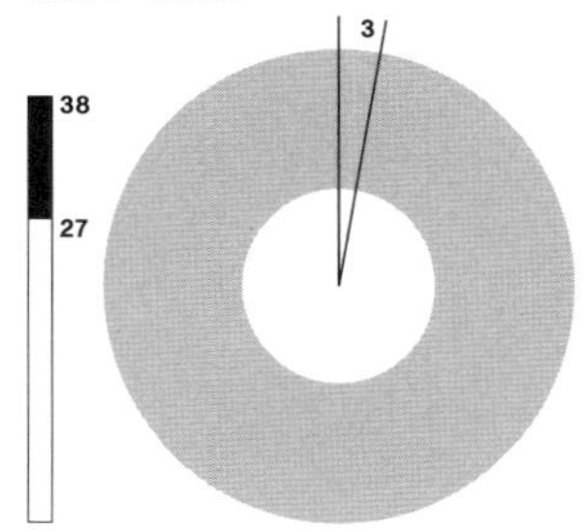

Tower Uniqueness Ratio: 3 %
Floor Area Ratio: 5.97
Tower Stories Range: 27–38

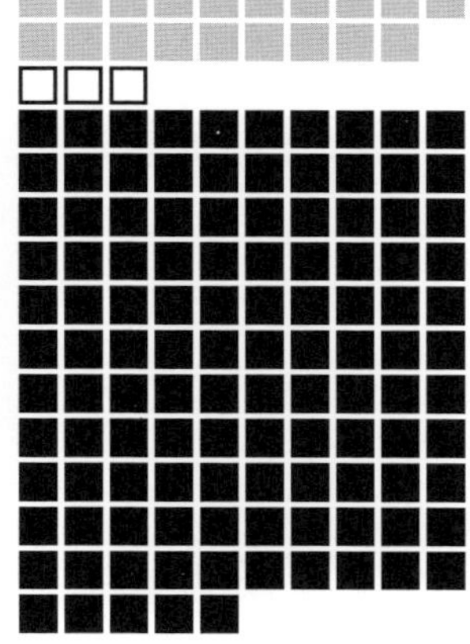

Site Area: 192,367 m²
Footprint Area: 33,565 m²
Built Area: 1,148,732 m²
Floor Area Ratio: 5.97
Plot Site Coverage: 17.4 %

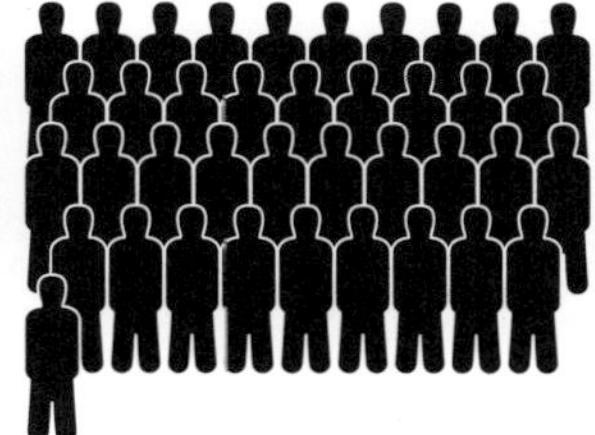

Population: 39,391 Residents

Site Area Density:
4.89 m²/Res

Built Area Density:
29.16 m²/Res

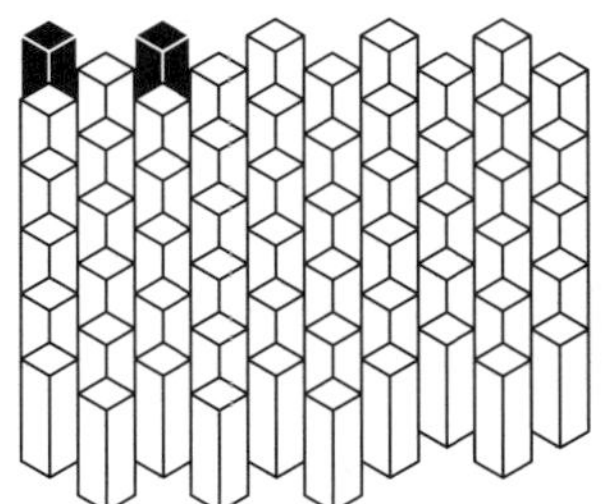

58 Blocks
2 Unique
56 Copies
Tower Uniqueness Ratio: 3 %

Standardization / Deviation

Mei Foo Sun Chuen 99 Blocks	Taikoo Shing 61 Blocks	Heng Fa Chuen 48 Blocks	Kornhill 32 Blocks	City One 52 Blocks
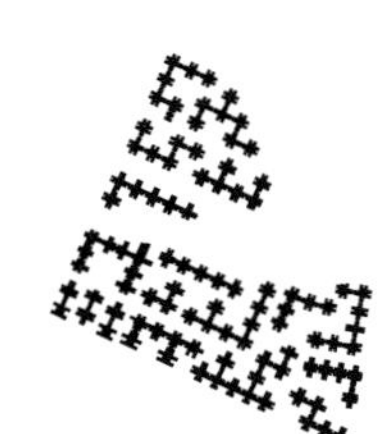	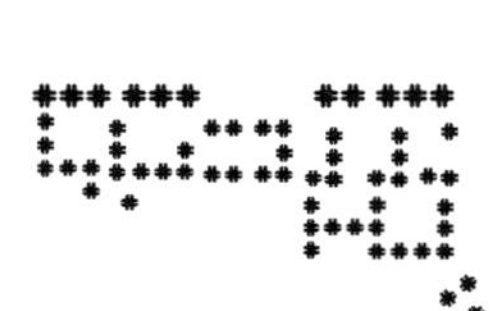	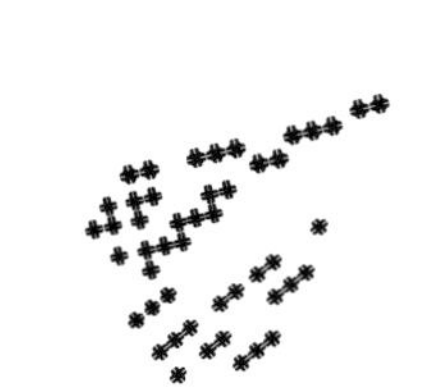	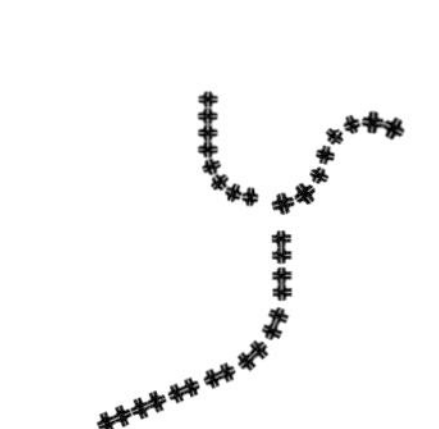	

A

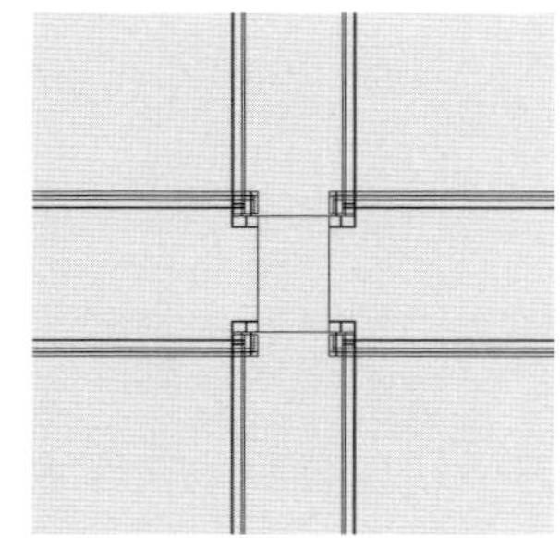 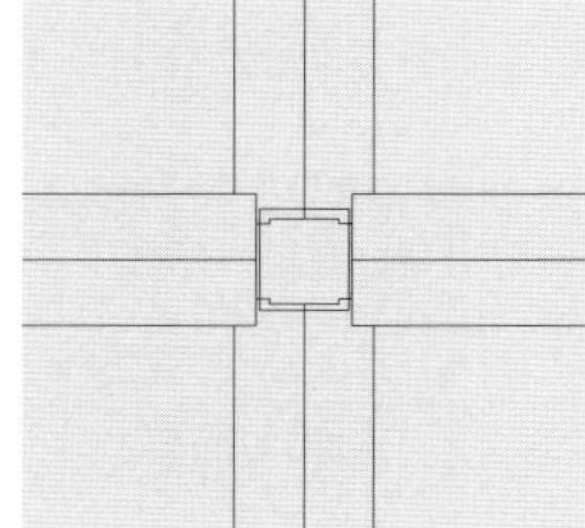 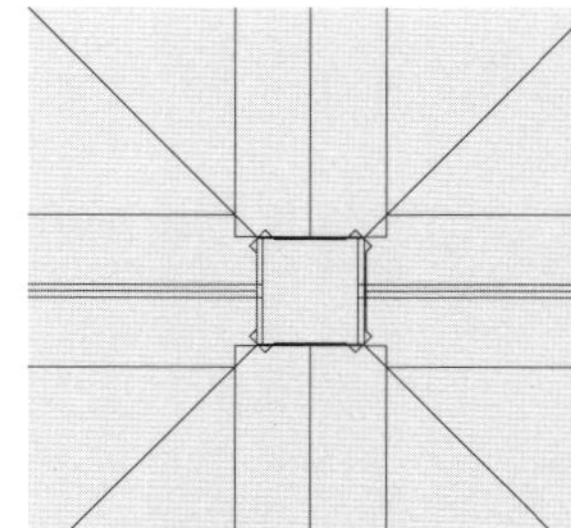 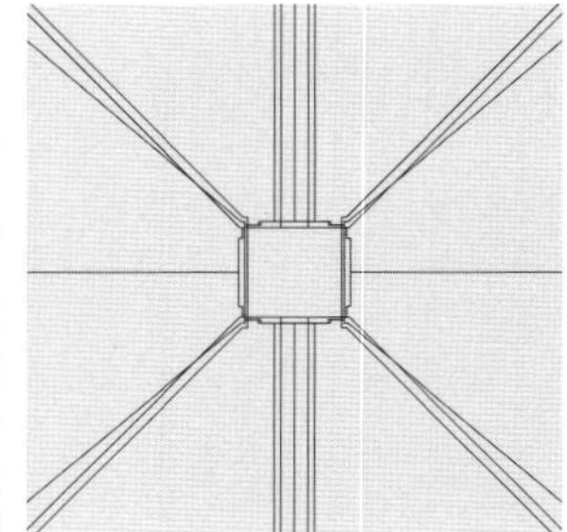 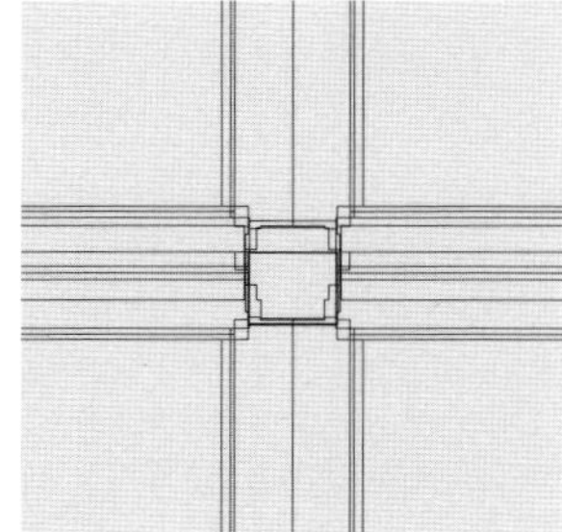

B

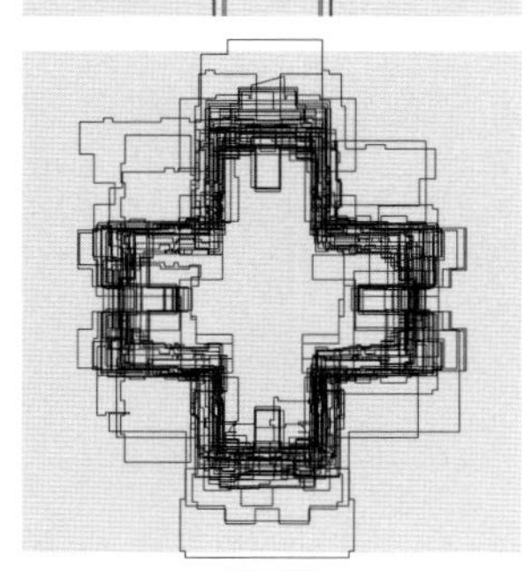 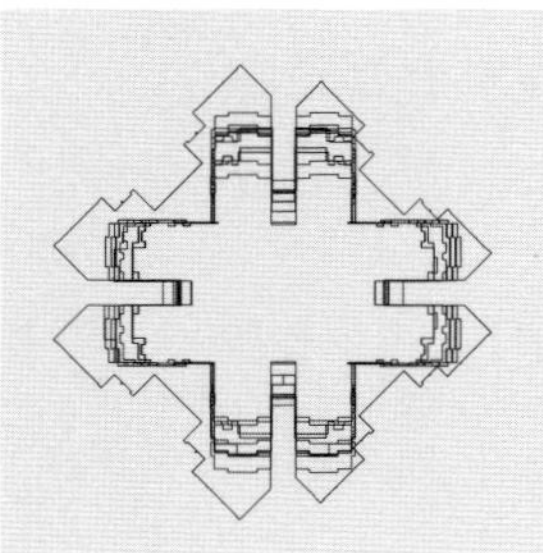 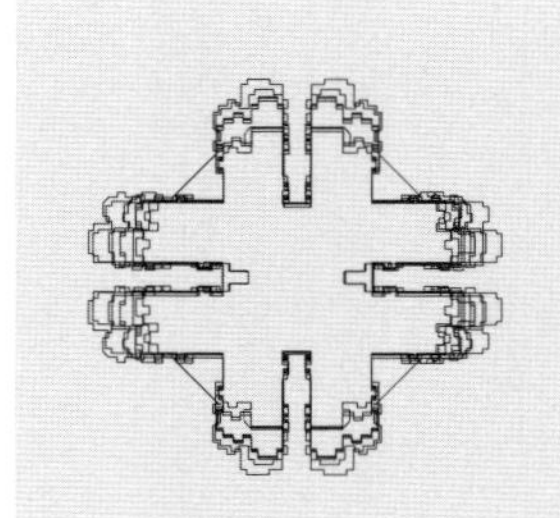 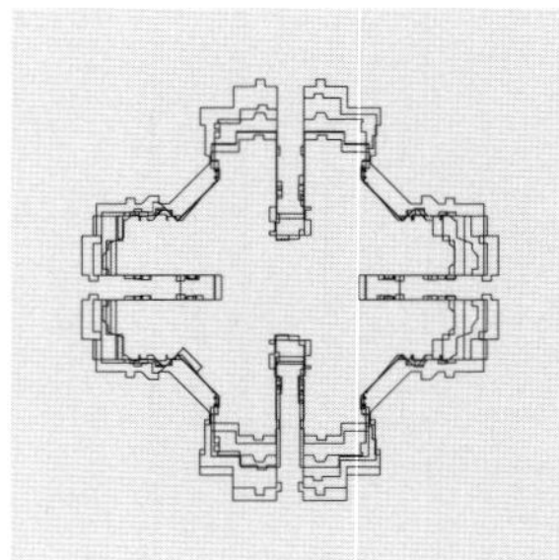 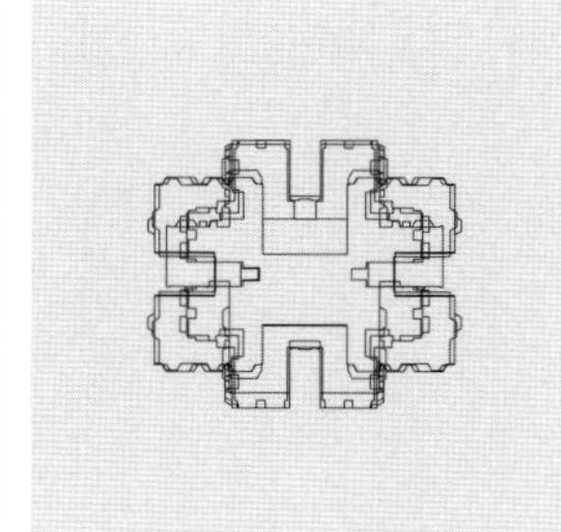

C

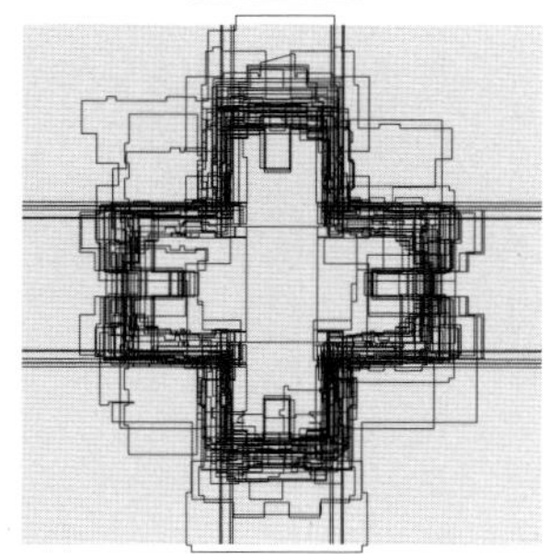 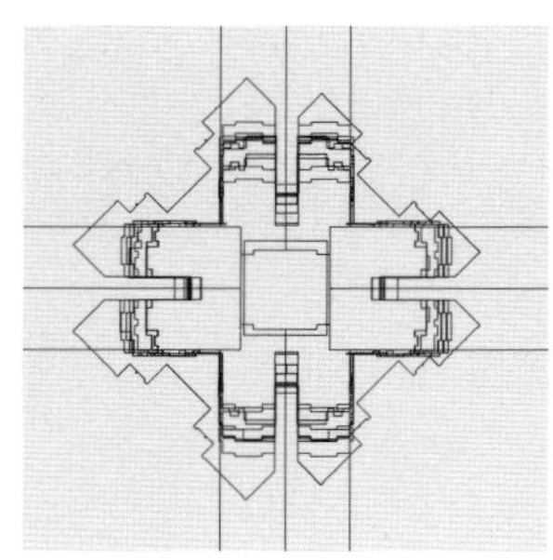 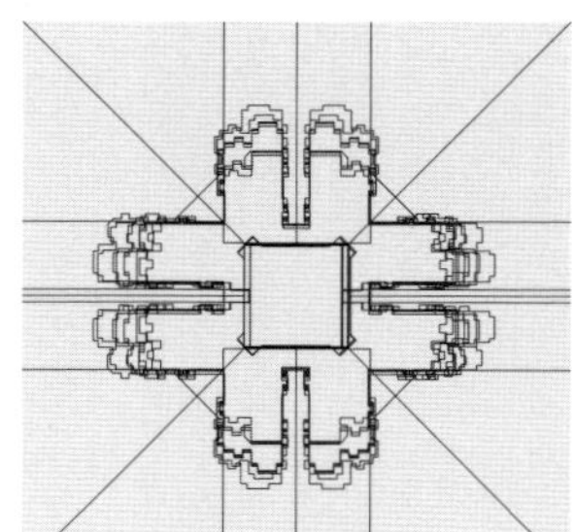 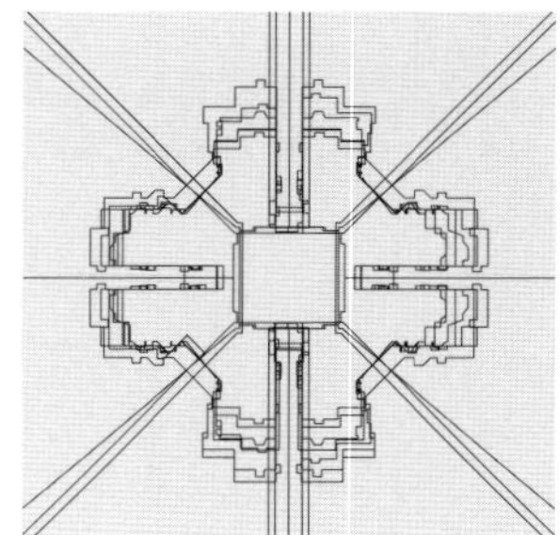 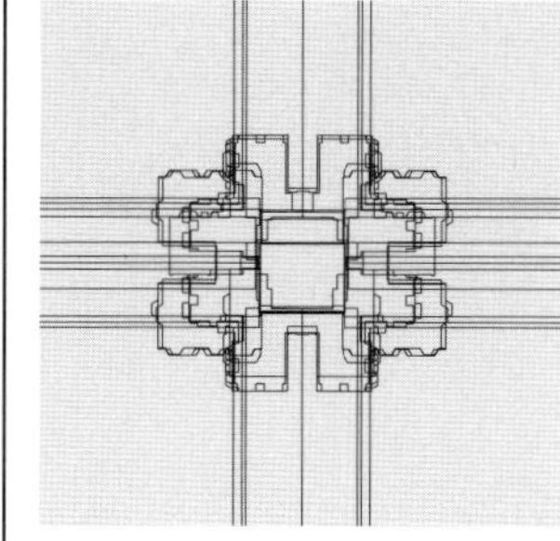

D

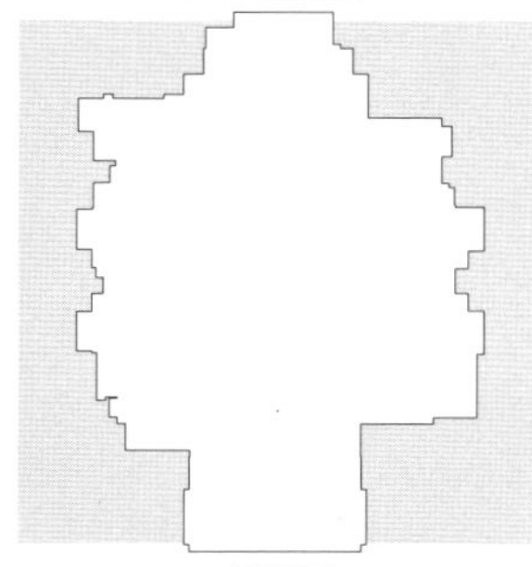 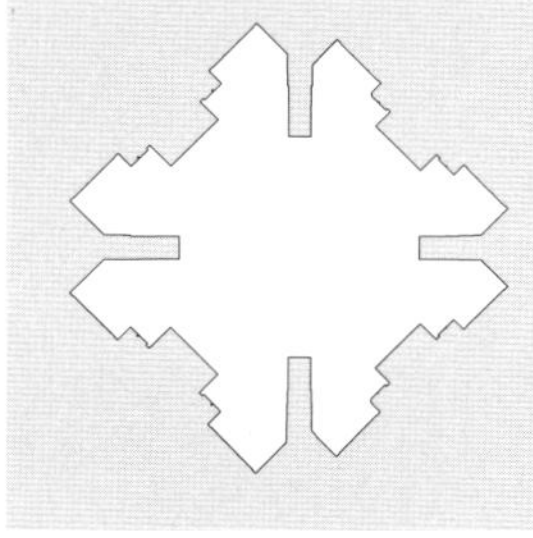 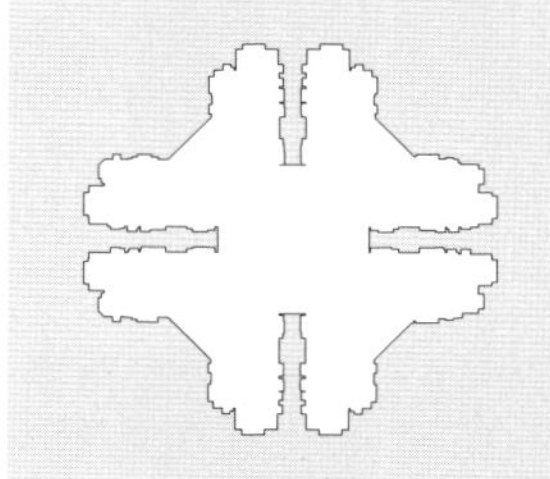 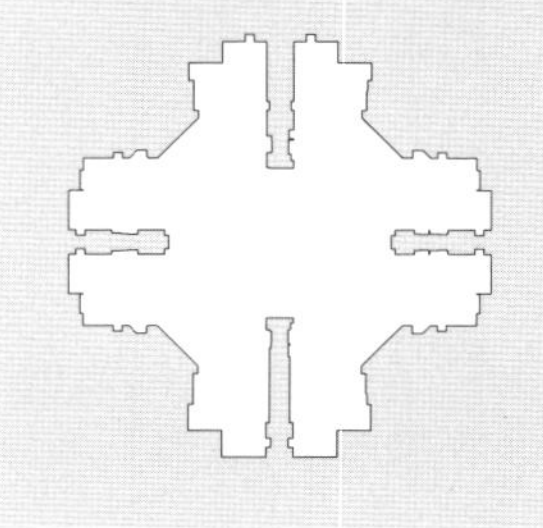 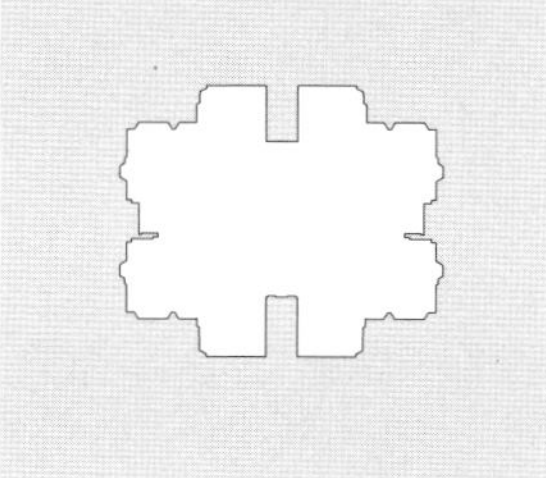

E

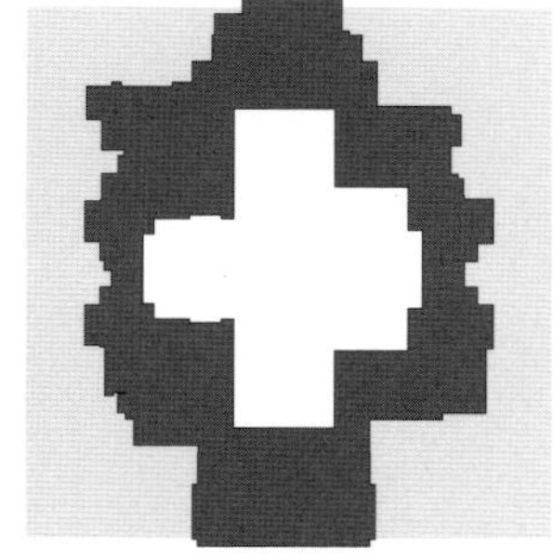 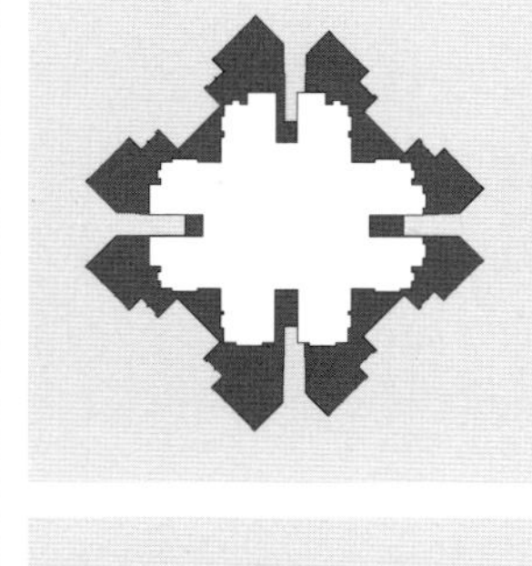 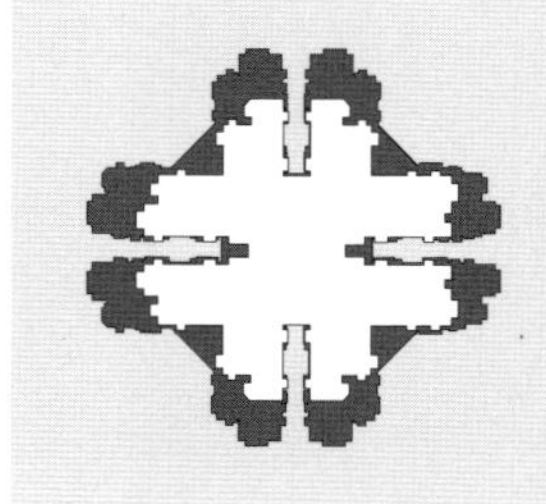 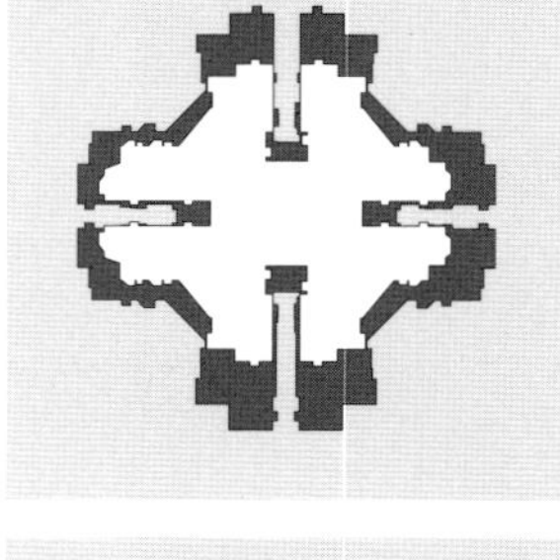 

F

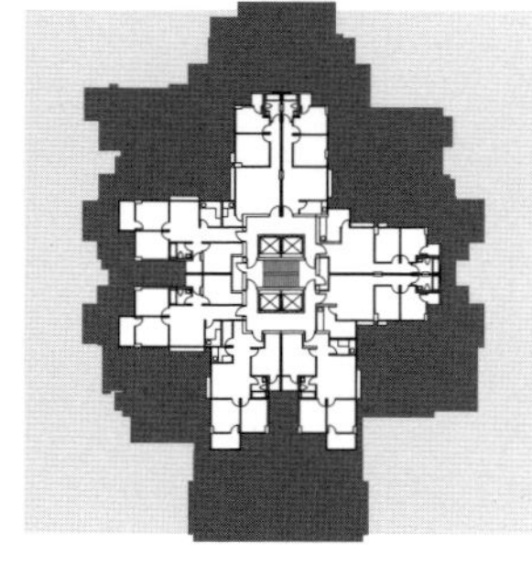 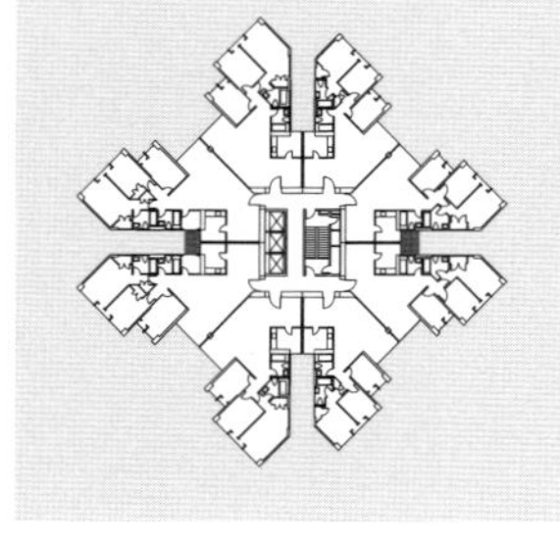 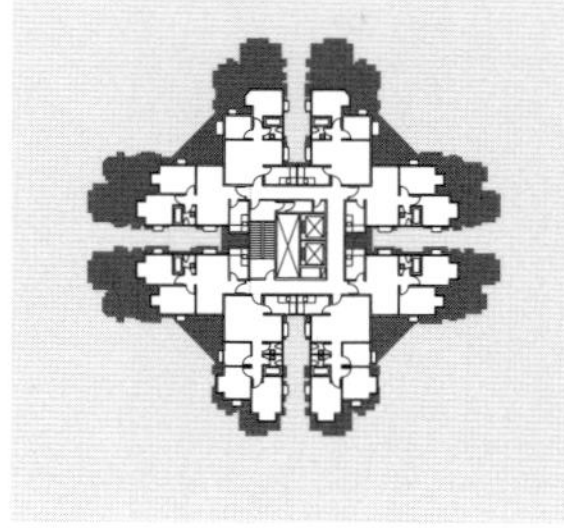 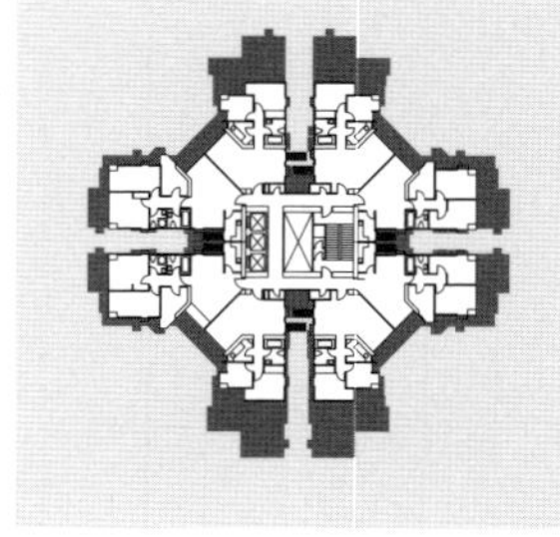 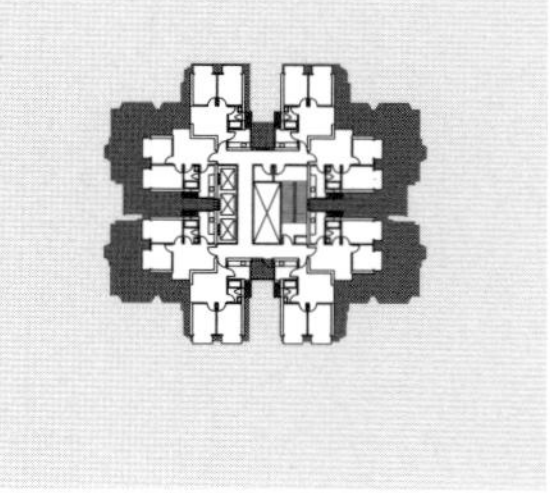

ROW A – Mirroring Lines
ROW B – Tower Superimposition
ROW C – Superimposed Towers with Mirroring Lines
ROW D – Composite Perimeter of all Towers
ROW E – Composite Perimeter with Smallest Towere
ROW F – Composite Perimeter with Dominant Tower

Whampoa Garden
88 Blocks

Belvedere Garden
19 Blocks

Laguna City
38 Blocks

South Horizons
34 Blocks

Kingswood Villas
58 Blocks

A

B

C

D

E

F

Dominant Tower Types

A B C D E F

Mei Foo Sun Chuen Type A 43

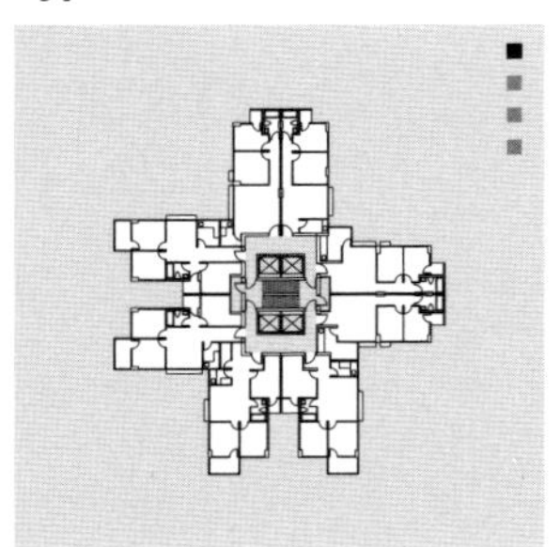

Block 90 3 Copies

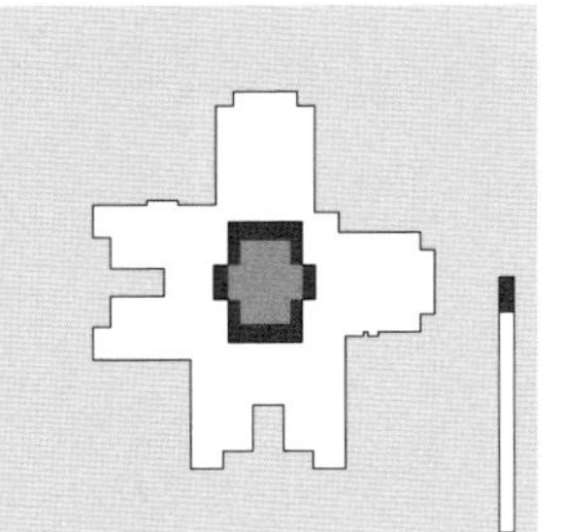

Gross Floor Area: 568.8 m^2
Core Area: 80.9 m^2
Saleable Area: 487.9 m^2
Efficiency: 85.8 %

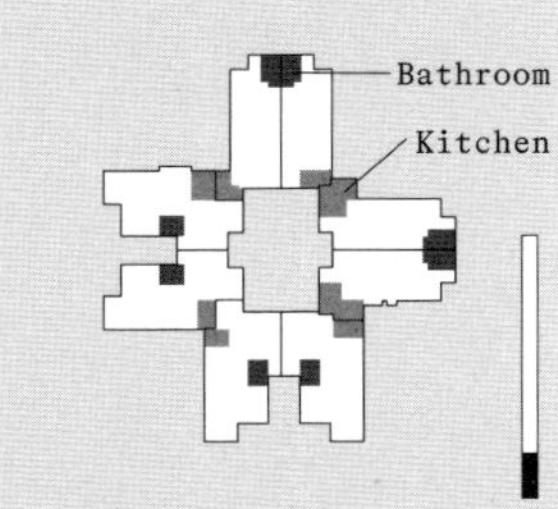

Avg. Unit Area Gross: 61.0 m^2
Avg. Kitchen Area: 5.6 m^2
Avg. Bathroom Area: 4.5 m^2
Service Area Ratio: 16.6 %

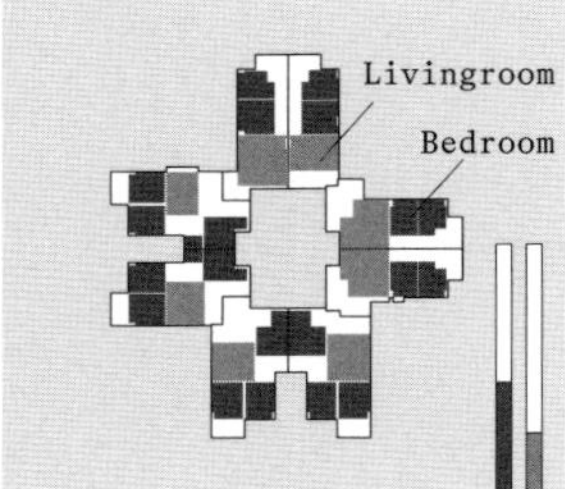

Avg. Bedroom Area: 25.1 m^2
Avg. Livingroom Area: 14.8 m^2
Bedroom Area Ratio: 46.2 %
Livingroom Area Ratio: 27.2 %

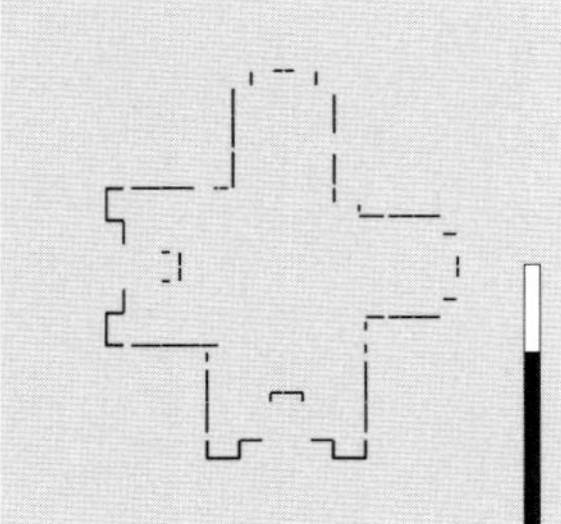

Window Length: 104.4 m
Perimeter Length: 157.0 m
Fenestration Ratio: 66 %

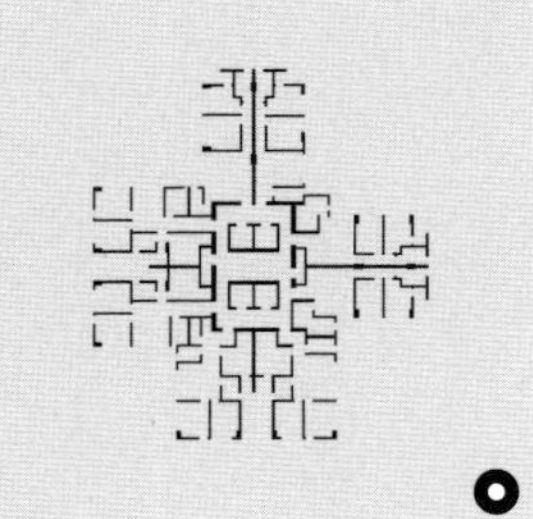

Partition Length: 357.5 m
Gross Floor Area: 568.8 m^2
Degree of Subdivision: 0.63

Taikoo Shing Type B1

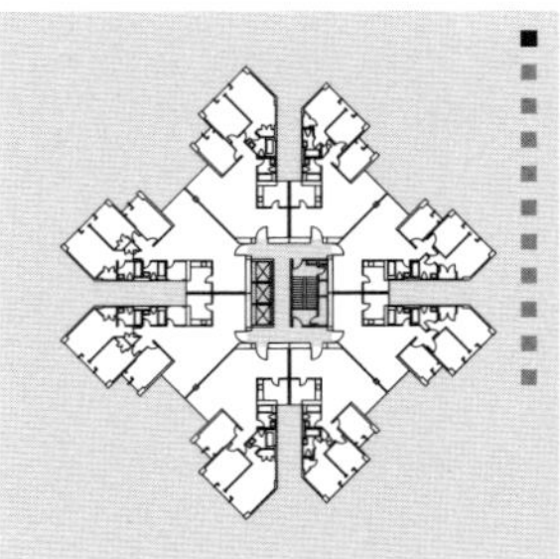

Block 33 10 Copies

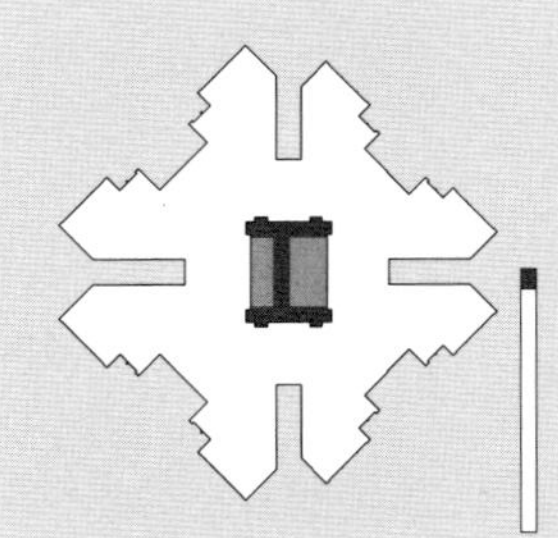

Gross Floor Area: 783.5 m^2
Core Area: 63.6 m^2
Saleable Area: 719.9 m^2
Efficiency: 91.9 %

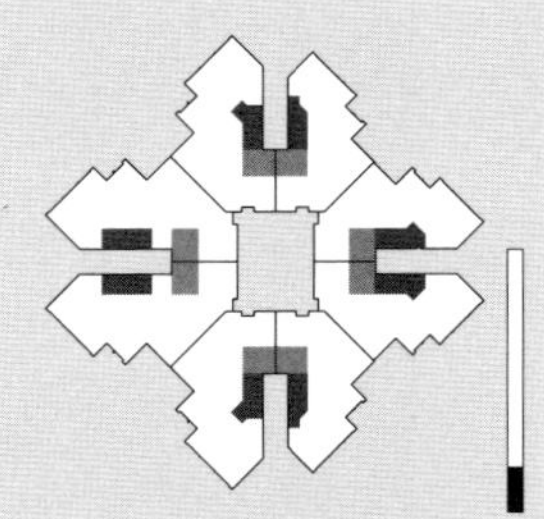

Avg. Unit Area Gross: 90.0 m^2
Avg. Kitchen Area: 6.9 m^2
Avg. Bathroom Area: 8.1 m^2
Service Area Ratio: 16.7 %

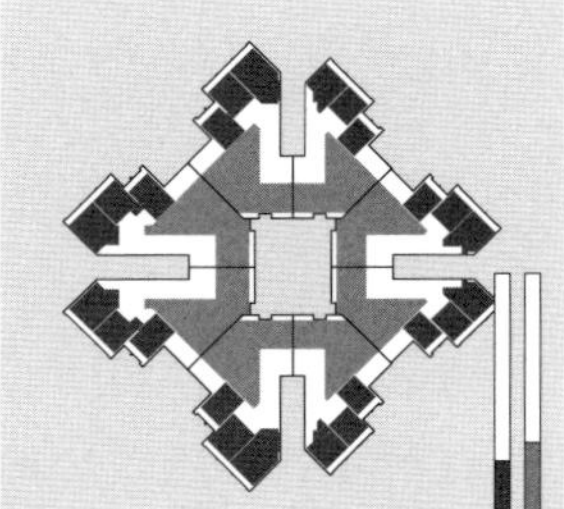

Avg. Bedroom Area: 24.7 m^2
Avg. Livingroom Area: 30.9 m^2
Bedroom Area Ratio: 27.5 %
Livingroom Area Ratio: 34.3 %

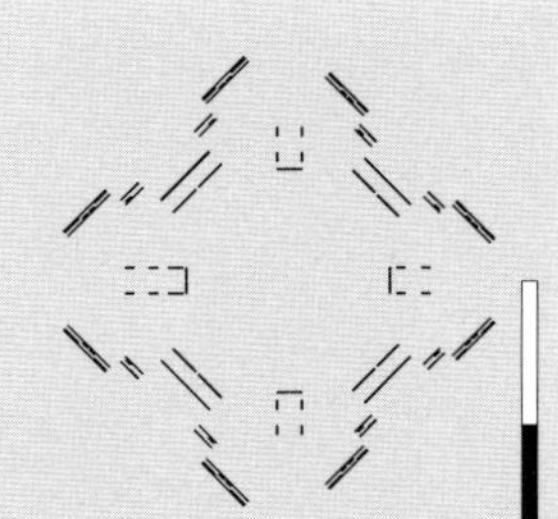

Window Length: 96.5 m
Perimeter Length: 221.1 m
Fenestration Ratio: 45 %

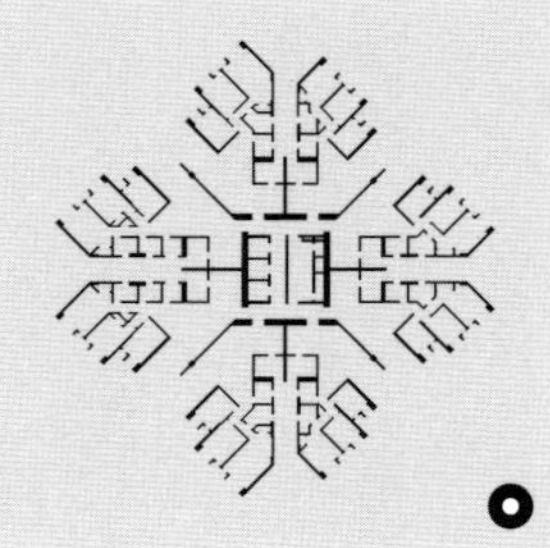

Partition Length: 498.4 m
Gross Floor Area: 783.5 m^2
Degree of Subdivision: 0.64

Heng Fa Chuen Type A3

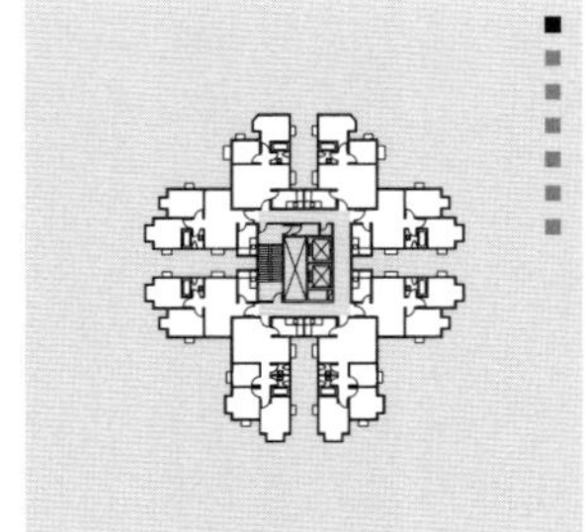

Block 3 6 Copies

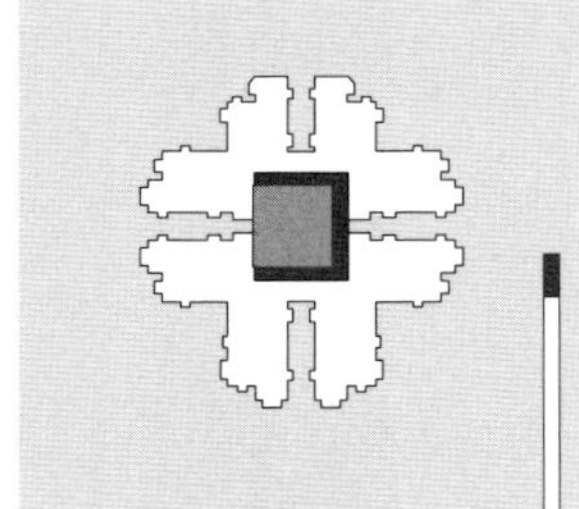

Gross Floor Area: 478.6 m^2
Core Area: 79.5 m^2
Saleable Area: 399.1 m^2
Efficiency: 83.4 %

Avg. Unit Area Gross: 49.9 m^2
Avg. Kitchen Area: 5.5 m^2
Avg. Bathroom Area: 4.3 m^2
Service Area Ratio: 19.8 %

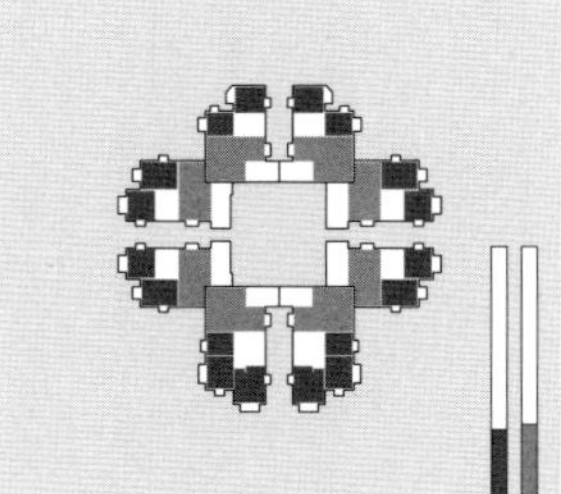

Avg. Bedroom Area: 14.5 m^2
Avg. Livingroom Area: 15.5 m^2
Bedroom Area Ratio: 29.1 %
Livingroom Area Ratio: 31.1 %

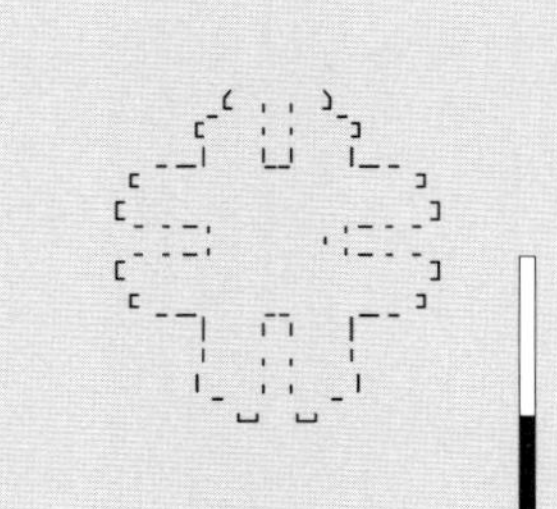

Window Length: 85.7 m
Perimeter Length: 223.51 m
Fenestration Ratio: 39 %

Partition Length: 452.4 m
Gross Floor Area: 478.6 m^2
Degree of Subdivision: 0.95

Kornhill Type A3

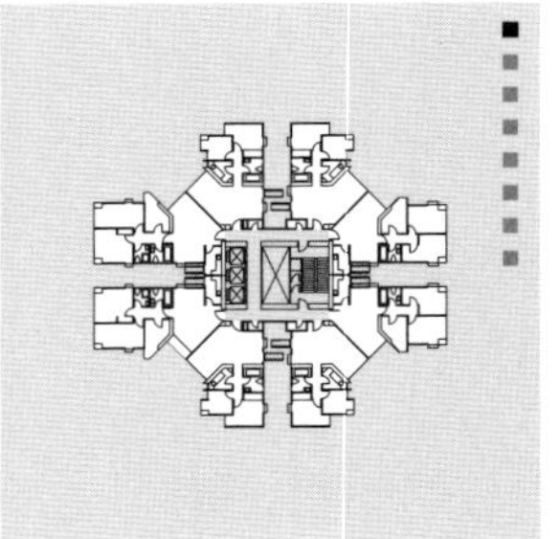

Block 9 7 Copies

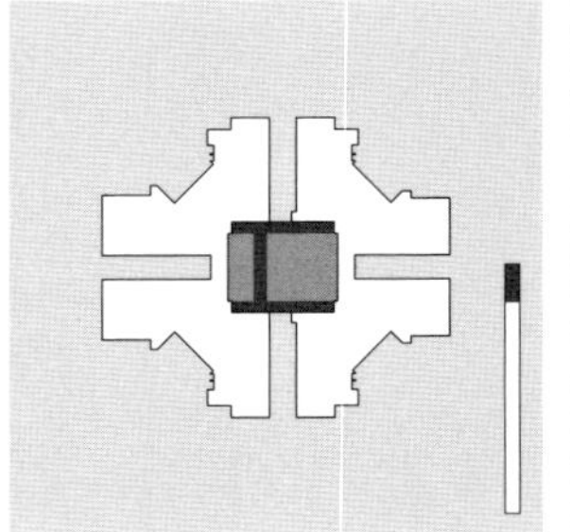

Gross Floor Area: 546.3 m^2
Core Area: 84.5 m^2
Saleable Area: 461.7 m^2
Efficiency: 84.5 %

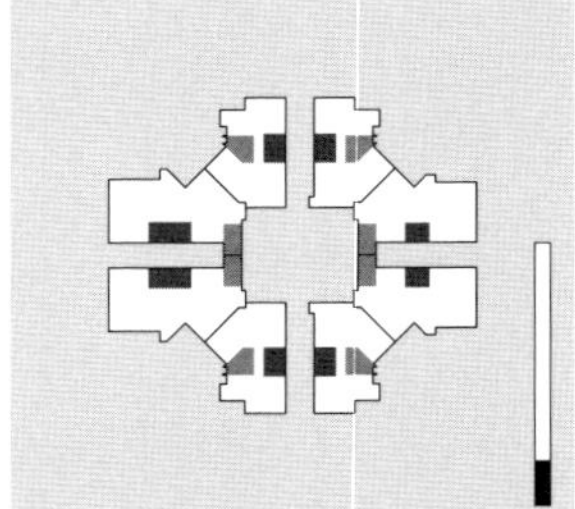

Avg. Unit Area Gross: 57.7 m^2
Avg. Kitchen Area: 4.6 m^2
Avg. Bathroom Area: 5.2 m^2
Service Area Ratio: 17.0 %

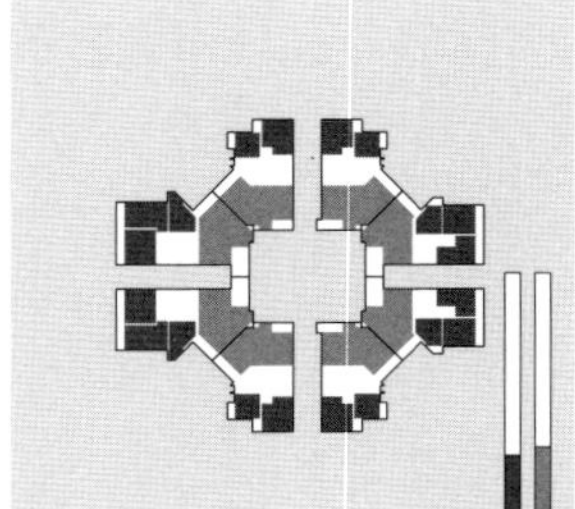

Avg. Bedroom Area: 16.9 m^2
Avg. Livingroom Area: 18.7 m^2
Bedroom AREA RATIO: 29.3 %
Livingroom AREA RATIO: 32.4 %

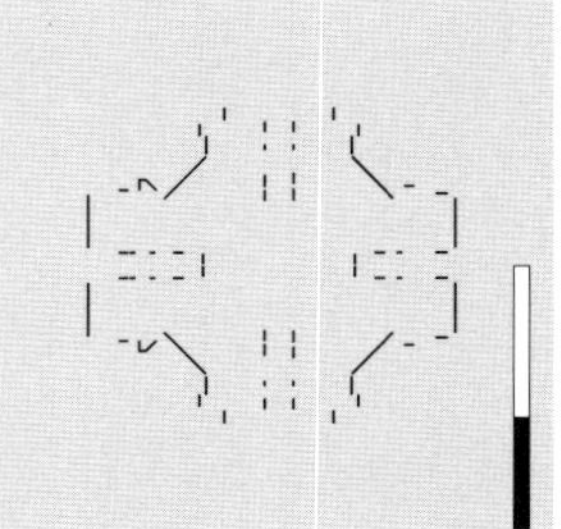

Window Length: 87.9 m
Perimeter Length: 206.7 m
Fenestration Ratio: 42 %

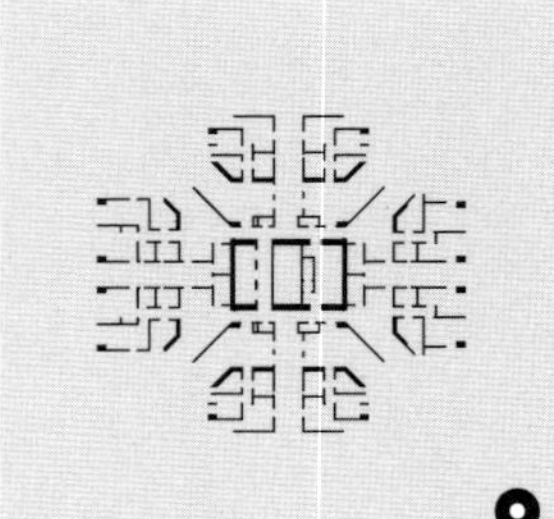

Partition Length: 367.8 m
Gross Floor Area: 546.3 m^2
Degree of Subdivision: 0.67

City One Type A5

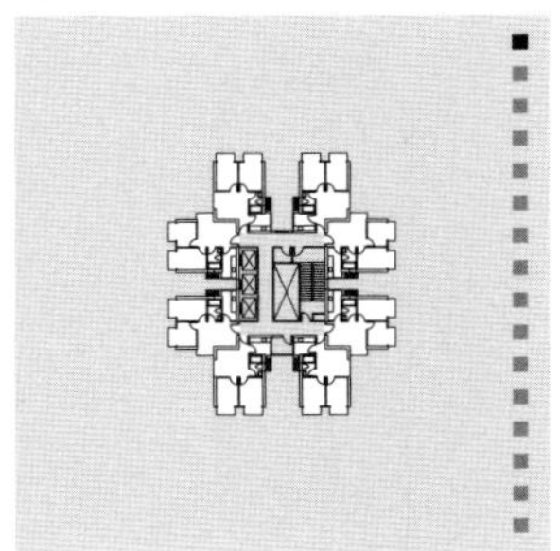

Block 37 15 Copies

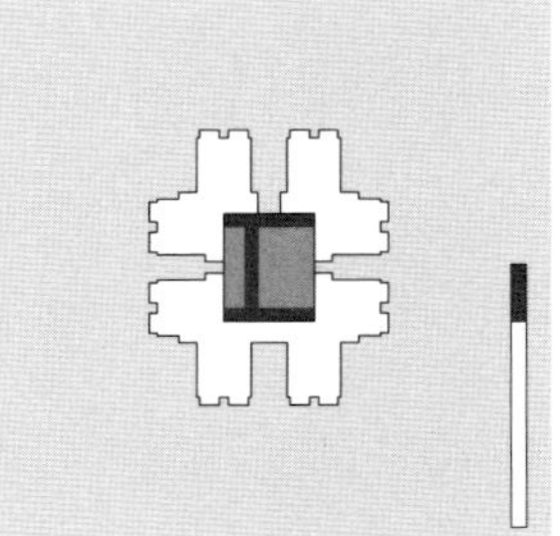

Gross Floor Area: 340.4 m^2
Core Area: 75.6 m^2
Saleable Area: 264.8 m^2
Efficiency: 77.8 %

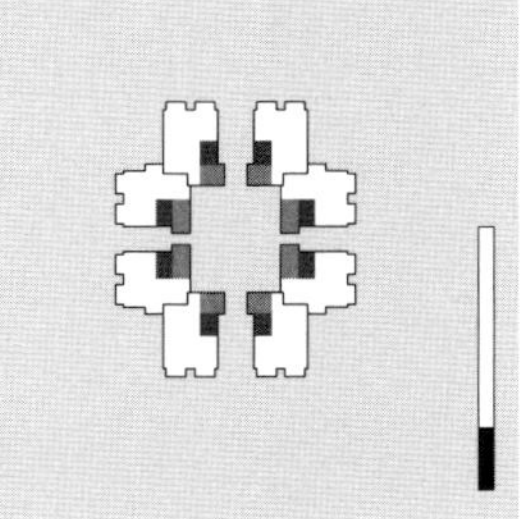

Avg. Unit Area Gross: 33.1 m^2
Avg. Kitchen Area: 4.3 m^2
Avg. Bathroom Area: 3.4 m^2
Service Area Ratio: 23.5 %

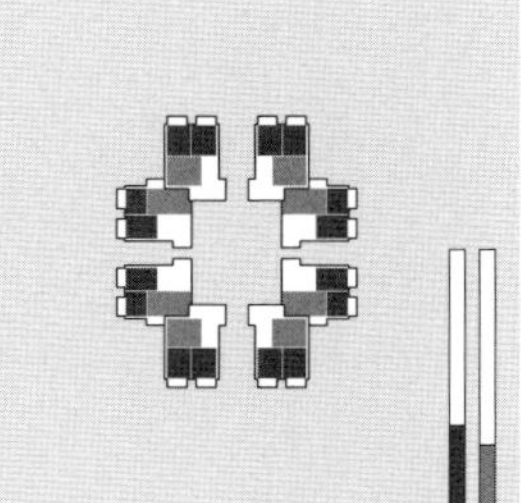

Avg. Bedroom Area: 10.5 m^2
Avg. Livingroom Area: 8.1 m^2
Bedroom Area Ratio: 31.7 %
Livingroom Area Ratio: 24.3 %

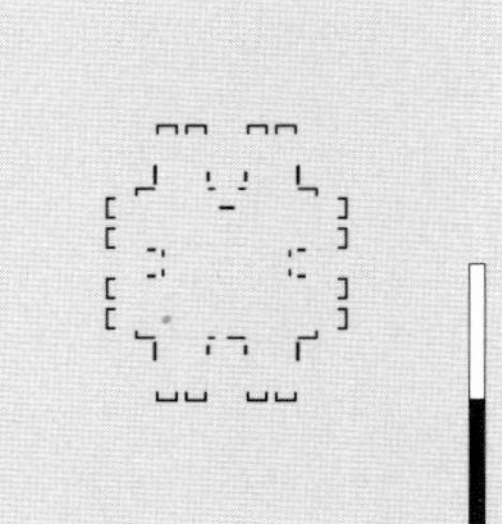

Window Length: 74.9 m
Perimeter Length: 154.5 m
Fenestration Ratio: 48 %

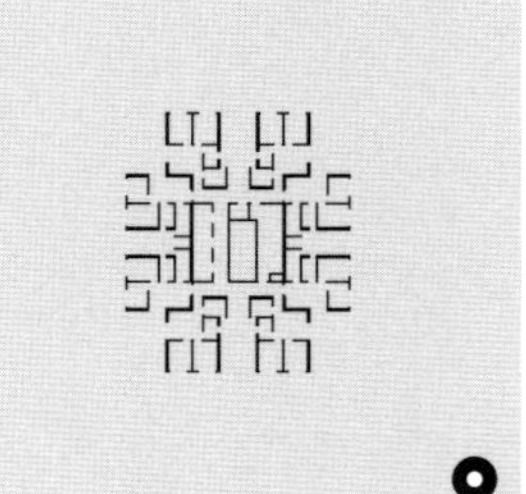

Partition Length: 226.2 m
Gross Floor Area: 344.2 m^2
Degree of Subdivision: 0.66

ROW A – Floor Plan
ROW B – Core + Saleable Are
ROW C – Kitchen + Bathroom Area
ROW D – Living + Bedroom Area
ROW E – Fenestration Ratio
ROW F – Degree of Subdivision

● – Degree of Subdivision

Whampoa Garden Type D1

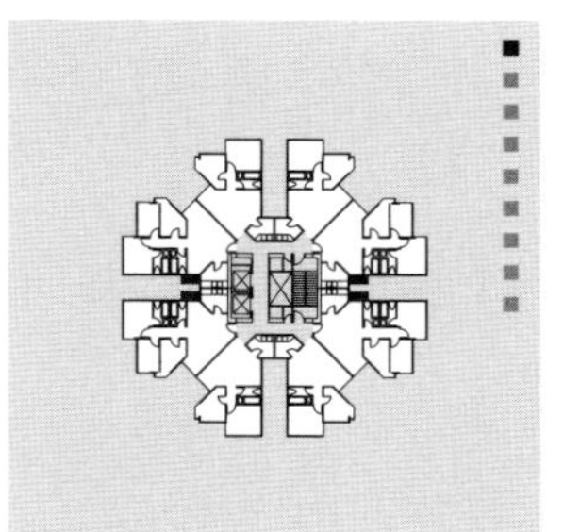

Block 38 8 Copies

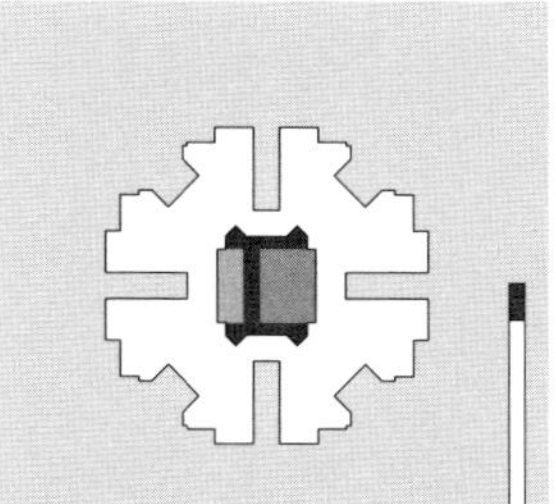

Gross Floor Area: 522.6 m²
Core Area: 75.6 m²
Saleable Area: 447.0 m²
Efficiency: 85.5 %

Avg. Unit Area Gross: 55.9 m²
Avg. Kitchen Area: 5.4 m²
Avg. Bathroom Area: 6.3 m²
Service Area Ratio: 21.0 %

Avg. Bedroom Area: 18.7 m²
Avg. Livingroom Area: 18.4 m²
Bedroom Area Ratio: 33.5 %
Livingroom Area Ratio: 32.9 %

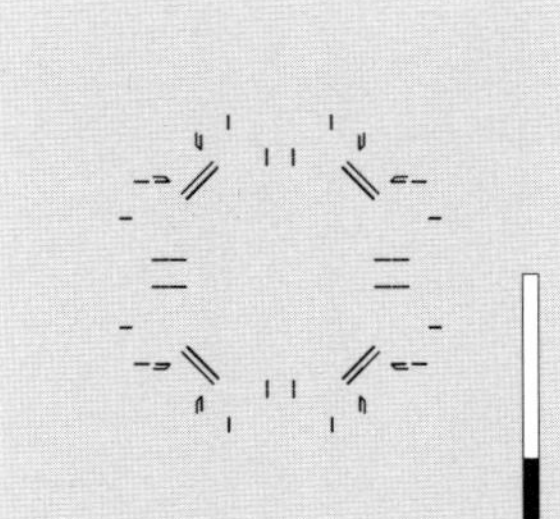

Window Length: 52.8 m
Perimeter Length: 181.0 m
Fenestration Ratio: 30 %

Partition Length: 373.5 m
Gross Floor Area: 522.6 m²
Degree of Subdivision: 0.71

Belvedere Garden Type A2

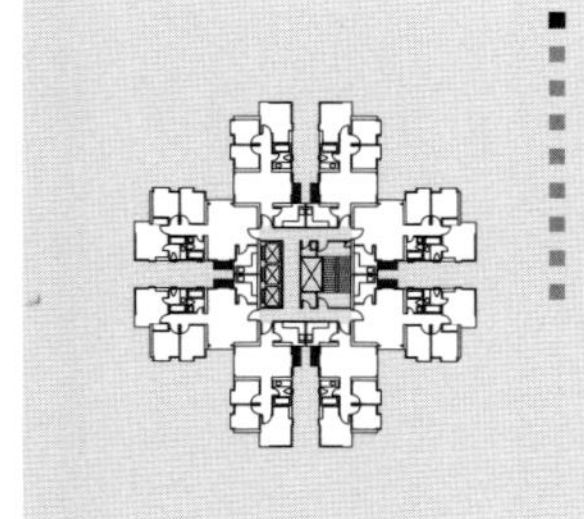

Block 4 8 Copies

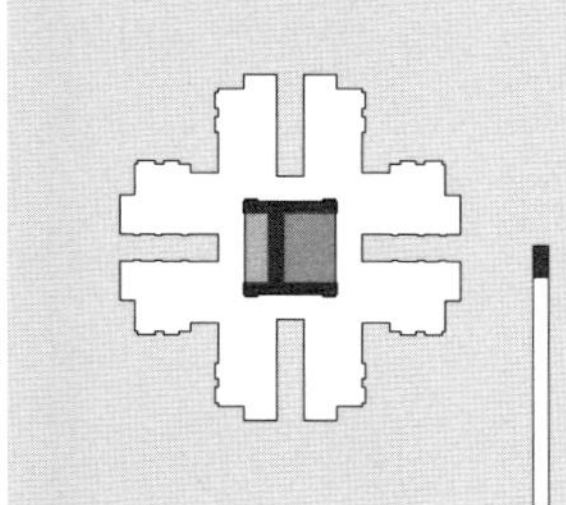

Gross Floor Area: 532.9 m²
Core Area: 67.0 m²
Saleable Area: 466.0 m²
Efficiency: 87.4 %

Avg. Unit Area Gross: 58.2 m²
Avg. Kitchen Area: 6.2 m²
Avg. Bathroom Area: 6.3 m²
Service Area Ratio: 21.4 %

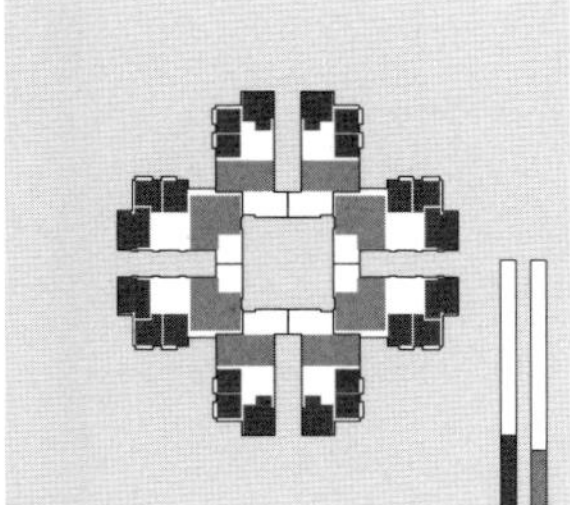

Avg. Bedroom Area: 19.2 m²
Avg. Livingroom Area: 15.9 m²
Bedroom Area Ratio: 32.9 %
Livingroom Area Ratio: 27.2 %

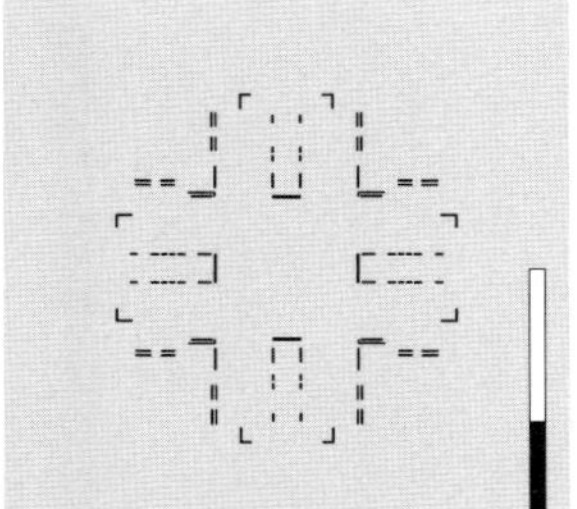

Window Length: 85.8 m
Perimeter Length: 209.8 m
Fenestration Ratio: 42 %

Partition Length: 342.0 m
Gross Floor Area: 533.0 m²
Degree of Subdivision: 0.64

Laguna City Type A2

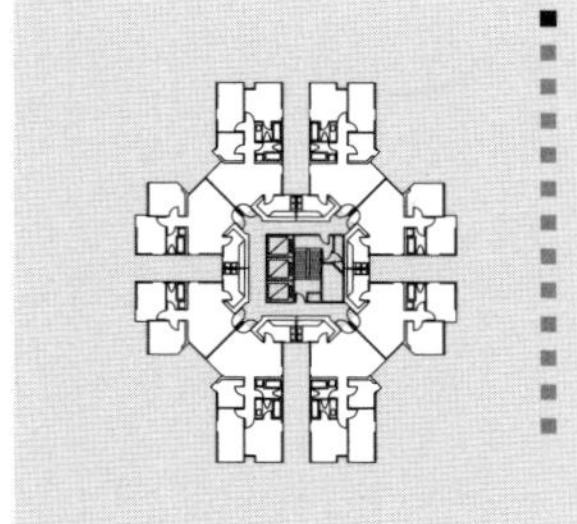

Block 2 12 Copies

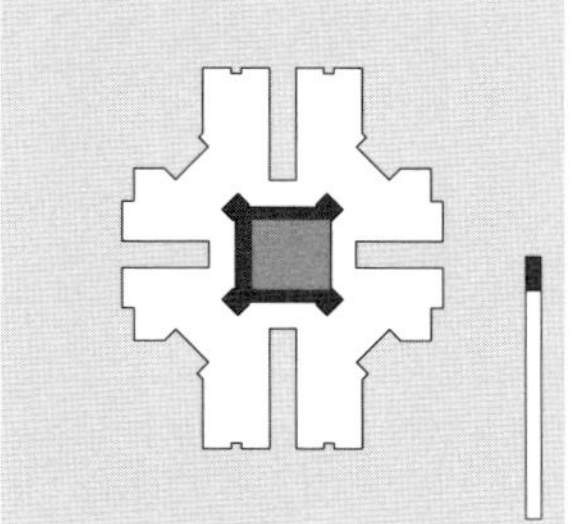

Gross Floor Area: 591.8 m²
Core Area: 80.0 m²
Saleable Area: 511.8 m²
Efficiency: 86.5 %

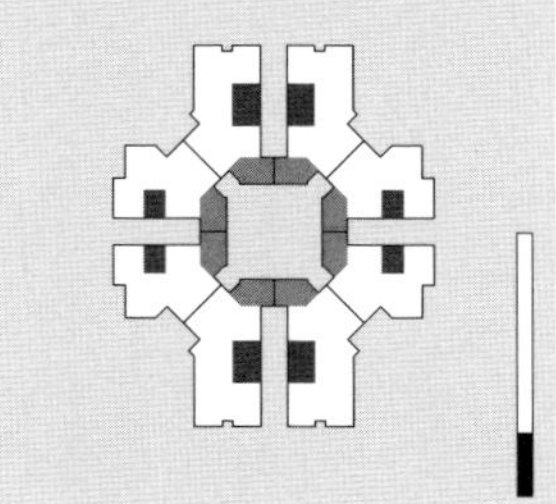

Avg. Unit Area Gross: 64.0 m²
Avg. Kitchen Area: 8.1 m²
Avg. Bathroom Area: 7.1 m²
Service Area Ratio: 23.8 %

Avg. Bedroom Area: 20.3 m²
Avg. Livingroom Area: 20.6 m²
Bedroom Area Ratio: 31.7 %
Livingroom Area Ratio: 32.3 %

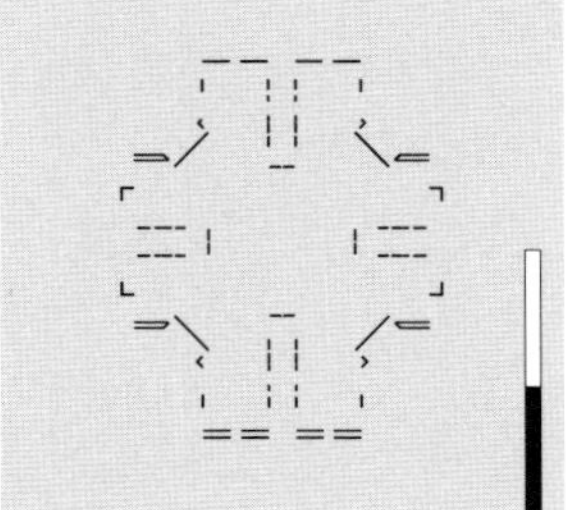

Window Length: 99.1 m
Perimeter Length: 202.4 m
Fenestration Ratio: 48 %

Partition Length: 357.1 m
Gross Floor Area: 202.5 m²
Degree of Subdivision: 0.60

South Horizons Type B3

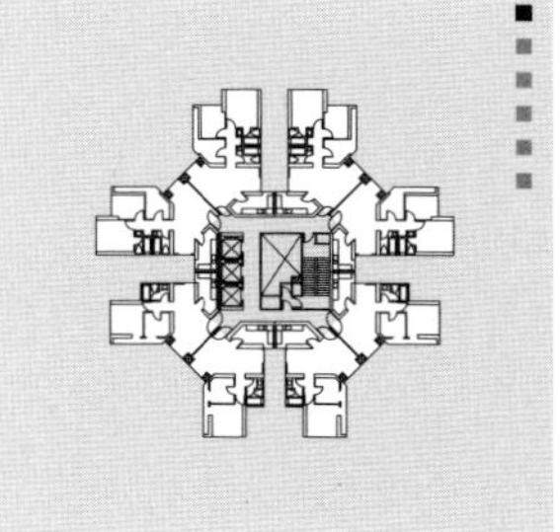

Block 29 5 Copies

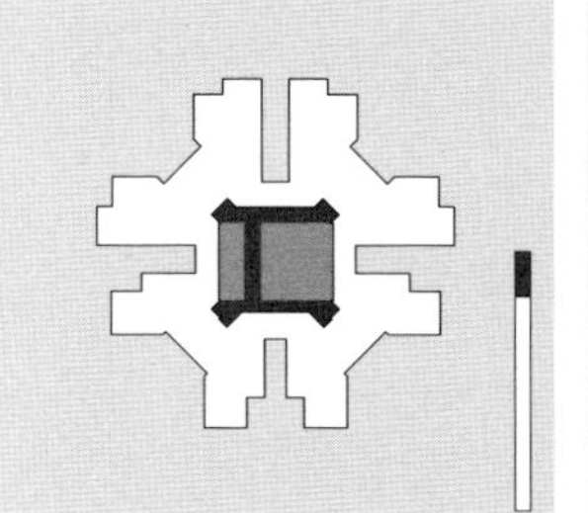

Gross Floor Area: 569.9 m²
Core Area: 101.5 m²
Saleable Area: 468.4 m²
Efficiency: 82.2 %

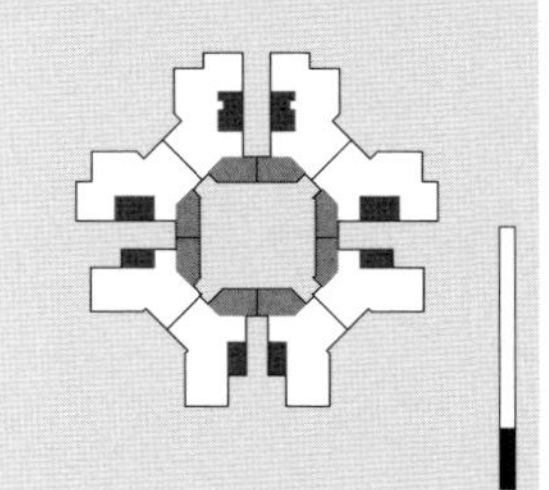

Avg. Unit Area Gross: 58.6 m²
Avg. Kitchen Area: 8.5 m²
Avg. Bathroom Area: 6.3 m²
Service Area Ratio: 25.4 %

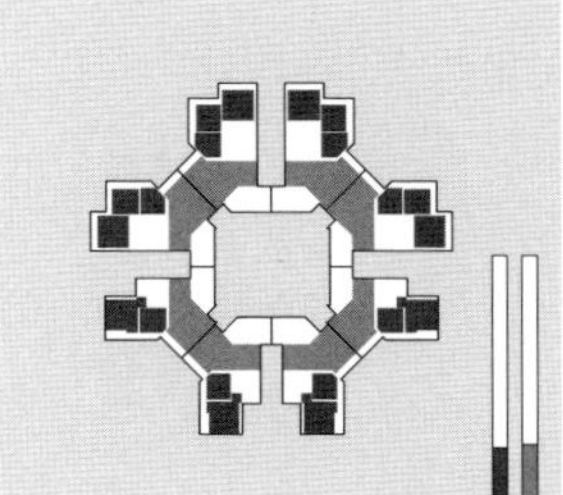

Avg. Bedroom Area: 15.5 m²
Avg. Livingroom Area: 16.2 m²
Bedroom AREA RATIO: 26.4 %
Livingroom AREA RATIO: 27.7 %

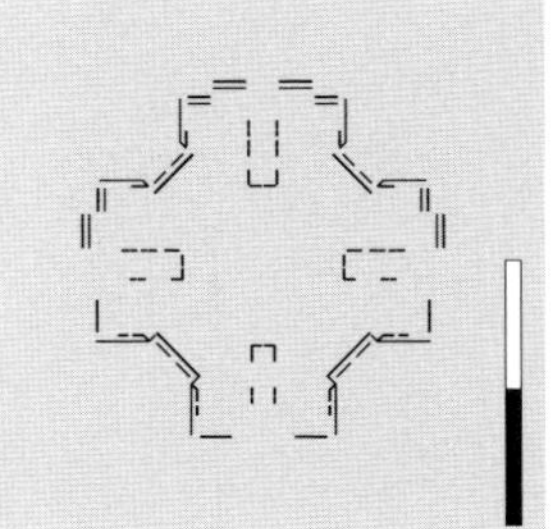

Window Length: 95.6 m
Perimeter Length: 187.4 m
Fenestration Ratio: 51 %

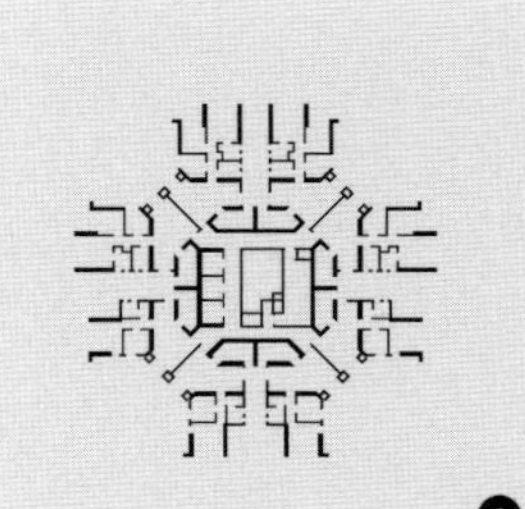

Partition Length: 372.3 m
Gross Floor Area: 570.0 m²
Degree of Subdivision: 0.65

Kingswood Villas Type A1

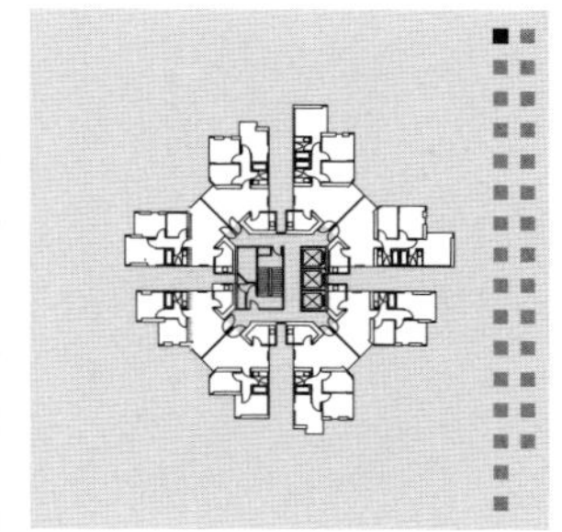

Block 1 29 Copies

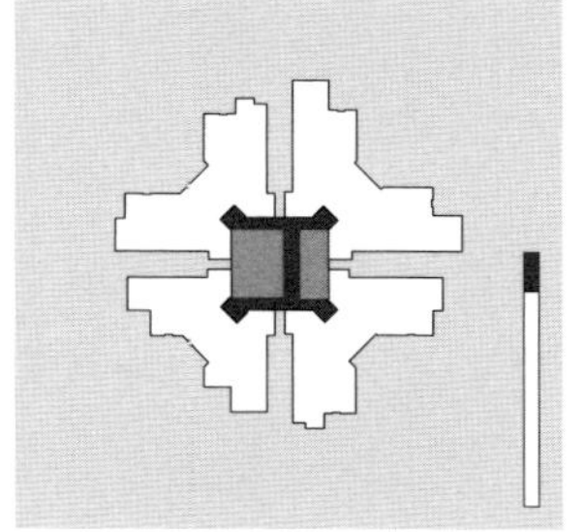

Gross Floor Area: 516.2 m²
Core Area: 81.7 m²
Saleable Area: 434.5 m²
Efficiency: 84.2 %

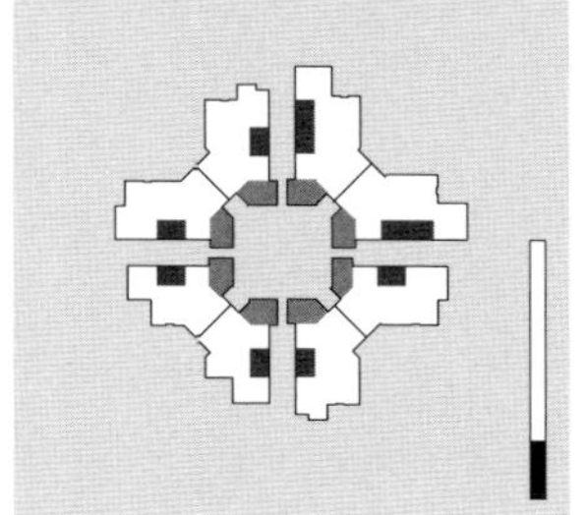

Avg. Unit Area Gross: 54.3 m²
Avg. Kitchen Area: 6.7 m²
Avg. Bathroom Area: 5.5 m²
Service Area Ratio: 22.4 %

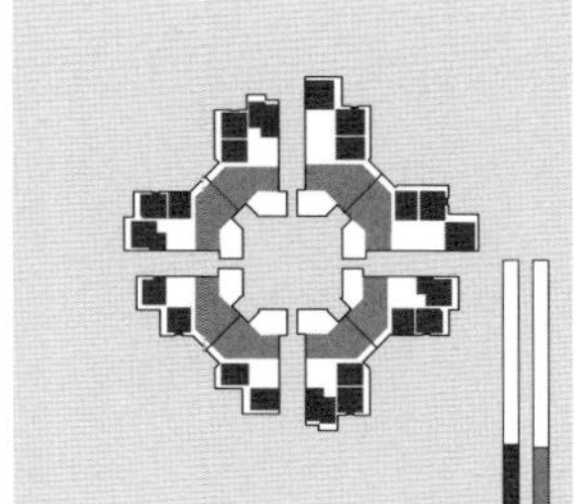

Avg. Bedroom Area: 15.7 m²
Avg. Livingroom Area: 15.0 m²
Bedroom Area Ratio: 28.8 %
Livingroom Area Ratio: 27.6 %

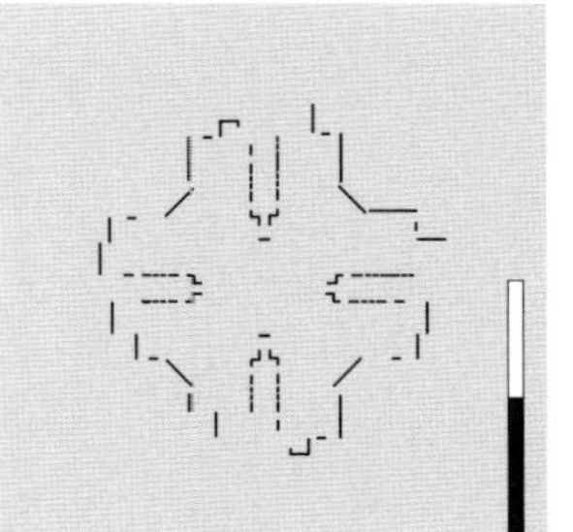

Window Length: 113.7 m
Perimeter Length: 205.5 m
Fenestration Ratio: 54 %

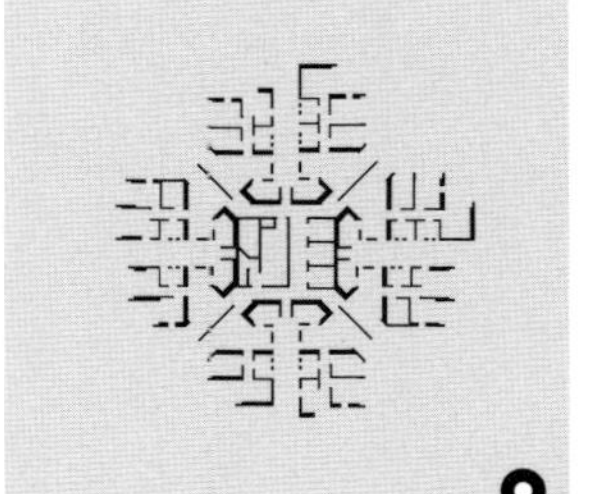

Partition Length: 316.9 m
Gross Floor Area: 516.2 m²
Degree of Subdivision: 0.61

A B C D E F

Dominant Unit Types

Mei Foo Sun Chuen
13,149 units

A

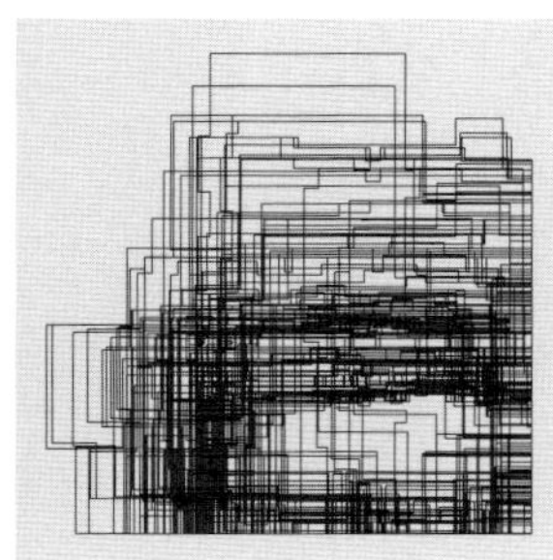

223 Unit Types

B

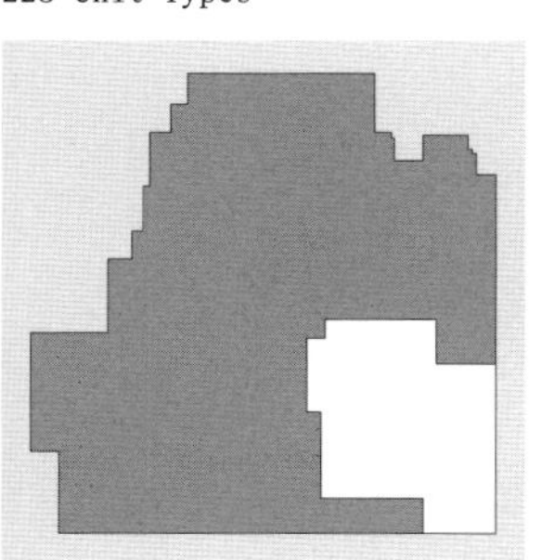

SMALLEST UNIT
Area: 40.61 m²

C

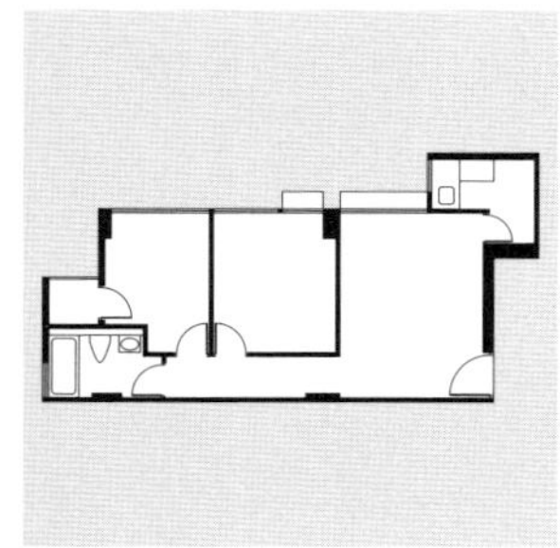

MOST REPETITIVE UNIT
2 Bed-Room Apartment

D

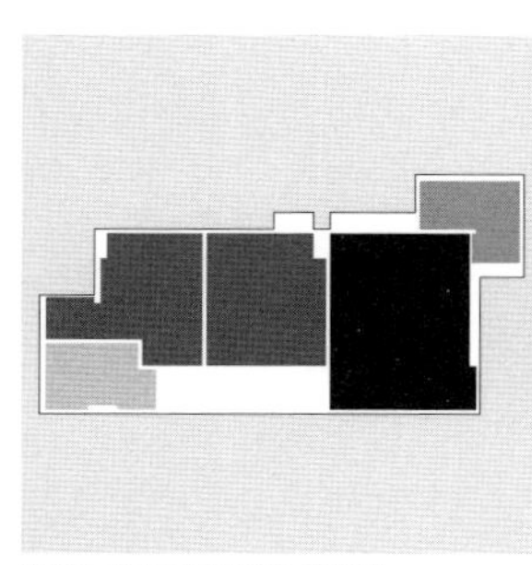

MOST REPETITIVE UNIT
Area: 51.18 m²

E

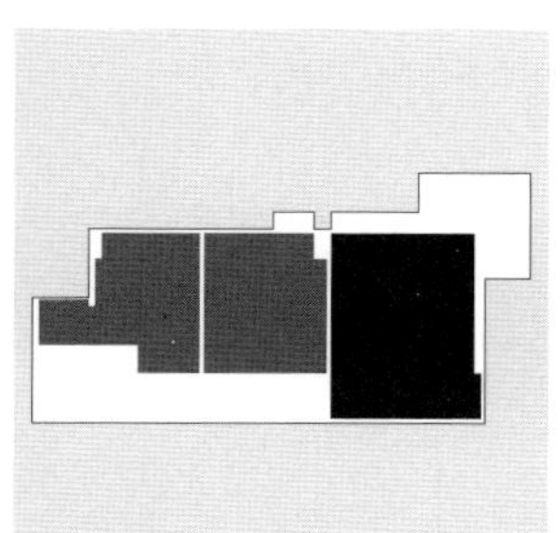

MOST REPETITIVE UNIT
Livingroom Area: 14.57 m²
Bedroom Area: 17.64 m²

F

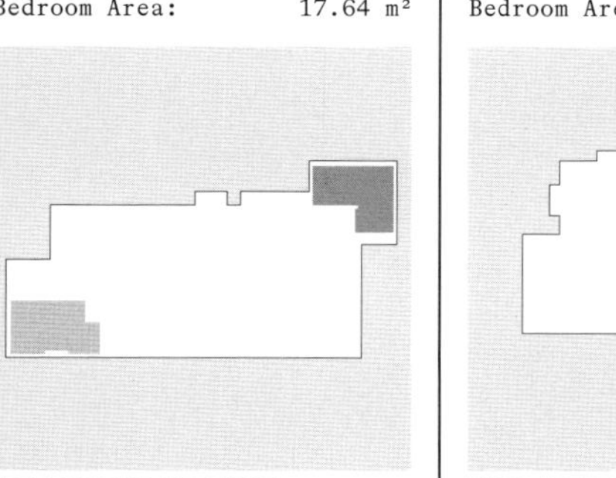

MOST REPETITIVE UNIT
Bathroom Area: 3.87 m²
Kitchen Area: 3.66 m²

Taikoo Shing
12,696 units

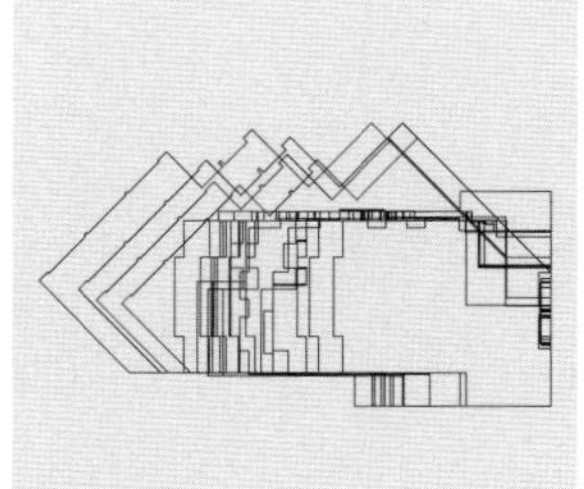

28 Unit Types

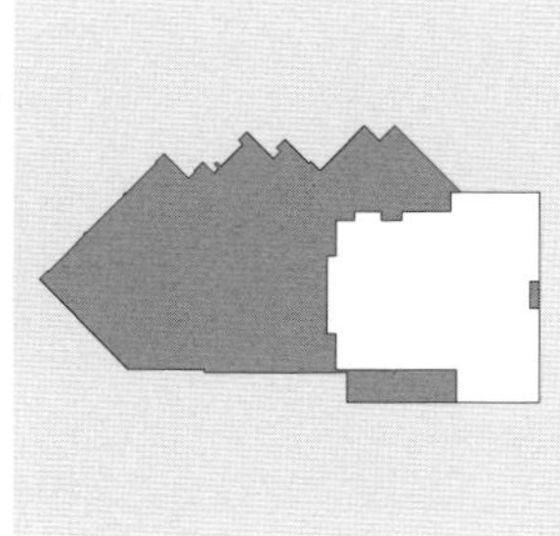

SMALLEST UNIT
Area: 41.69 m²

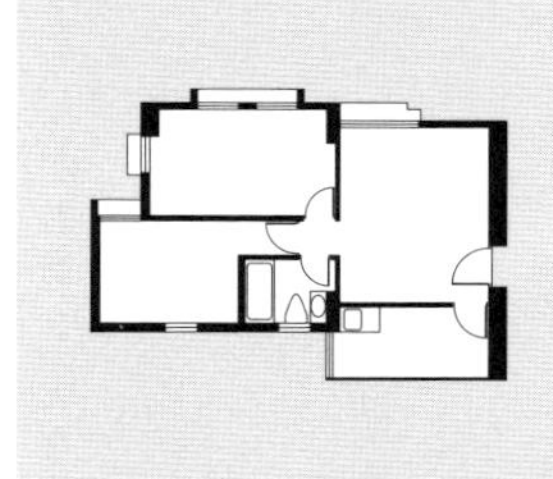

MOST REPETITIVE UNIT
2 Bed-Room Apartment

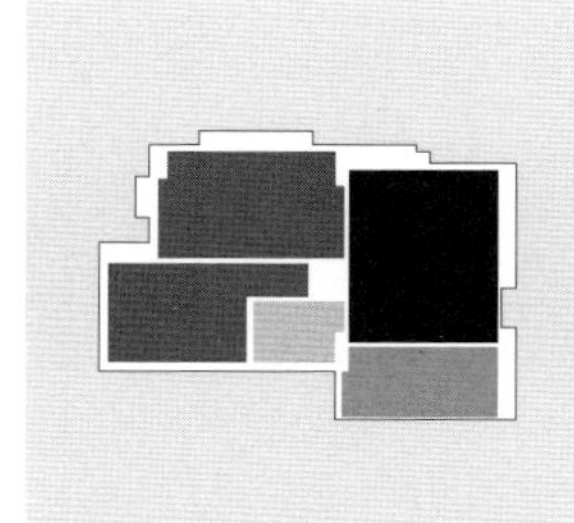

MOST REPETITIVE UNIT
Area: 48.40 m²

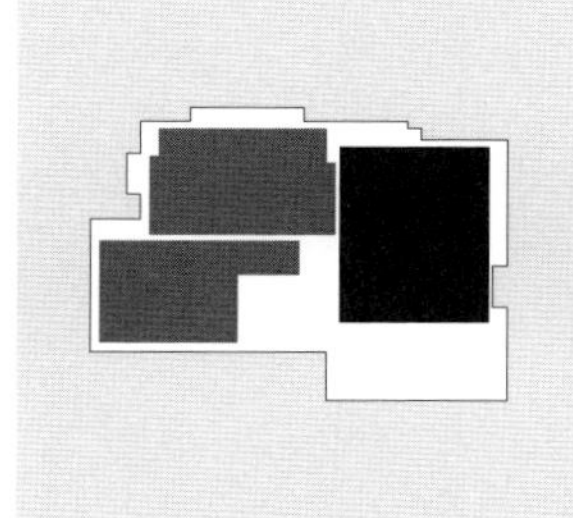

MOST REPETITIVE UNIT
Livingroom Area: 12.77 m²
Bedroom Area: 17.24 m²

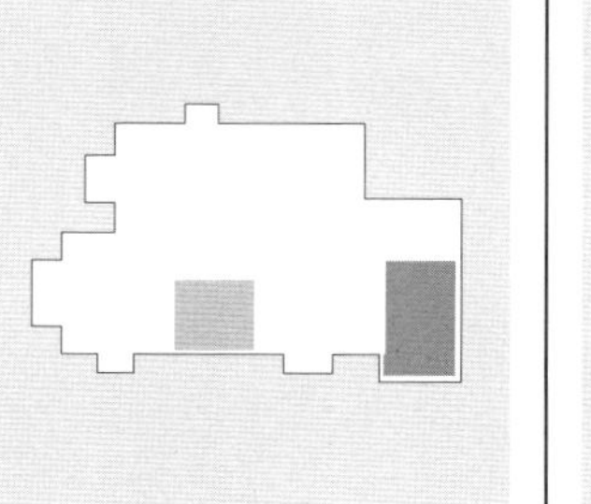

MOST REPETITIVE UNIT
Bathroom Area: 2.66 m²
Kitchen Area: 5.26 m²

Heng Fa Chuen
6,505 units

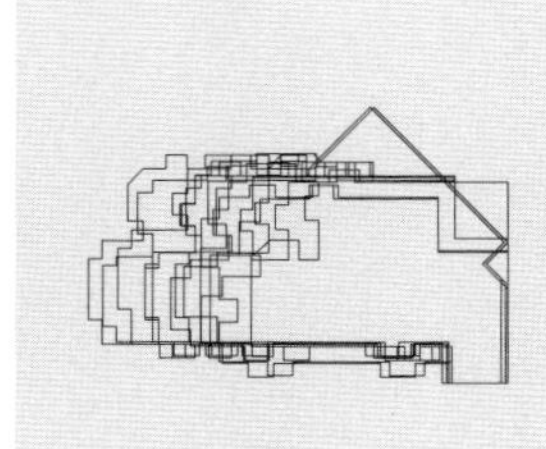

15 Unit Types

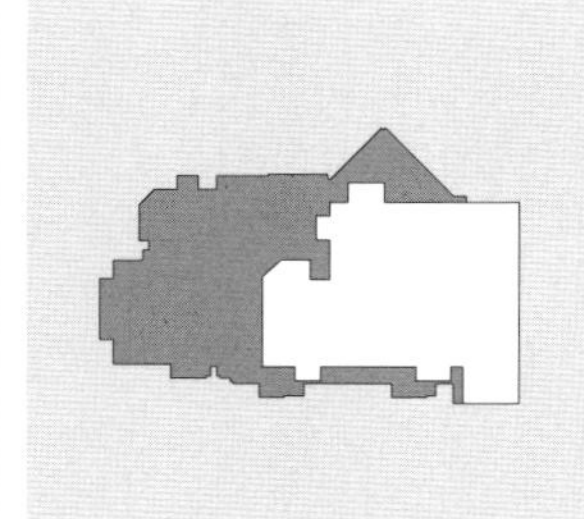

SMALLEST UNIT
Area: 45.29 m²

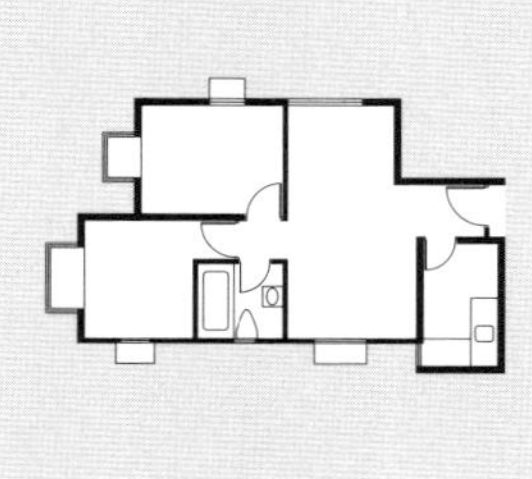

MOST REPETITIVE UNIT
2 Bed-Room Apartment

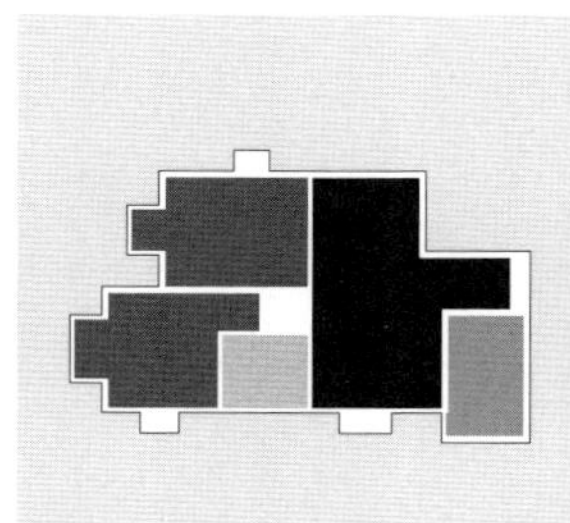

MOST REPETITIVE UNIT
Area: 48.20 m²

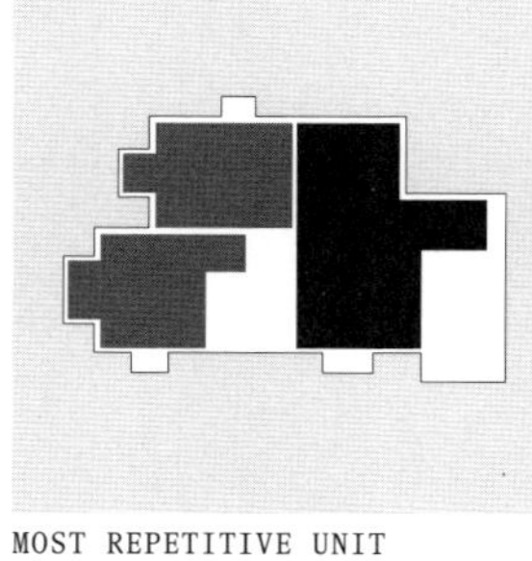

MOST REPETITIVE UNIT
Livingroom Area: 15.52 m²
Bedroom Area: 16.27 m²

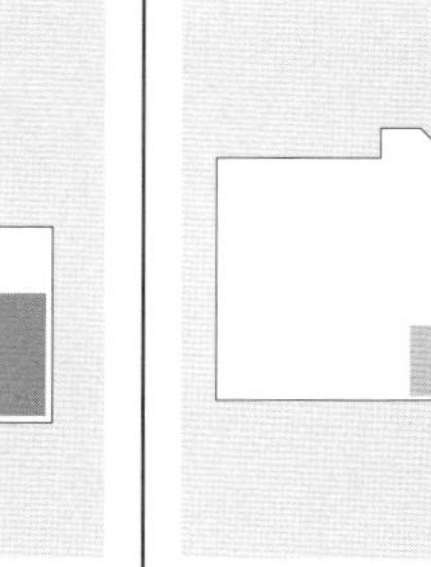

MOST REPETITIVE UNIT
Bathroom Area: 3.13 m²
Kitchen Area: 4.48 m²

Kornhill
6,651 units

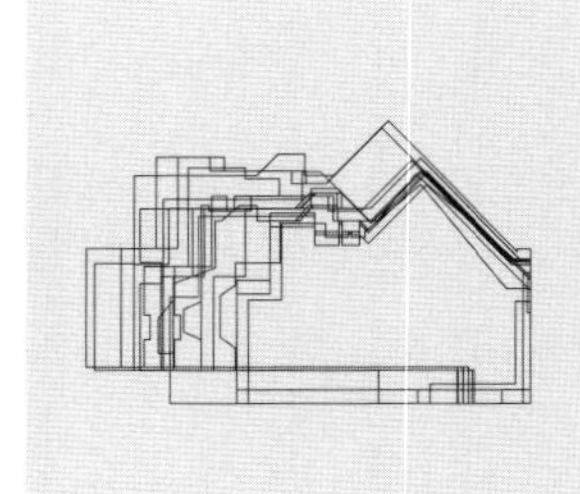

29 Unit Types

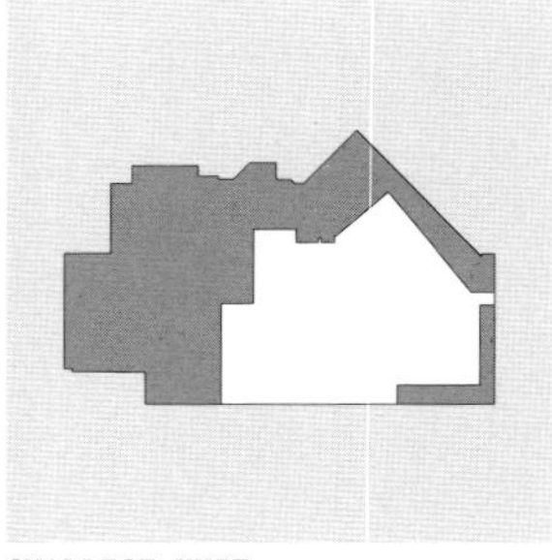

SMALLEST UNIT
Area: 46.39 m²

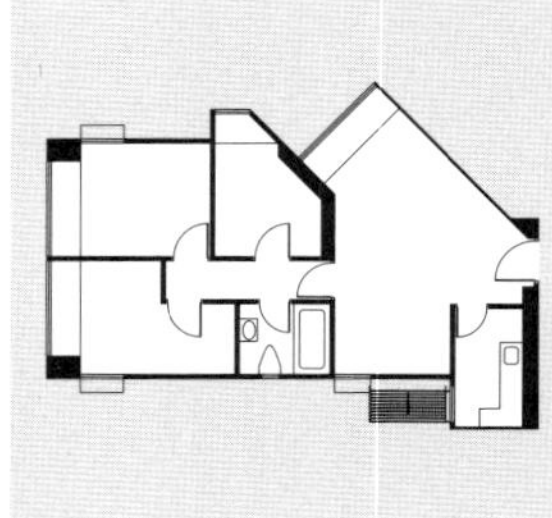

MOST REPETITIVE UNIT
2 Bed-Room Apartment

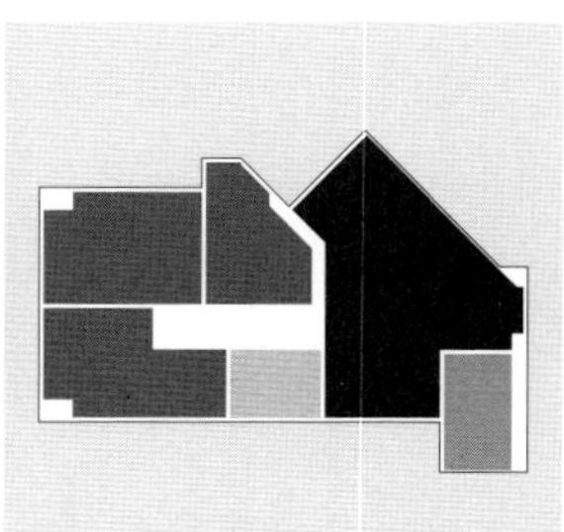

MOST REPETITIVE UNIT
Area: 59.49 m²

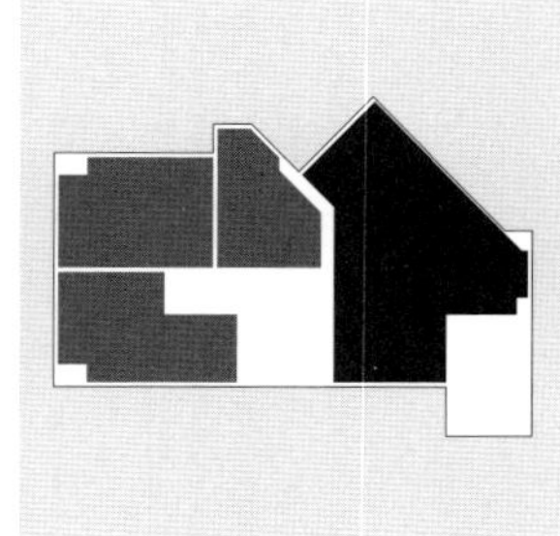

MOST REPETITIVE UNIT
Livingroom Area: 19.16 m²
Bedroom Area: 22.74 m²

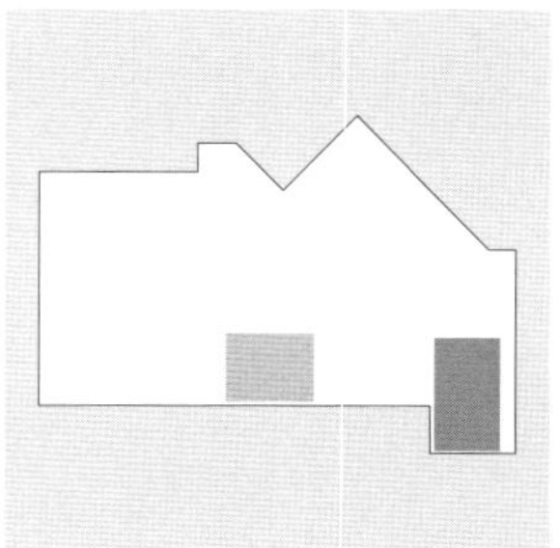

MOST REPETITIVE UNIT
Bathroom Area: 3.10 m²
Kitchen Area: 3.89 m²

City One
10,643 units

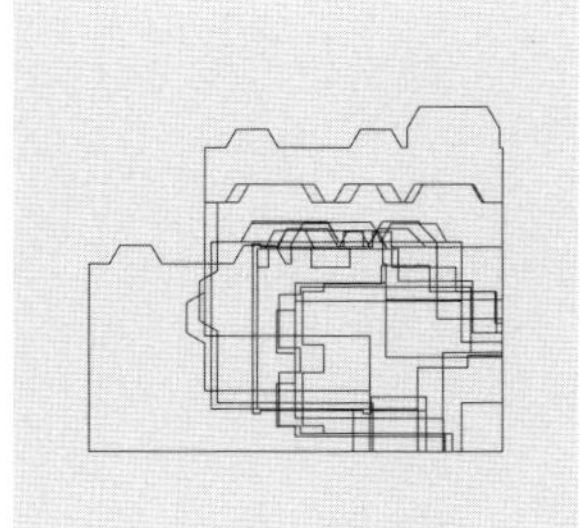

18 Unit Types

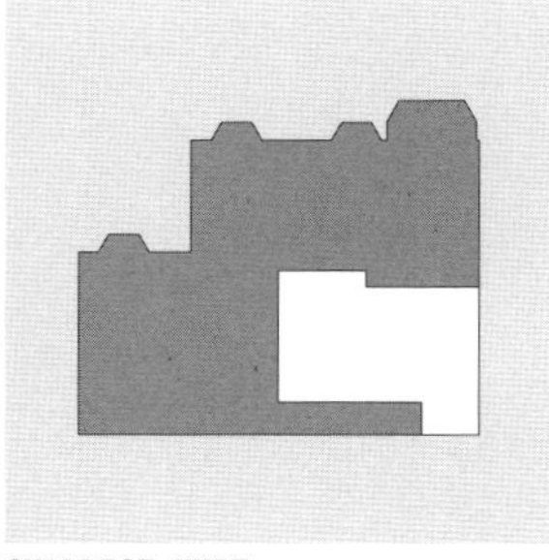

SMALLEST UNIT
Area: 30.55 m²

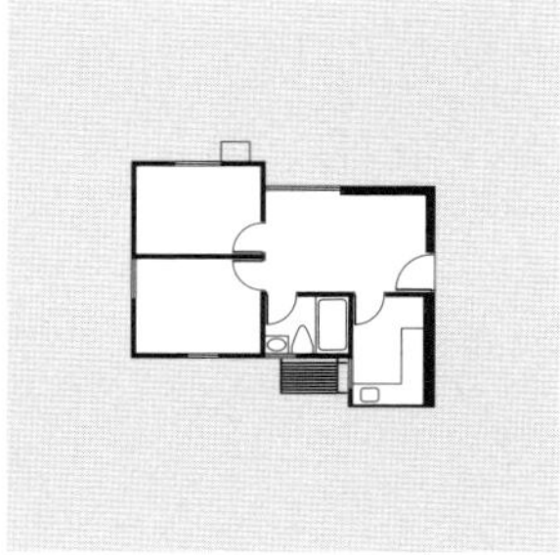

MOST REPETITIVE UNIT
2 Bed-Room Apartment

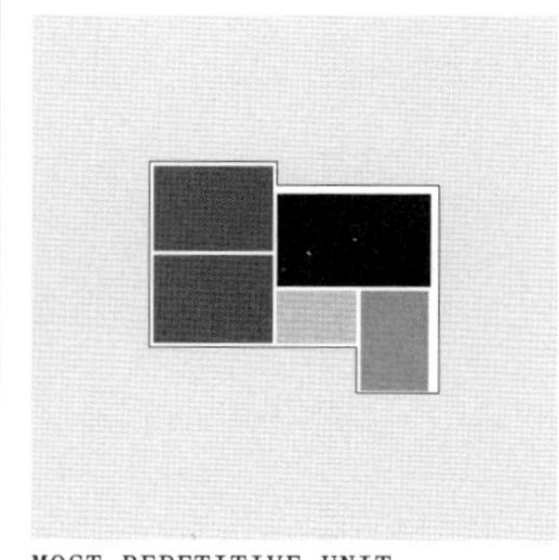

MOST REPETITIVE UNIT
Area: 30.55 m²

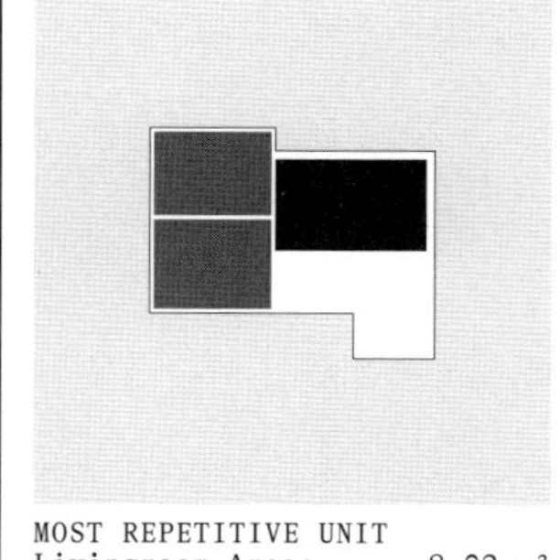

MOST REPETITIVE UNIT
Livingroom Area: 8.03 m²
Bedroom Area: 11.68 m²

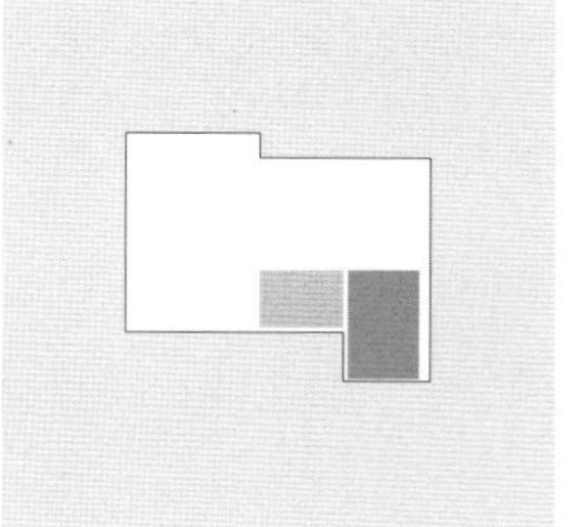

MOST REPETITIVE UNIT
Bathroom Area: 2.45 m²
Kitchen Area: 3.90 m²

ROW A – Unit Superimposition
ROW B – Composite Perimeter with Smallest Unit
ROW C – Most Repetitive Unit
ROW D – Program Distribution
ROW E – Living Areas and Bedrooms
ROW F – Kitchens and Bathrooms

Whampoa Garden
10,441 units

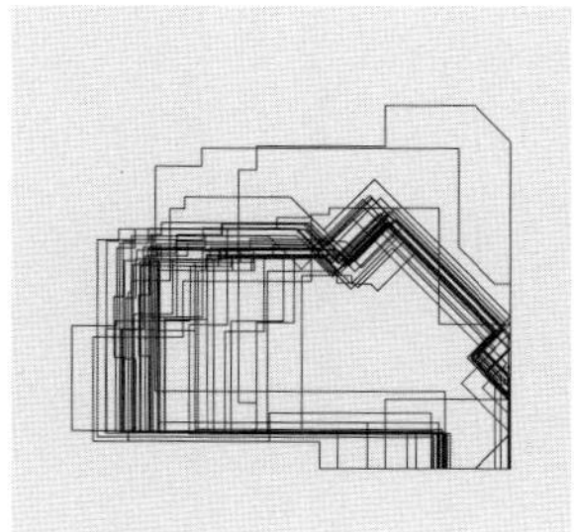

49 Unit Types

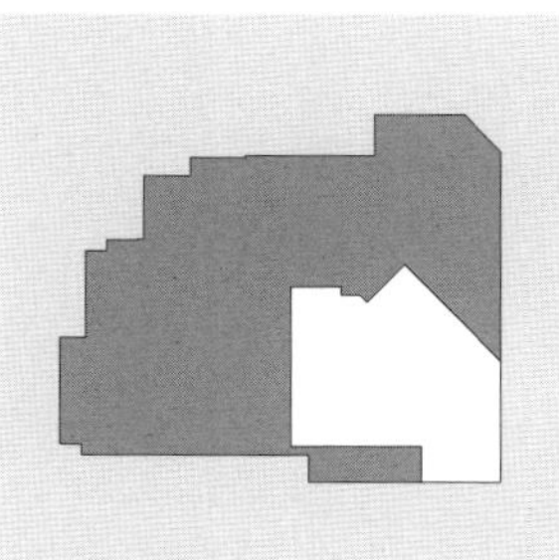

SMALLEST UNIT
Area: 36.97 m²

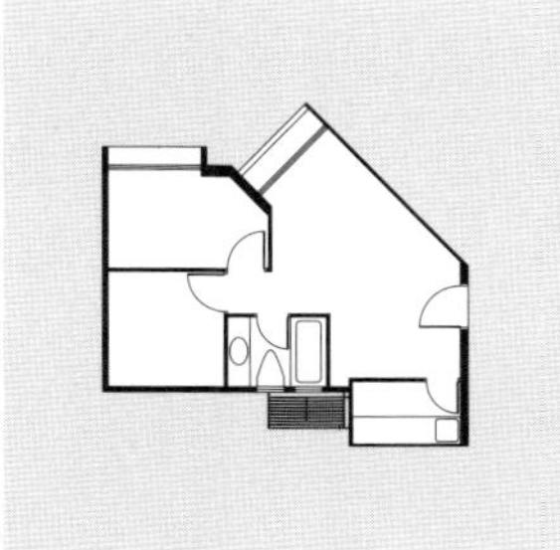

MOST REPETITIVE UNIT
2 Bed-Room Apartment

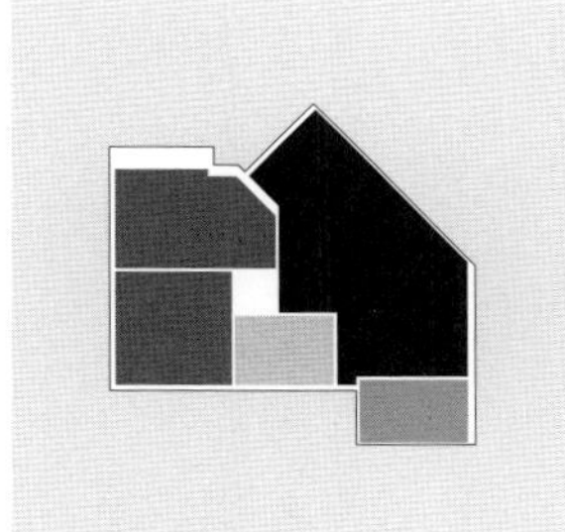

MOST REPETITIVE UNIT
Area: 44.19 m²

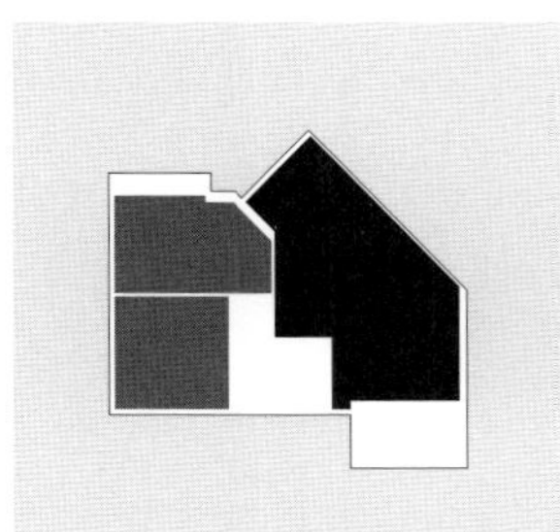

MOST REPETITIVE UNIT
Livingroom Area: 17.53 m²
Bedroom Area: 13.72 m²

MOST REPETITIVE UNIT
Bathroom Area: 17.53 m²
Kitchen Area: 13.72 m²

Belvedere Garden
6,016 units

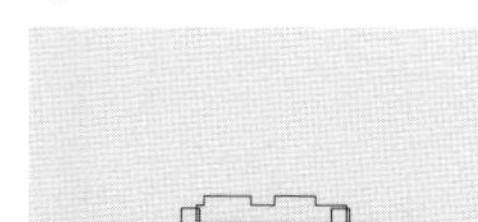

8 Unit Types

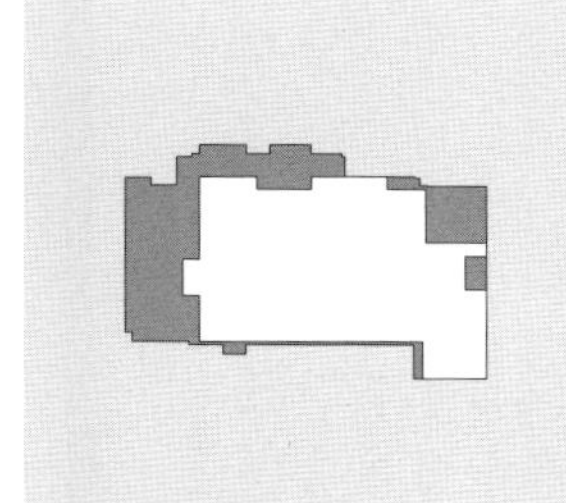

SMALLEST UNIT
Area: 48.17 m²

MOST REPETITIVE UNIT
2 Bed-Room Apartment

MOST REPETITIVE UNIT
Area: 64.06 m²

MOST REPETITIVE UNIT
Livingroom Area: 18.55 m²
Bedroom Area: 21.85 m²

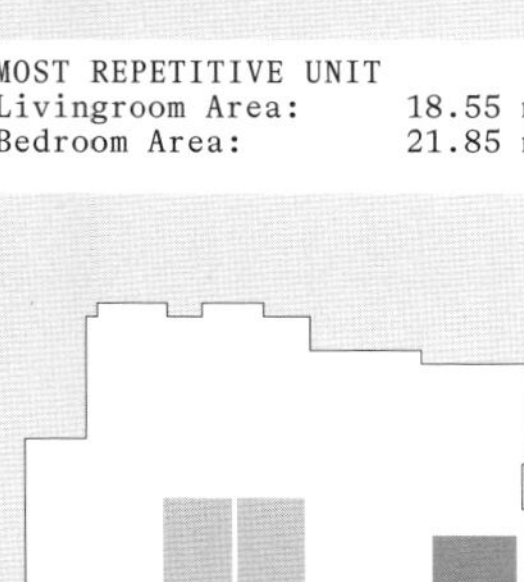

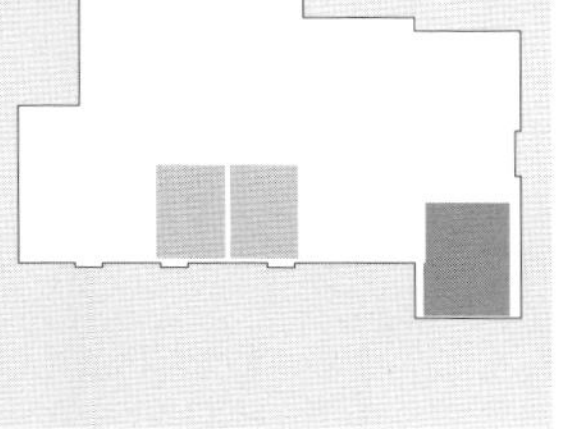

MOST REPETITIVE UNIT
Bathroom Area: 18.55 m²
Kitchen Area: 21.85 m²

Laguna City
8,072 units

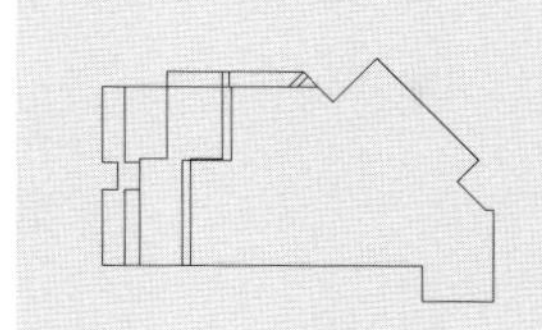

4 Unit Types

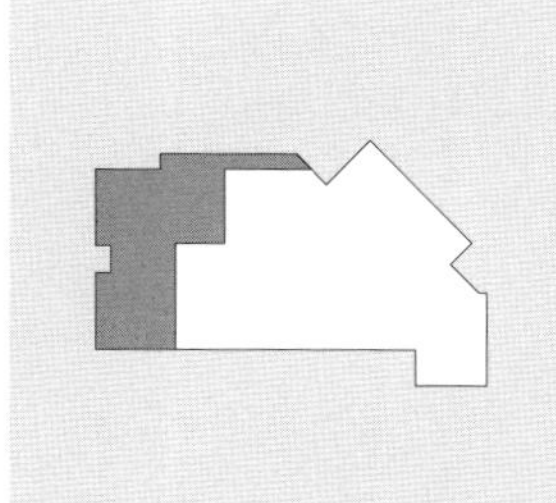

SMALLEST UNIT
Area: 54.90 m²

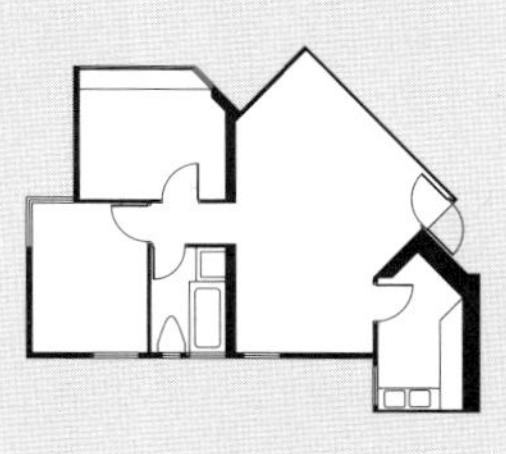

MOST REPETITIVE UNIT
2 Bed-Room Apartment

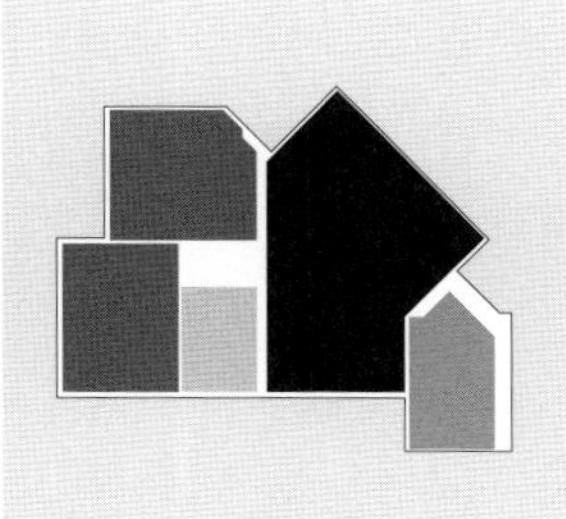

MOST REPETITIVE UNIT
Area: 56.19 m²

MOST REPETITIVE UNIT
Livingroom Area: 21.01 m²
Bedroom Area: 17.21 m²

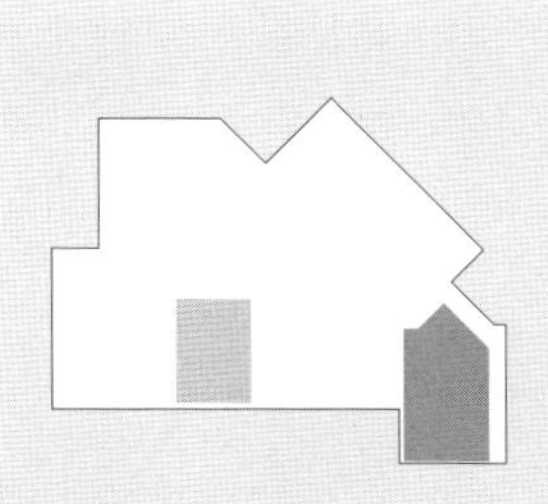

MOST REPETITIVE UNIT
Bathroom Area: 21.01 m²
Kitchen Area: 17.21 m²

South Horizons
9,813 units

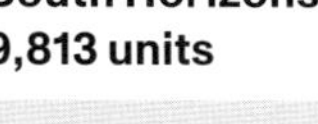

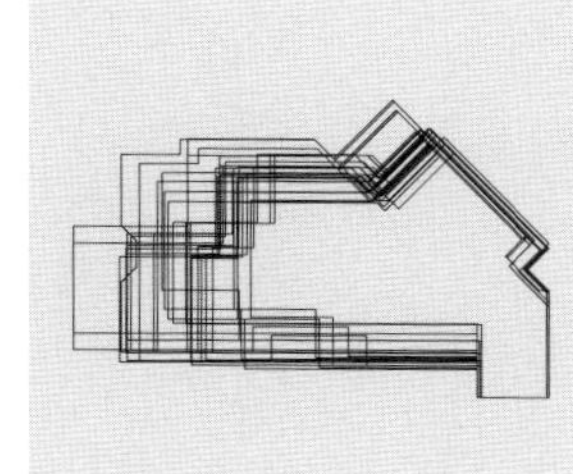

31 Unit Types

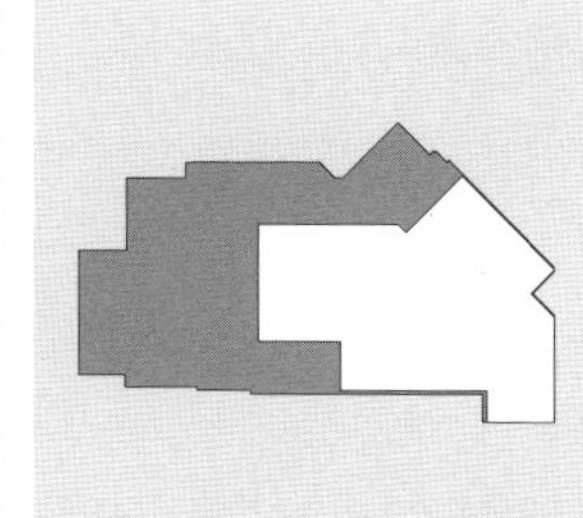

SMALLEST UNIT
Area: 52.21 m²

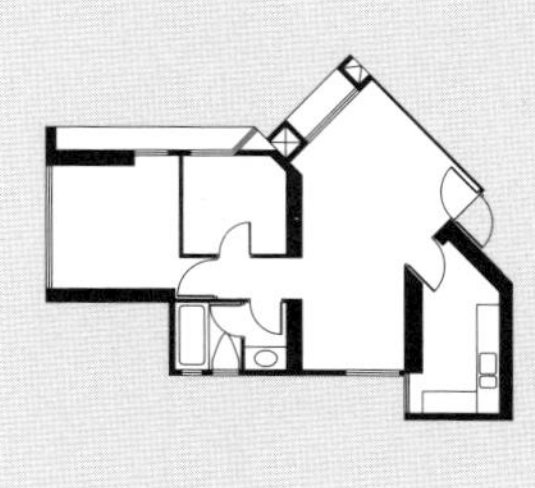

MOST REPETITIVE UNIT
2 Bed-Room Apartment

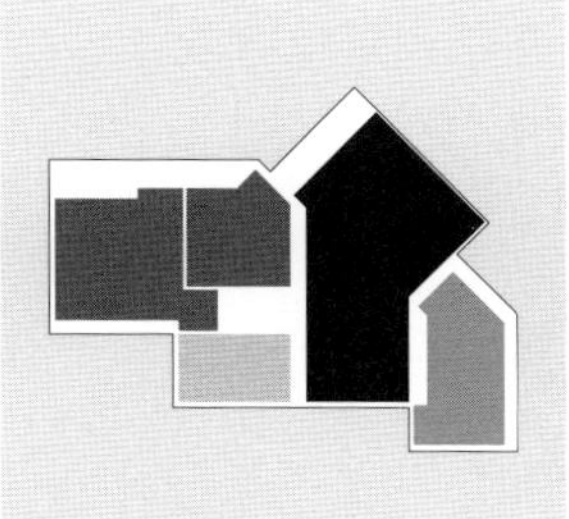

MOST REPETITIVE UNIT
Area: 53.34 m²

MOST REPETITIVE UNIT
Livingroom Area: 15.83 m²
Bedroom Area: 13.65 m²

MOST REPETITIVE UNIT
Bathroom Area: 15.83 m²
Kitchen Area: 13.65 m²

Kingswood Villas
15,927 units

15 Unit Types

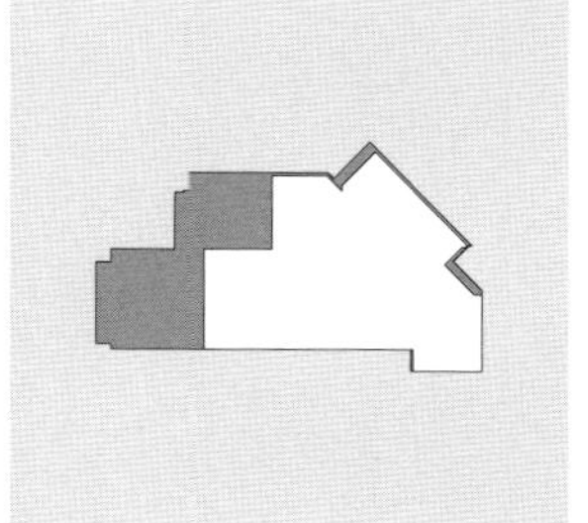

SMALLEST UNIT
Area: 44.89 m²

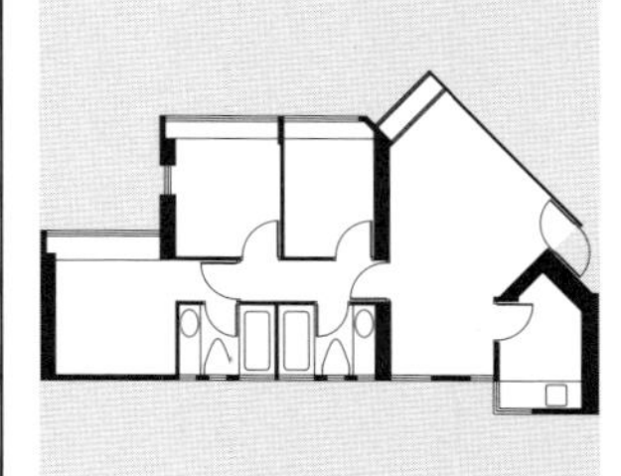

MOST REPETITIVE UNIT
2 Bed-Room Apartment

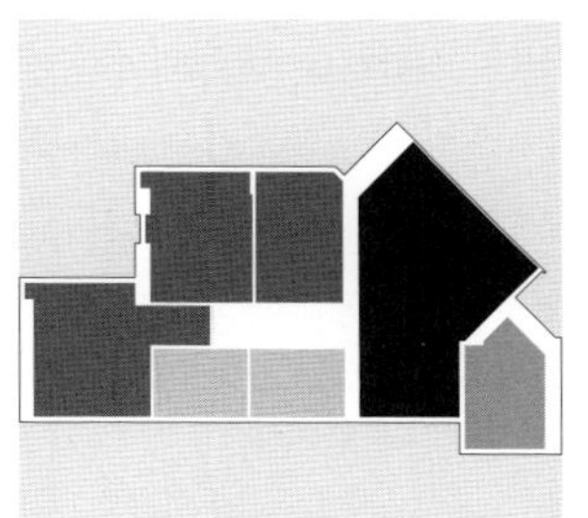

MOST REPETITIVE UNIT
Area: 63.81 m²

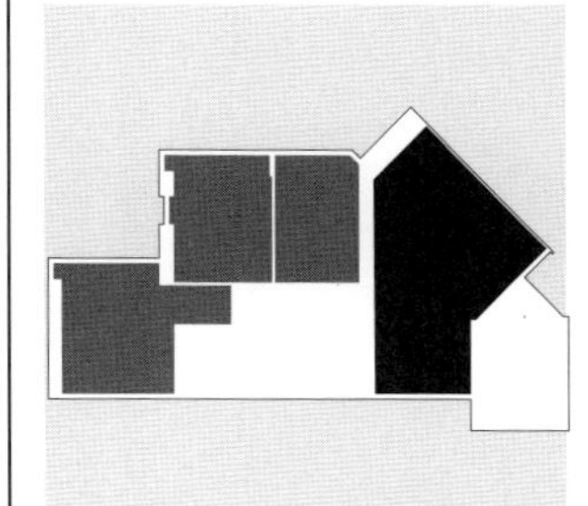

MOST REPETITIVE UNIT
Livingroom Area: 16.46 m²
Bedroom Area: 22.44 m²

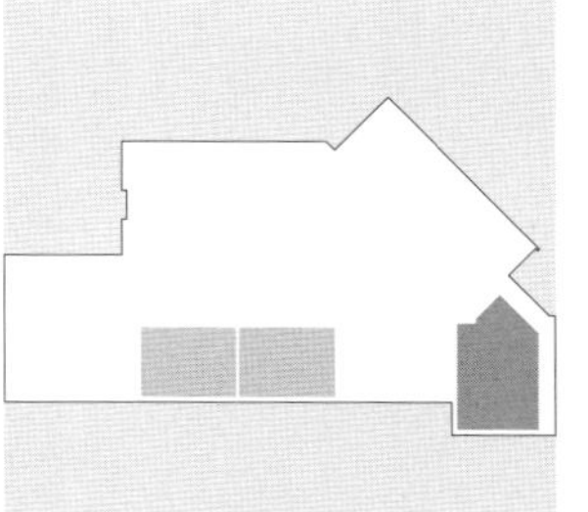

MOST REPETITIVE UNIT
Bathroom Area: 16.46 m²
Kitchen Area: 22.44 m²

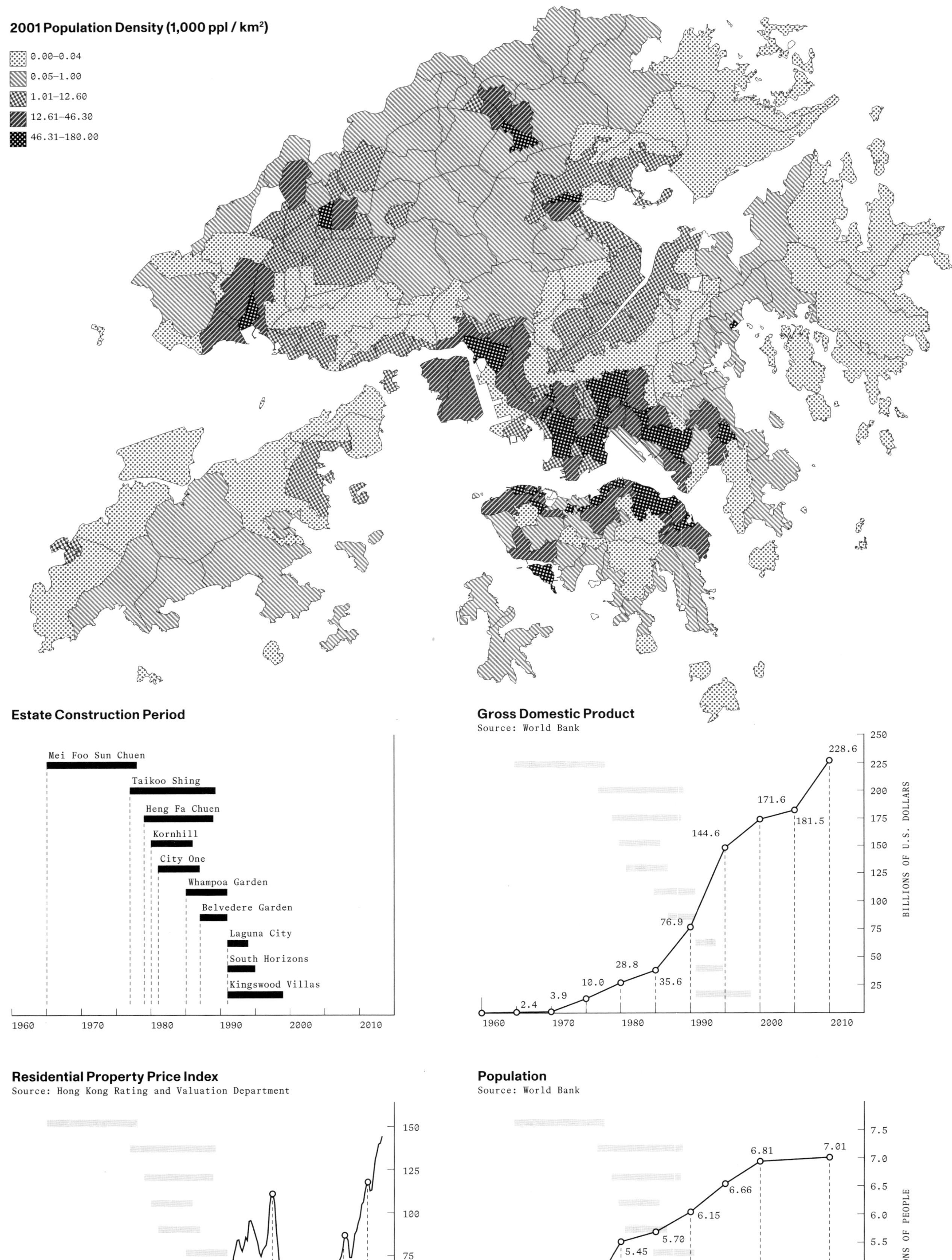

Residential Property Price Index

Source: Hong Kong Rating and Valuation Department

Population

Source: World Bank

Interview with Carolin Fong

The authors of *Cities of Repetition* (CoR) conducted the following interview with Carolin Fong to discuss the challenges of and strategies for housing design in Hong Kong from an architect's point of view. At the time of interview, Fong was Director and Head of Sustainable Development of DLN Architects Ltd. She is an architect by profession and an Authorized Person in Hong Kong. Her experience includes the design of over fifteen large-scale housing projects in Hong Kong and East Asia.

CoR With its beginnings in European modernist architecture there began an approach to high-density housing where a single tower type is developed and repeated. This strategy was applied to large scale, modernist housing developments throughout the world and many, as history has shown us, have been condemned for societal problems associated with an architectural type. Hong Kong presents a different case as high-rise housing has evolved as the prominent housing type for citizens across the socio-economic spectrum without any dramatic societal problems. How do you view the difference between high-rise dwelling in Hong Kong versus other places in the world?

Fong I think Hong Kongers are used to high-rise housing estates. The city is so dense that people often don't have a choice. Everybody who is born here has accepted the fact that they will live on the 19th floor in flat B, for example. You don't necessarily know who lives next-door, but it's totally okay for us. So basically, I think it's the culture of Hong Kong that is different from other places in the world. Typically in many Western countries with more space, there is more choice in terms of housing. Hong Kong is accustomed to dense living and citizens do not challenge it.

CoR How do you approach the design of high-rise and high-density buildings as an architect?

Fong Because domestic spaces are typically very small, you have to make sure that they work. For example, the minimum size for a bedroom is 2.1×2.05 meters. That is the smallest size possible that can fit a bed. And the living room must have adequate wall space so that one can place his or her TV against it opposite the sofa. All these considerations come down to "typical unit plans" for each type of unit—two-bedroom, three-bedroom, etc. Repeating these typical unit plans to become a floor plan is a way of designing. And if the developer accepts the repetitive nature of the projects, then all the focus goes into making the prototype perfect.

When an architect had designed one building in perfect compliance with the minimum requirements of the building code, and all the dimensions have been perfected, then the focus could shift to the public spaces: lobbies, the hardscape, the gardens, and the penthouse. Hong Kong architects tended to do this rather than trying to make every tower different. Usually developers don't like making every tower different because it raises the overall cost of the project. Having all the towers repeated is obviously more cost effective for the developers and more manageable for the architects. This is not a very nice thing to say about architects, but it is true. But, to be honest, replicating the design of towers in a project is becoming more and more rare nowadays.

CoR How do you explain the typical cruciform shaped floor plan that so many towers in Hong Kong follow?

Fong The cruciform shape of the plan was very popular and common until the end of the millennium. It was the optimum result of an equation comprising of the types of units that developers sought during that period based on efficiency and structural considerations, and used to determine the layout of two- and three-bedroom units in a highly-efficient cruciform plan with minimal circulation space, which eats up valuable floor area. The cruciform plan is also structurally efficient and therefore reduces construction costs.

If you need to provide very small units, like one-bedroom or studio units, it is not possible to have a cruciform plan, which would become very inefficient. Over the past four or five years, the price of housing has grown so high that developers have calculated that most people cannot afford two-bedroom units anymore.

CoR Thus tower form is changing based on real estate value and reduced unit size?

Fong Yes, many people today are probably in the market for a 200-square-foot studio, which will cost about four million Hong Kong dollars in a mid-range residential district.

CoR So the unit types that were preferable to developers in the 1960s, '70s, and '80s are no longer marketable and that has changed typical building form?

Fong Nowadays, the brief that I get from a developer is more complicated than in past years. Within a typical brief, it is not uncommon to have over seventy percent of units being small—one- or two-bedrooms or studio units. And the footage of these units is also becoming smaller and smaller. Buildings are increasingly being built with very, very small units. The remaining twenty to thirty percent will be three-bedroom units with a handful of very luxurious four- or five-bedroom units.

Compared with many of Hong Kong's older housing estates, the types of units in a residential project are comparatively more varied nowadays. With a broader range of unit types and smaller unit sizes, cruciform plans no longer work for today's typical brief.

CoR What makes building form so central to housing design in Hong Kong?

Fong The fundamental issue behind housing design in Hong Kong is the efficiency of the floor plan. You are only allowed to build a certain total square footage on a plot of land. The common areas cannot be sold, therefore, you need to squeeze the core, which is the common area. Making the vertical circulation core smaller means you make the salable area larger. In that sense, with today's market prices and preferred unit sizes, a cruciform plan would be highly inefficient. Today's typical small one- or two-bedroom units cannot wrap a typical core designed for a project like Mei Foo. They have to be arranged along a corridor, usually with over thirteen units per floor to share the common facilities like lifts and stairs, otherwise the plan will be too inefficient.

CoR How would you describe this attitude towards housing design in an economic context as extreme as Hong Kong where real-estate prices are consistently among the highest in the world?

Fong Form follows finance. It's really about finance—everything revolves around finance.

CoR So would you argue, then, that the key to architecture in this city is about maximizing profit?

Fong It's about "optimizing the site's potential in the financial sense" or "maximizing the profit for the client." But this is also about designing for what the buyers want.

I must add that many developers look for more in a project—environmental performance, contribution to the neighborhood, advances in technologies, and innovation in design and construction, etc. However, the financial return of the project is still the fundamental purpose of the project. And it is only within the framework of profit optimization that architects and building professionals have to strive for these additional values in their designs.

CoR Can you provide an example of buyer's desire trumping developer's profit?

Fong For instance, many of the housing estates from the 1990s were built with a triangular, or what they call the "diamond-shaped" living room. Back then, that floor plan design was very popular because this layout suits the cruciform plan of the tower, and the spatial efficiency is very high. And the angular shaped living room avoids the "overlooking problem." Before the diamond-shaped living room came along, windows of rectangular living rooms of adjacent units in a cruciform plan used to be adjacent to each other at a ninety-degree angle, so one could look directly into a neighbor's flat. With the diamond shape design, the living room windows of adjacent units were aligned, and the improvement was acceptable. But people soon found that it is very difficult to put furniture into a diamond-shaped space. Residents are always looking at their televisions from a slanted angle because the living room has diagonal walls. So nowadays, we always use a rectilinear layout based on buyer preferences, even when an angular layout can yield higher floor efficiency or increase structural stability.

CoR Some of the estates documented in our research have more repetitive towers than others. From an architect's point of view, what are the driving motivations for design when confronted with the task of designing fifty or more towers on a single site?

Fong When designing large sites for a private residential project, the primary concern is the optimization of views for the units, if you ask me. Other than that, the main concern is to create an integrated and large outdoor open space for the enjoyment of the residents, which creates view values for the units. It is all about creating values for the residential units.

Of course there are also other subjective site constraints—most common being traffic noise—that needs to be considered in the planning. Bad views, like facing a cemetery or funeral house, must also be avoided.

When a unit has an open view or a sea view, its unit price will increase. In our submission of schematic designs, we always have a sea view/garden view unit ratio. This is one way a developer judges a design.

CoR So in a sense there is a formulaic set of criteria that a developer would use to value a design scheme, such as maximum planning efficiency, percentage of units with sea views, and, therefore, a large part of the architect's task is reduced to number crunching?

Fong In a sense, yes. It is number crunching through the design, though we also have to figure out how the developers value the views. Some believe that the units must have a full sea view in order to have a higher selling price. Some believe that even a partial view is a good enough reason to raise the price. Different developers have different ways of judging and marketing. We have to understand how the developers view these values. Then, as architects, we have to exercise our skill to provide a design that answers to these values. Because, for example, if you want a lot of sea views, your corridor will need to be longer which will then jeopardize your efficiency. How do you balance that? How can you achieve the most sea views without designing a very long corridor and wasting space? How can you create more view values for your units through the planning of outdoor landscape spaces? These are the key challenges to design in this context, when it comes to getting a design approved by the client or winning a design competition for a large housing project.

CoR How does it typically work? Does the developer only approach one architect at a time?

Fong No, they typically approach quite a few.

CoR And then they hire the designer who gives the cheapest bid?

Fong Not necessarily. Some developers value the ability of the firm, their track record, and their committed resources for the projects. Some consider the cost as the most important factor. But the one who can produce the best, or most profitable design always has a better chance to get the job.

CoR The plan that achieves the most efficiency and achieves the most views?

Fong Yes, that will be the winning design—for private residential projects. It does not have to be the "prettiest," but it has to achieve the value for the investment. I think the quality of the design—in terms of its spatial quality, architecture, and its relationship with the city, all of the things that we learn in design school—come in later. Architects debate those issues during a design project but they are not really what gets us the job. After all, a housing estate is an investment and we exert our skills to achieve the best results for the developers. After we get the job, we go along to think about those values, and with a lot of efforts we try to integrate them into the project to be realized.

CoR After analyzing the hundred biggest privately developed estates in Hong Kong, we realized that most of them reveal three major urban patterns: the Cartesian grid, the staggered grid, or the linear aggregation. Do you have an answer in how these came into place, and if so what are the advantages or disadvantages?

Fong I think the driving force for the patterns, or the master plan that you see, is not because the architect wanted to achieve master plan based on mere form; it is really a consideration of multiple factors including finances. The units are arranged so that they are looking over each other as little as possible. This is the first thing—when units look over each other, you go to your window and you immediately see another unit. This leads to a reduced sale price. Architects strive to design for the best views for all units. And, of course, the shape of the site also comes into play.

CoR You would argue that the urban form comes into place from the desire to maximize privacy, maximizing views, and maximizing efficiency on site. So these are the driving forces of how an urban form comes into place?

Fong Yes. I think an urban form grows out of the logic of the building typology.

CoR From the developers' point of view, what other criteria are important to the development of housing in Hong Kong today?

Fong After efficiency and views comes the ease of construction, maintenance in relation to the low construction costs, and the time it takes for construction. Is your design very difficult to construct? Does it take a lot of time? Can your design be separated into phases such that sales of units can happen in phases also? Does your design satisfy the code easily? All of these are important criteria to the financial success of a project.

Ultimately, it is whether the product is attractive to the potential buyers. You need to provide a punchline for the developer. That's one point. Sometimes the developer requires us to help them to market their development, and if our design carries a theme—architecturally—it really helps.

CoR So an architectural narrative is also important to a large-scale housing project?

Fong Yes. It would be really great and satisfying if we create a poetic design for the project, and still be able to satisfy the efficiency and the views.

CoR How does governmental regulation affect architectural design and development?

Fong The government sets the game rules. They are concerned with basic issues such as the safety, hygiene, and structural soundness of the building. They are concerned, since 2010, about the ventilation of the neighborhood of the development. They are also concerned about ensuring fairness to the rights of people who spend their life's savings to purchase a flat. All these are affecting the design of residential projects.

One example of government regulation to increase the transparency of the private residential market was the Residential Properties (Firsthand Sale) Ordinance, which was put into place five years ago. Firsthand sale means that if people want to sell their units for the first time, they have to abide by a certain set of rules. You wouldn't believe that these rules did not exist before 2013. Before that time, if you bought a flat that the developer said was 800 square feet, you wouldn't know where those 800 feet were because your house is not actually 800 square feet. Home buyers were paying for space outside their flats. The regulators figured that they should do something about this. And if you ask me, that change affected the entire market. Now, all sales must be done by saleable area. The saleable area is something well defined and tangible that buyers can more easily comprehend.

CoR Even though the salable area still includes something like building structure?

Fong It does, by definition. The area of any walls adjacent to or enclosing your unit are included in the saleable area. You can imagine how this definition of saleable area drives the detailed planning of the building structures—structural walls, thin or thick, are always planned to be adjacent and to enclose residential units as far as possible. This is a rule that architects and developers have followed since saleable area was adopted.

CoR That dramatically affects the entire floor plan layout and structural design.

Fong Yes, it affects the design, but these kinds of subtle operations make an enormous difference in the overall efficiency of the project. They can make a three or four percent difference in efficiency, which has dramatic effects on the bottom line of a project, financially speaking.

CoR While the Firsthand Sale Ordinance was designed to be more fair to the homeowner, the result is that architects can creatively implement new design tactics to use the regulations for the financial benefit of the developer?

Fong The answer to your question is yes, we do. But I would say it is much better than before when one does not even know where the footage they are buying comes from.

As architects, we do what is needed to optimize the profit of our clients. I'm not ashamed of it. It's the ecology of things in the private sector. For the buyers, of course, it may sound off-putting, but the Firsthand Sales regulation also requires very clear representation of the layout in a sales brochure, including thickness of walls etc., and there are absolutely no hidden tricks in terms of area or information about the unit being sold.

For the developer, in an environment with extremely high land values, it is natural that they would endeavor to ensure optimum return of their investments.

CoR Hong Kong is known for its strict building regulations, and some say the city is a direct imprint of the code. How strict is the building code in respect to the development of estates in Hong Kong and in what ways has this shaped your own practice as an architect designing housing in the city?

Fong We must abide by the code in the most efficient way. The code is strict, and it is becoming more and more difficult and brain draining to satisfy them in the design.

In the old days, with the cruciform plan, there were a lot of reentrances into the facade into which you could tuck building services. The floor plan perimeter was very long in comparison to the floor area in those housing types, and each unit usually had at least two sides of external wall frontages; providing all the required windows and fitting in all the external drainage pipe works, outdoor AC units, flue apertures, etc., was relatively easy. Today, the same set of codes still applies. But with very small units arranged along a corridor, each unit only has one side of exterior frontage. And that limited frontage needs to serve as windows of rooms behind, outdoor air conditioner units, balconies, utility platforms, flue apertures, vertical drainage pipes for kitchens, AC platforms, etc.—everything is competing for every inch of external facade.

In Hong Kong, the code governing hot water heaters' flue aperture provision is very strict because we used to have accidents in which people died from carbon monoxide poisoning due to poorly installed gas water heaters. So by law, each residential unit must have a flue aperture complying with very specific dimensional requirements, even if an electric water heater is used for that unit. With units that are so closely packed together in residential towers, it is always a time and effort draining exercise to find a place to put the flue. A similar problem arises with the air conditioner. Nearly each project comes down to the struggle of millimeters when dealing with elevation design.

To answer your question, it's not about the strictness of the code, it's the circumstances in which the architects need to comply with the code requirements that is making life very difficult.

CoR Therefore, through a hyper-competitive economic environment and a set of strict building regulations, architectural creativity is defined differently in Hong Kong than in other regions of the world. Are there any other regulations that you would attribute to changing the urban form of the city through architectural building code?

Fong Yes. There are other issues such as area exemptions that complicate matters even further. I think the rules surrounding balconies and utility platforms have had a large impact on urban form. Before 2000, you could build apartment buildings with balconies and utility platforms, but, back then, they ate into your permitted floor area ratio, so few developers used them. Since 2000, these areas can be exempted from the permitted gross floor area, and it has become a "rule" that each residential unit must have a balcony and a utility platform.

Bay windows are another example. Since the '80s and until 2010, bay windows projecting less than 0.5 meters were exempted from permitted floor areas, so all developers used them in their estates during those years. In 2009, there was a public complaint about developers earning too much from the sale of units and buyers not knowing how the measurement of area was calculated—this problem was described as "inflated areas." At the same time, the public became aware of the bulkiness of buildings and began to consider that different types of GFA exempted features, bay window being one of them, should no longer be allowed for exemption from the permitted area calculations, such that the building bulk can be reduced. Around the same time, when the First-hand Sales Ordinance came into force, the depth of a bay window's projection that can be GFA exempted has been drastically reduced from 0.5×0.1 meters. In reality, bay windows are no longer an effective design choice. Getting rid of the bay window, if you ask me, did not benefit the public. Units are already very small. The bedroom is only 2.1×2.05 meters. When we could use the bay window, the room would feel bigger, at least, and you could put your books on it. Some people would even put their bed on the bay view window, and the room would look bigger. Now, all new units are without any bay windows, so entering a typical Hong Kong bedroom today really feels like walking into a box.

The concerns back in 2009/10 about the "bulkiness" of buildings and the district ventilation problem could have been dealt with by other means—today we have a very detailed set of rules to ensure building separation codes are met for district ventilation. Under the Firsthand Sales rule, bay windows cannot inflate the salable area in any way. It is really sad that this practical and unique feature of HK residential units can no longer exist for the benefit of the occupants.

CoR When looking at the historical development of the public sector in Hong Kong, we realized that the architects experimented with many different typologies, such as the H-Shaped plan of the Mark I housing, the linked twin-towers with vertical courtyards seen in the Wah Fu Housing Estate or the Y-Shaped plans used in Hong Kong's Trident blocks. In the private sector, the cruciform plan shape, which was used since the 1960s in the Mei Foo estate, has become the predominant housing type across the territory.

Fong And architects and developers have carried that on for decades, until unit sizes were reduced due to high sales prices.

CoR Do you think the development of public housing in Hong Kong affected the private sector at all? And in your opinion, why did the private sector not produce more variations in residential building typology?

Fong Yes, developers of private housing do not want their developments to look like public housing because the people buying their own homes do not want their homes to look like public housing. One characteristic of public housing is that they typically have long, wide corridors. I once had a project in Tseung Kwan O for which we designed quite a long corridor, such that the units could be arranged to maximize the views. The developer actually took a long time to decide whether this was okay. Because the corridor was relatively long, the developer feared that potential buyers would think it resembled a public housing estate. In the end, they decided the benefits of the views outweighed that risk, and they decided they would put in additional money and really dressed up the hallways to avoid any resemblance with public housing. So, if you ask me what the effect of public housing on private housing is, it is that the two must not resemble each other.

And then, if you ask me why public estates are more experimental with typologies, I would say that they have more rooms and reasons to experiment. Each typology of public housing design is implemented a lot of times and in many different locations. The government remains the owner of all these housing estates even after they are occupied. Government housing design is very concerned with the construction costs (large quantities to be built), maintenance (perpetual maintenance responsibility lies with the government), and possible complaints from occupants (the occupants are "customers" renting the place from government, not owners responsible for the premises). Furthermore, views are far less important to the design of public housing units in comparison to private sector housing.

CoR In recent years, many Hong Kong citizens have critically commented on the increasing-price and shrinking size of apartment units. What have been the most significant trends in the design of housing over recent decades? What are the forces that have shaped these trends from your perspective?

Fong Well, the price. Private housing in Hong Kong is always about the money. As I said, the units are getting smaller because of the total price of the flat itself. Based on market trends, the sales departments of developers usually decide a threshold of the sizes of different types of units such that the potential buyers of such type of units can afford them. For example, the salable area of a two-bedroom unit cannot be more than, say, 400 square feet—different developers have different figures. It could be a calculation to match gross sales price with the estimated affordability of young working couples with a newborn and a domestic helper with the financial assistance from their parents and grandparents. As architects, we do not know the math, but we have to design to fit the developer's strategy. Sometimes we have to shrink by a few square feet in order to meet their demands. So architects for private housing projects really have to keep the criteria of the developers in mind. And buildings requiring smaller and smaller units have changed the whole picture—the typology—if you ask me. As I said, the cruciform is extinct. Really, I don't have one single plan in my office that uses this. Because everyone is going for much smaller units now, and it is simply not efficient to wrap eight, small units around a vertical core.

CoR What are the fundamental concerns of the typical HK real estate developer in the provision of housing versus the fundamental concerns of the residents of the housing you have designed?

Fong Well, I guess I have answered the first part of the question already. The fundamental concern for the developer is always the investment return. This is before any aspirations to achieve environmental friendly design, creating benefits for the community, innovations in technology, etc., which more and more developers are concerned with. For architects in Hong Kong working in private residential housing developments, I think 70 to 75 percent of our efforts are spent, not on architecture in the aesthetic sense, but in achieving efficiency, marketability, and profitability for the developer and resolving issues with the building codes and various government departments.

Consider the concept of pre-sale consent, the permission to start the sales of the residential units before the buildings are completed. You can imagine how important this is financially for the developers. So preparation for the application of pre-sale consent is a very important task for architects working on private residential developments. Under these circumstances the architect's efforts are geared to getting the plans approved, getting the construction going, getting through all the hurdles to achieve this as quickly as possible, and according to the developer's schedule. So, this is really the administrative side of the architect's work—this is a very large part of our work.

CoR And in terms of the residents' concerns?

Fong If you ask me, the resident's concern is also the developer's concern. If the residents are not happy, they will simply not buy the units of that development. The residents' concern is, of course, the quality of the building. If you ask me whether the residents are concerned that a building is a cruciform, or whether they have a nice exterior elevation, I don't think they are. Their concern is whether the unit they are buying is good to live in—whether they can fit their furniture, whether there are enough electrical sockets and the switches are in the right place, whether they can configure their wardrobe in an appropriate way. If it is right, it will be a nice home for them and, more importantly, a good investment of their life savings.

CoR Perhaps we can come back to the concept of repetition. Based on our research, there are older estates which have a vast number of different unit types. For example, Mei Foo, developed in the 1960s, has 223 different types of units, which were all designed without computers and drawn by hand. Whereas in Laguna City, developed in the 1990s, there are only four unit types. Is it a concern for the architect or for the developer to offer diversity in flat types and sizes? Does the developer still want to provide choice to potential buyers?

Fong The differences must have their own specific reasons of their time, but I think the developers always want to give choices. If the buyers have a choice then they will choose what's best for them. In the old days, if you ask me, buyers of flats were very happy to be able to own a flat already and, therefore, they didn't ask for too much. For example, the handover standards of kitchens and bathrooms were very simple back then. Life was easy for developers and architects.

What type of units and what facilities will be provided in a development is always the decision of the developers. For architects, we must follow the brief. The developers decide what is the best strategy, how many, and what type of units to build. Architects can give their suggestions, but at the end of the day, it is never an architect's decision. We get the brief and we design according to it.

So when you ask if I am concerned about the repetition observed in some of the older developments, of course I am concerned. When all towers look the same, and when one must rely only on a sign to know the way home, it is very disheartening. Sometimes this is a result of commercial decision, but more often, the regulations or planning control play a part also. For example, we seldom have any design flexibility to create height variations for the towers in a project—to make some towers taller and others shorter—because of a height restriction control. One might think that architects can always create the variation by making some towers shorter than the allowable height limit, but the reality is, if any of the towers are shorter than the maximum allowable height then we cannot provide all the permitted areas within the site.

One such example is the TKO (Tseung Kwan O) areas near the MTR stations. Most of the sites have a height restriction of 168 mPD, and, as you can see, nearly all the towers are of the same height—168 mPD.

CoR What do you mean by mPD?

Fong Meters above Principle Datum. Sometimes the building height restriction is described in this way. Sometimes it is in the actual building height from street level. It depends on which document is stating the height limit.

And the number of refuge floors—for fire escape—also controls the building height. For example, if the building is taller than fifty-stories, two refuge floors, instead of one, will be required.

CoR So the whole estate is capped at forty-nine stories.

Fong Forty-nine stories above ground at 168 mPD. It's a perfect fit.

CoR Within the context of Hong Kong, to what extent can an architect push a developer to pursue a particular design intent that potentially runs counter to efficiency or profitability?

Fong Now, that is the crux of the problem for architects here. If he or she only pursues architectural values without understanding why that project has existed in the first place—as an investments for financial return—then it is really of no use, and he or she will certainly be very disappointed. To be able to create good architecture will require much more diligence—and the basic starting point is to be able to deliver what your client wants most.

CoR He or she is not going to design for the private housing market.

Fong No. He or she should go for projects like museums or schools. I would never say these building types are any less challenging, but the focus is far less on the financial return aspects of the design as compared to private residential developments.

CoR So money comes first and then ...

Fong The drive or the challenge architecturally for private residential projects is that we have to fulfill the brief and at the same time also create something more. I think that is what we are always trying to do. It makes things difficult for us actually. You could have planned the building in such a way so that it is less difficult to resolve problems, but because of all these things you want to do—you want your building form to respect a certain urban axis, or to break down the scale of a very long façade when all units are stretched out to face the sea—there are added challenges, and things could have been easier if you just stop when the brief is fulfilled. And that's where more problems come in, practical problems about how you're going to implement the design come in.

CoR When looking at Singapore, one of Hong Kong's competitors, you see a lot more experimentation in the public and private housing sector. Buildings such as the recent Pinnacle at Duxton or Paul Rudolph's Colonnade portray a very different approach to housing. Why is it so difficult in Hong Kong to produce more unique buildings?

Fong Our building codes and the push for salable area efficiency leave little room for creativity. To be creative in building design, you need to have some area footage at the architect's disposal. Architects here spend great amounts of effort to convince the authorities to grant exemption to exclude areas of design features from GFA calculation, since any footage spent on these features will be a direct reduction from salable area in the project. Nearly all clients I met respect the design of architects and do not mind spending the little extra on construction costs for design features, but when it comes to reduction of salable area, it is a different story.

But coming back to your comparison, to be fair, we do have very creative residential projects in HK, but they are usually of much smaller scale, like a single tower or single group of towers.

CoR Working with a firm that has major projects across east Asia, to what extent has the typology of housing in Hong Kong influenced other cities, regions, and countries? In particular, mainland China?

Fong I would say it is the way we optimize the land potential, in terms of building more on that land rather than spreading everything out. All cities have requirements about the plot ratio, the area that you are allowed to build on a piece of land divided by the site area.

In China, some might say that their plot ratio is still very low, around two or three. But in China there are deep setback requirements from the site boundary, so in reality the actual plot ratio can be as high as five. In HK, the plot ratio for very large sites is always five. For smaller urban redevelopment sites, it can be up to ten, depending on the characteristic of the site.

CoR Will HK continue to densify and use standardization as a tool for providing reasonably priced housing?

Fong No. I do not believe the design can make the costs of private housing in HK reasonable. Design can enhance the value of the property, but design cannot lower the price of housing in HK.

CoR Standardization has nothing to do with providing reasonably priced housing?

Fong To give an example by just quoting some figures from newspapers, the land price of a recently sold private housing site in the Cheung Sha Wan area, which is a mid-range housing area, is approximately HK$17,000/ft^2 GFA. The construction costs for high quality housing is approximately HK$4,500/ft^2 GFA. There are, of course, other financial costs and fees. On the other hand, the sales price based on the current market is approximately HK$30,000/ft^2 GFA. You can see that some reduction of construction costs by standardization will not help much in terms of the price of housing.

Even if the construction costs can be lowered by five to ten percent through standardization, we do not know whether this savings can be translated into actual reduction of sales price of housing. The value of the housing is not really related to the costs. It is more related to the land price and market value.

The price of housing has very little to do with design. A good design will create more value and the price will reflect that. But in Hong Kong the price of housing depends on the land price, and the market. That is what is fundamental.

I once had a residential project in early 2000s when the market was going down, as we were doing the design. Due to the pessimistic atmosphere of the residential market at that time, we were instructed to carry out cost saving exercises to lower the construction costs. During that period, the sales price was approximately HK$2,500 to HK$3,500/ft^2 GFA. But by the time the building was completed and went into sales, the market turned around and sales prices reached over HK$4,000/ft^2 GFA. But the construction cost saving exercise was already implemented. So you can see how little construction costs will affect the sales price of units.

So, to answer your question, a good design can elevate the value but the sales price is unrelated, really. If you ask me, in Hong Kong, in this twisted environment, it's always the land price and the market.

CoR What are the lessons to be learned from the past fifty years of private housing development in Hong Kong? Is there any foreseeable trend for the future?

Fong I think the lesson we learned from the past is that the repetitiousness, the monotony, is something that should be avoided. The repetition in estates like Whampoa Garden, Laguna City, or Telford Garden is monotonous—all the towers look the same. Either because architects and designers are actively trying to avoid it, or because of the change in the size of units that created a totally different battlefield for the design of housing, I do not think repetitions of the same tower within a large residential project is possible anymore. To optimize the value of the flats, and to resolve all sorts of practical problems with inadequate façade length for units behind, all of us must become very site- and location-specific in terms of residential tower design.

In a context as extreme as HK, I don't think repetition or standardization of building form in private residential development can or will happen anymore. On a more micro level, repetition and standardization of unit designs, and especially its elevation arrangement, will continue and become more evident, as their designs becomes more "perfect" in terms of efficiency of resolving all practical and code requirements.

Photographs

Visualizing Repetition

18
19
19

IMPOUNDING ZONE
NON-EMERGENCY AMBULANCE TRANSFER SERVICE
非緊急救護運送服務
HOSPITAL AUTHORITY

GIVE WAY
讓
LOOK LEFT

Photographic Survey

Mei Foo Sun Chuen

Taikoo Shing

Heng Fa Chuen

Kornhill

City One

Whampoa Garden

Belvedere Garden

Laguna City

South Horizons

Kingswood Villas

The Authors

Jason F. Carlow and Christian J. Lange are architects and educators. As colleagues teaching design studios together at the University of Hong Kong they developed a keen interest in housing, high density urbanism and the relationship between building code and architectural form. Carlow is Associate Professor and Head of the Department of Architecture at American University of Sharjah in the United Arab Emirates. Lange is Associate Professor in the Department of Architecture at the University of Hong Kong, where he serves as Director of the Fabrication and Material Technologies Lab and leads the Robotic Fabrication Lab.

Acknowledgements

There are a number of institutions and academic units that have supported us as authors, teachers, researchers and designers during the past decade. We would like to thank the University of Hong Kong (HKU), the HKU Faculty of Architecture, the HKU Shanghai Study Center and the HKU Department of Architecture for their significant financial and academic support. The American University of Sharjah (AUS), the AUS College of Architecture, Art & Design, and the Department of Architecture at AUS have also provided important resources and support for this endeavor.

There are also many individuals from the University of Hong Kong community, without whom our research project and publication would have not been possible. We must recognize the late Faculty of Architecture Dean, Ralph Lerner, for bringing each of us to Hong Kong in the first place. His energy, leadership, and encouragement made HKU an amazing place to teach, research, innovate, and collaborate with peers. We thank former Dean, Professor David Lung, for his support and assistance in securing funding for initial research and production. Thanks also to Chris Webster and Eric Schuldenfrei, for their generous support of the project.

Thanks to past and present HKU colleagues Weijen Wang, Andersen Lee, WS Wong, Jonathan Solomon, Juan Du, Tom Verebes, David Erdman, Wilson Liu, Kelvin Wong, and Shirley Hung for their support, advice, and guidance during the life of the project. Our gratitude goes to our contributor, Cole Roskam, for reviewing our texts, offering advice when needed, and for his historical framing of the project through his foreward. Thanks to Carolin Fong for her interview herein and many hours spent with us explaining the tricks, spatial strategies, regulatory loopholes, and economic incentives in the design of Hong Kong's housing. Thanks to Eyad Houssami for his editorial contributions.

Thanks to the small army of former students and colleagues who contributed to the research and production of the graphics for the book and the exhibition over the years, including Jan Henao, Benjamin Hollberg, Joni Low Ee Harn, Mono Tung, Dominic Co, Kelsi Su, Cherene Hui, Diamond Chan, Peony Tang, Theo Tong, Joel Wong, Garkay Wong, Lisa Lai, and Mariam-Al Hachami. Special thanks to Ching Ying Ngan for her organizational skills and keen graphic eye.

Thanks to Gordon Goff, Jake Andersen, and the team at ORO/AR+D for their editorial guidance, advice, and patience as we completely changed and redesigned the book from front to back. And thanks to our invaluable graphic designer, Kay Bachmann. Given the graphic nature of this volume and the importance of conveying information clearly through visual means, Kay made a critical editorial and creative contribution to this project, for which we are immensely grateful.

Finally, thanks go to our respective partners in life, Julia and Alice, for their tolerance of this never-ending project and their encouragement, support, and patience along the way. Despite the fact that our young children, Lika, Letta, Lukas, Oskar, and Helen, did everything in their power to slow down and delay the production of this volume, we thank them regardless for providing much needed joy and distraction, and for making our lives much less standard and repetitive.

Legal Notice

Published by Applied Research and Design Publishing, an imprint of ORO Editions.

Gordon Goff: Publisher

www.appliedresearchanddesign.com
info@appliedresearchanddesign.com

Authors
Jason F. Carlow, Christian J. Lange

Foreword
Cole Roskam

Photographs
Christian J. Lange (p. 121–142)

Book Design
Kay Bachmann (OfficeKB)

Project Manager
Jake Anderson

10 9 8 7 6 5 4 3 2 1 First Edition

ISBN: 978-1-939621-65-8

Prepress and Print work by ORO Editions Inc.
Printed in Hong Kong

AR+D Publishing makes a continuous effort to minimize the overall carbon footprint of its publications. As part of this goal, AR+D, in association with Global ReLeaf, arranges to plant trees to replace those used in the manufacturing of the paper produced for its books. Global ReLeaf is an international campaign run by American Forests, one of the world's oldest nonprofit conservation organizations. Global ReLeaf is American Forests' education and action program that helps individuals, organizations, agencies, and corporations improve the local and global environment by planting and caring for trees.

ORO EDITIONS
Attn: Editorial Dept
31 Commercial Blvd, Suite F
Novato, CA 94949

SAN FRANCISCO
31 Commercial Blvd. Suite F
Novato, CA 94949 – USA
T 1.415.883.3300
F 1.415.883.3309

LOS ANGELES
1705 Clark Lane, Suite 2
Redondo Beach, CA 90278
USA
T 310.318.5186

MONTREAL
180 Chemin Danis
Grenville PQ, J0V 1B0
Quebec, Canada
T 1.415.233.1944

SINGAPORE
2 Venture Drive
#11–15 Vision Exchange
Singapore 608526
T 65.66.2206

SHENZHEN
Room 15E, Building 7,
Ying Jun Nian Hua Garden,
Dan Zhu Tou, Shenhui Road, Buji,
Longgang district,
Shenzhen, China 518114w
T 86.1372.4392.704

BUENOS AIRES
Juramento 3115
Buenos Aires C1428DOC
Argentina
T 54.911.6861.2543